D0569269

Health Impact Assessment

Health Impact Assessment

Principles and Practice

Martin Birley

publishing for a sustainable future

London • New York

First published 2011
by Earthscan
2 Park Square, Milton Park, Abingdon, Oxon OX14 4RN

Simultaneously published in the USA and Canada
by Earthscan
711 Third Avenue, New York, NY 10017

Earthscan is an imprint of the Taylor & Francis Group, an informa business

© 2011 Martin Birley

The right of Martin Birley to be identified as author of this work has been asserted by him in accordance with sections 77 and 78 of the Copyright, Designs and Patents Act 1988.

All rights reserved. No part of this book may be reprinted or reproduced or utilised in any form or by any electronic, mechanical, or other means, now known or hereafter invented, including photocopying and recording, or in any information storage or retrieval system, without permission in writing from the publishers.

Earthscan publishes in association with the International Institute for Environment and Development

Trademark notice: Product or corporate names may be trademarks or registered trademarks, and are used only for identification and explanation without intent to infringe.

British Library Cataloguing in Publication Data
A catalogue record for this book is available from the British Library

Library of Congress Cataloging in Publication Data
Birley, Martin H.
   Health impact assessment : principles and practice / Martin Birley.
       p. ; cm.
   Includes bibliographical references and index.
   ISBN 978-1-84971-276-7 (hardback) — ISBN 978-1-84971-277-4 (pbk.) 1. Environmental health. 2. Health risk assessment. 3. Environmental impact analysis. 4. Medical policy. I. Title.
   [DNLM: 1. Public Health. 2. Environmental Health. 3. Health Policy. 4. International Cooperation. WA 100]
   RA565.B57 2011
   362.196'98—dc22

                                   2011000737

ISBN: 978-1-84971-276-7 (hbk)
ISBN: 978-1-84971-277-4 (pbk)

Typeset in New Baskerville
by JS typesetting Ltd
Cover design by Susanne Harris and Veronica Birley, all photos supplied by www.tropix.co.uk
Printed and bound in the UK by MPG Books Group
The paper used is FSC certified.

With gratitude to all the people who have helped to grow IIIA as a vital instrument for improving human welfare

Contents

List of figures, tables and boxes

FIGURES

TABLES

BOXES

Foreword by Professor Sir Michael Marmot

A key insight, going back at least to Hippocrates but which is regularly forgotten, is that health of populations is profoundly influenced by what happens outside the health (care) sector. The World Health Organization (WHO) Commission on Social Determinants of Health summarized this by saying that health, and health inequalities, are influenced by the circumstances in which people are born, grow, live, work and age, and the structural drivers of those circumstances – the social determinants of health. The WHO Commission on Social Determinants of Health, and the English Review of Health Inequalities that I chaired, called for an examination of the impact on health equity of all policies.

But how to do it? That is where this much needed book comes in. The health inequalities of the future are born out of today's policies, programmes, plans and projects: and most of these, in fact, lie outside the health sector. Health impact assessment (HIA) is a powerful instrument to protect and promote future health equality.

Practical HIA is a complex process, as this text makes abundantly clear. To many people, health impact assessment is a new concept requiring new skills. A pioneer and global expert in the subject, Dr Birley has been grappling successfully with this challenge for many decades. His book reveals extraordinary experience and insight, providing depth and context to existing HIA guidelines, invaluable to those who are trying to use them. There has long been a need for a clear and illustrative introductory text such as this, which enriches the knowledge and expertise of the practitioner.

This important book should appeal to a wide audience, touching, as it does, on a range of interests from the academic to the corporate, from the legal and ethical to the sustainable and the political. It succeeds in showing the points where these are contiguous. Dr Birley has taken a complex and emotive subject and addressed it in a comprehensive manner; thereby producing what constitutes an authoritative work of reference for one person, and a textbook for another.

The book provides valuable insights to readers from a wide range of professions who care about public health, health equality and health impact. It will be an important text for university courses that deal with these themes.

Sir Michael Marmot is Professor of Epidemiology and Public Health at University College London and chaired the Commission on Social Determinants of Health and the Strategic Review of Health Inequalities in England post-2010 (Marmot Review)

Foreword by Dr Robert Goodland

The evolution of social responsibility by public bodies, multilateral financial institutions and transnational corporations has shown slow but steady progress over the past 30 years, with some reversals. Initially the focus was on reducing environmental damage. Environmental impact assessment was born in the 1970s and spread rapidly around the globe, creating a body of practitioners and professional organizations, such as the International Association for Impact Assessment. In my role in the World Bank, the environmental focus was international development. We recognized the hidden costs of development; we struggled with external costs, and the need to protect vulnerable ecosystems and communities. While environmental and social impact assessment is now well developed and understood, there is still much confusion about health impact assessment (HIA) and a severe lack of global capacity. In the World Bank, we were acutely aware that the construction of big dams in warm climates could lead to the proliferation of tropical diseases and we took steps to prevent that. But most of the other health impacts of dams and, indeed, other projects went unheeded. There was a gap in our understanding that took time to fill. The health community was largely focused on health care, rather than the impacts of other development policies on health. There were a few notable lone voices in WHO and elsewhere but they received relatively little support. The author of this book, Martin Birley, is one of those and was well ahead of his time. I have known of his work for many years and watched as HIA was slowly pushed into the mainstream led by his efforts.

Development depends on the cooperation of many disciplines from engineering to sociologists to spatial planning. Professionals of all sorts need some awareness of the unintended impacts of their work. This should be incorporated in their professional training. Until now, there has been no introductory textbook on HIA to support that training, which provides a detailed account not just of the theory, but also of the practice. A book that includes not just the public sector, but also industry; that encompasses not just the high-income economies, but also the low-income ones.

I am delighted that Martin has taken on the challenge of recording his decades of hands-on, muddy-booted experience, telling us both what he has found that works and what doesn't. The early practitioners of any field have few, if any, teachers to learn from and must discover it for themselves. Later practitioners can learn from their mistakes. There is a lot of misunderstanding, worldwide about HIA, what it entails and how it integrates with the other forms of impact assessment. I am glad to see that Martin has tackled this.

Finally there are the issues of cumulative impact, especially those associated with our excess consumption of fossil fuels and water, and damage to the natural environment of soils, forests, fisheries and whole ecosystems. This is one of the cutting edges of impact assessment thought. Here again Martin is beating new paths, seeking to understand the implications of climate change and energy scarcity for the practice of HIA.

I warmly commend this book. Apart from being by far the most useful work on how to operationalize HIA, the book is brilliantly comprehensive and – dare I say it? – a profoundly wise distillation of the author's unparalleled experience worldwide. It is about much more than HIA: implementation of its judicious recommendations will help both the public and private sector to boost their effectiveness at low cost. It is required reading for all development agencies and those wanting to promote health and well-being related to industrial and infrastructure projects.

Dr Robert Goodland is former senior environmental adviser to the World Bank Group in Washington, DC and former president of the International Association for Impact Assessment

I have watched health impact assessment (HIA) grow from an off-the-wall, cranky idea to a mainstream pursuit. I am reminded of Schopenhauer: 'All truth passes through three stages: first it is ridiculed; then it is violently opposed; finally it is accepted as being self-evident.' In the early days it was considered absurd that health should look at other sectors; later I was directed to stop developing HIA; now I understand that health impact assessment has become self-evident. I am delighted to have been a central part of its story.

I wrote this introductory textbook because there wasn't one. The material grew from my personal experience. Some of it derives from teaching and training materials that I have used and tested in many HIA lectures and courses. I have observed the bewilderment of experts when presented with the question 'What is health?'. I have had trainees with health backgrounds baffled to discover that health is largely determined by decisions made outside the health sector.

My audience for those courses is drawn from many cultures and includes postgraduate students and professionals, among others. About half have some form of health background such as medicine, nursing, occupational safety or community health. Many of our training courses are intersectoral – in fact, we encourage as good a mix as we can get. But experts from different disciplines are not always skilled in communicating with their colleagues from other sectors. I hope I have included enough explanation and examples to make my meaning clear to readers from many different walks of life.

HIAs are now conducted in such a range of settings and contexts that they seem to be speaking a different language from one another and have very different aims and objectives. Finding the common thread is getting harder. For example, there are civil servants in the UK who look to maximize the health gain of local public sector policies. On the other hand, there are managers of transnational corporations who want to safeguard the reputation of their company while implementing a large-scale project in a rural area of a low-income country. Then there are policy-makers in the European Commission who seek to ensure that new social policy conforms to treaty agreements. And so on.

I did not intend that this book would cover all, or even most, of the vast field of health impact assessment. I have recorded my knowledge and experience, both what I knew and what I didn't know. There are still huge and unexplored territories in the area of expertise covered by HIA. I try to map out some of the issues that have yet to be resolved. These await discovery by HIA practitioners. If that means you, I wish you well!

The author, Dr Martin Birley, has been at the forefront of global health impact assessment for over 25 years. He was involved in development of the discipline since its early days in tropical medicine, to its later emergence as an instrument both for healthy public policy in the UK and Europe, and for community safeguards by international development banks. He has personal experience of many sectors including water resource development, agriculture, oil and gas, housing and planning. He has written guidelines for health agencies, development banks and transnational corporations. He continues to work all across the globe as well as in the UK. He has been a university academic and a corporate adviser, and runs training courses for institutions worldwide. He currently heads up a London-based consultancy, Birley HIA. Martin's depth of experience places him in an ideal position to provide an introduction to the rapidly expanding field of health impact assessment, for the benefit of newcomers and experienced practitioners alike

Acknowledgements

The list of people who have inspired or helped with this book is very long and I fear that I may not mention them all. First and foremost, Veronica Birley, my wife and partner, who has been part of my journey for more than 40 years. Veronica encouraged, edited and debated with me about all aspects of this book, and also researched and provided the photography. Special thanks to our talented daughter Bethany Birley for additional picture research; our son, Roland Birley, for his continued interest and encouragement; also to our good friend Gemma Hutchinson for a brief but inspired contribution. I am indebted to the authors of the forewords, Sir Michael Marmot and Dr Robert Goodland, your kind words do me a great honour. I am sincerely grateful to those who put in long hours of labour reading the manuscript, picking up my errors and challenging me to explain more clearly: Alan Bond, Lea den Broeder, Margaret Douglas, Eva Elliott, Peter Furu, Robert Goodland, Mark McCarthy, Rob Quigley, Alex Scott-Samuel, Francesca Viliani, Salim Vohra, Aaron Wernham and Colleen Williams. I wish to thank those old friends and colleagues whose constant support encouraged me to proceed: Robert Bos, Ben Cave, Andy Dannenberg and Ben Harris-Roxas.

I would like to acknowledge the influence on my thinking of colleagues not acknowledged elsewhere: Balsam Ahmed, Debbie Abrahams, Kate Ardern, Hugh Annett, Dick Ashford, Steven Ault, David Bradley, Andrew Buroni, Joanna Cochrane, Mark Divall, Hilary Dreaves, Mike Eastwood, Charles Engel, Debbie Fox, Liz Green, Amir Hassan, Ralph Hendrickse, John Jewsbury, Bill Jobin, Geert de Jong, John Kemm, Rob Keulemans, Flemming Konradsen, Karen Lock, Jenni Mindell, David Nabarro, Ken Newell, Marla Orenstein, Andy Pennington, Eddie Potts, Mike Service, Margaret Whitehead and to apologize to many others who are unnamed. Many others are acknowledged through citations in the book.

I must also acknowledge colleagues who presented papers and asked questions at innumerable conferences, workshops or training courses. You have all contributed to my understanding of this subject. My editor and publisher, Tim Hardwick, and his team at Earthscan have been helpful throughout. Last, but by no means least, I offer my kind regards and blessings to all the people and communities that I had the privilege to visit and learn from over the past 30 years: in Africa, Asia, Latin America, North America, the Middle East and Europe. I thank you.

DISCLAIMER

I must add a disclaimer: the opinions expressed remain those of the author alone. Throughout this book, tables and bulleted lists are provided as examples and illustrations. For clarity, they are deliberately selective. They are not intended as checklists. Each HIA is necessarily different from the last. The impacts described do not occur in every case and there may often be additional impacts.

This introductory chapter will inform you:

- What health impact assessment (HIA) is about and why it is needed.
- Who this book is for.
- What the drivers are for HIA nationally and internationally, publicly and privately.
- The different contexts in which HIA takes place.
- Some other forms of assessment and how they overlap with HIA.
- Some of the challenges.

1.1 WHAT IS HEALTH IMPACT ASSESSMENT?

In this introductory text, health impact assessment (HIA) is an instrument for preparing justified recommendations for the management of the future health impacts of proposals. The proposals may be policies, plans, programmes or projects. HIA is primarily used before the proposal is activated and examines both positive and negative consequences. The intention of HIA is to make recommendations supported by evidence that modify the proposal in order to safeguard and enhance population health. The sentences in this opening paragraph are necessarily dense and incomplete and the remainder of the book is required in order to explain them.

Example of health impacts

A government may decide to build a road from the periphery to the centre of a city. The aim is to enable goods and people to be transported rapidly. The unintended impacts could include: obesity, social severance, road traffic accidents, respiratory disease, loss of well-being and psychosocial disorder.

Well-planned transport could improve fitness and well-being while reducing accident rates.

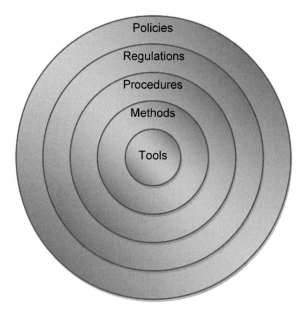

Figure 1.1 *An overview of HIA*

HIA itself is guided by policies, regulations, procedure, methods and tools: see Figure 1.1. The policies provide the enabling environment for HIA. When effective, this often leads to regulations to ensure that an organization implements the agreed policies. The procedure consists of deciding when an HIA is required, what it should include, who should draft it and who should be consulted, and how to manage reports, evaluation, acceptance and implementation. The methods of HIA include literature review, health profiling, systematic classification, analysis, community consultation, prioritization and development of recommendations. Tools refer to the various ways in which the methods are applied. When I first started to think about HIA, as a practising scientist, I naturally assumed that the method was the most important component. In practice this is not the case and this book will explain why.

There are a number of definitions of impact assessment, such as the following (IAIA, undated):

> Impact assessment, simply defined, is the process of identifying the future consequences of a current or proposed action.

There have also been a number of definitions of HIA. A popular one was the product of a meeting in 1999 that is known as the Gothenburg consensus (WHO European Centre for Health Policy, 1999). This was modified in 2006 by the International Association for Impact Assessment (IAIA) to include a final sentence that explains why the assessment is being conducted (Quigley et al, 2006):

> *A combination of procedures, methods, and tools by which a policy, program or project may be judged as to its potential effects on the health of a population, and the*

distribution of those effects within the population. HIA identifies appropriate actions to manage those effects.

The distribution of effects acknowledges that different communities will be affected in different ways, and this is also referred to as health inequalities. There is much additional detail that cannot be captured in a definition, for example the role of public participation and transparency.

1.1.1 Policies, plans, programmes and projects

This book is about the health impact of policies, plans, programmes and projects. The generic term that will be used for these is 'proposal'. A proposal is defined as a plan or suggestion, especially a formal or written one, put forward for consideration by others (OUP, 2010).

One characteristic of proposals is that they have aims and objectives that are stated in advance. For example, the aims might be to move people from A to B as quickly as possible, to extract minerals from a deposit, to improve educational status, or to provide a better medical service. All proposals have both intended and unintended consequences. The intended consequences are stated in the aims. The unintended consequences may include positive and negative impacts on human health.

Projects are tangible and specific. They often consist of specific infrastructure to be built in specific locations. Programmes are arrays of projects that may proceed sequentially or concurrently. Plans have a multitude of meanings. Policies are sets of decisions that are oriented towards a long-term purpose. When policies are made by governments they are often embodied in legislation and apply to a country as a whole. Much of the discussion in this book is focused on project-level HIA, but applies to a greater or lesser extent to all proposals. In some contexts, programmes and plans have a legal meaning. For example, the UNECE Protocol on Strategic Environmental Assessment applies to programmes and plans that are defined in a specific way (UNECE, 2010).

1.1.2 Timing of impact assessment

The IAIA definition above refers to proposed or current impact assessments. Sometimes people distinguish the following kinds of impact assessment:

- **Prospective** – undertaken before the proposal is implemented or construction has commenced; undertaken to safeguard the health of the affected communities in the future.
- **Concurrent** – undertaken during proposal implementation, construction or early operation. This is often a last-minute attempt to assess impacts, even though there is little room to change the design or operation.
- **Retrospective** – undertaken after the proposal has become operational; used to gather evidence about how proposals have affected health and to modify an existing operation. Arguably, this is actually evaluation, not impact assessment.

In this text, HIA refers primarily to prospective impact assessment, which is the most effective stage for safeguarding health.

1.2 WHO IS THIS BOOK FOR?

Despite nearly three decades of global capacity building, HIA remains a relatively unknown subject that does not form a large part of university curricula and is not invoked systematically in proposal design. It is often introduced at postgraduate degree level in fields such as public health, environmental epidemiology and environmental science. Professionals may first encounter it when they are managing a new proposal, as part of their duties as a health, safety, security and environment (HSSE) officer or sustainability officer, or when including health in an environmental impact assessment (EIA), social impact assessment (SIA) or strategic environmental assessment (SEA). Demand for HIA is increasing and a number of short training courses are available worldwide.

The audience for this introductory book is therefore very diverse and includes:

- university undergraduates and postgraduates;
- professionals seeking an introduction to the subject;
- participants in short courses;
- project stakeholders;
- informed or concerned members of the public, such as communities who wish to undertake an HIA themselves;
- HSSE managers in private corporations;
- occupational health and safety specialists;
- public health practitioners;
- policy-makers;
- environmental health practitioners;
- town, land-use and spatial planners;
- geographers;
- social scientists;
- development bank advisers;
- international development practitioners;
- national and international non-governmental organizations;
- water and sanitation engineers;
- agricultural development officers;
- miners and oilmen;
- road builders;
- landlords;
- EIA, SIA or SEA practitioners.

The audience may be located in any country in the world, because proposals are made everywhere. In many countries, professionals will be encountering HIA for the first time. So the needs of the audience will vary. Some will need a brief overview of what they must manage. Others will need to participate in an HIA as a member of a new team. Some will need in-depth knowledge and skills and will be seeking pointers to the wider literature. Many, but not all, of the skills required are generic. Some of the generic skills will be familiar to those trained in public health, EIA or SEA.

This book is intended as a personal account in which general principles are illustrated by practical experience for the benefit of newcomers to the subject. The orientation is project rather than policy, and includes low-income as well as high-income economies. It is not intended as an exhaustive review of the literature as there are already a number of excellent reviews and guides. See, for example, the HIA Gateway (APHO, undated).

One of the challenges in writing for a diverse audience is that the material will be too simple for some and too advanced for others. Whenever possible, I hope to err on the side of being clear and simple.

1.3 WHY IS HIA NEEDED?

As stated above, all proposals can have unintended positive and negative impacts. The impacts affect the environment, society and human health. Environmental impact assessment (EIA) was formulated in the 1970s in order to manage environmental impacts by encouraging transparency and the supply of information to the public. This included some, but by no means all, health impacts. For example, EIA can be a good tool for managing the health impacts of pollution, providing that an acceptable threshold has been agreed elsewhere for the concentration of a pollutant at the point of contact with a susceptible human. But there are many more health impacts that EIA cannot identify or resolve (Harris et al, 2009).

It is the duty of the health sector within national and local government to safeguard and improve community health. However, many major health improvements are made through other sectors and health is sometimes one of the least resourced of the government sectors. In a low-income economy, it may receive less than 2–5 per cent of the annual government budget (UNDP, 2008). In some of the more developed economies it may receive 8–15 per cent. Whatever the actual percentage, it is clear that far more money is spent by government in all the other sectors combined, than in the health sector itself. This is illustrated in Figure 1.2. The question then arises, which expenditure is likely to have the most influence on health: the small amount spent in the health sector, or the very much larger amount spent in all the other sectors? On the basis of expenditure alone, it can be argued that non-health sectors have far more influence on health than the health sector does. Some of these other expenditures are focused directly on human health: these include the supply of clean drinking water, sanitation, food safety and security. There are many more expenditures which do not include human health as a direct objective, but which affect it nevertheless. Transport, energy and education are examples.

Newcomers to health issues may believe that the major improvements that have occurred in public health status over time are the result of activities by the health sector. However, this is not generally the case. For example, infant mortality rates from communicable diseases in Europe fell rapidly before the introduction of effective medical care. What caused that? Apparently it was general improvements in the standard of living, the environment and the wealth of the community (McKeown,

Figure 1.2 *The share of the health sector in a national budget*

1979). With the introduction of modern vaccination, childhood mortality rates from communicable diseases fell further. While important, the health sector contribution added to a general trend that was already established.

Policies of non-health sectors, both public and private, have enormous influence on public health and well-being. For example, reducing pollution of air, water and food prevents poisoning and chronic disease. Social policies can reduce stress and mental illness. Transport policies can reduce road traffic accidents, obesity and heart disease.

HIA is primarily concerned with health in other sectors. The sectors include mining, oil production, agriculture, transport, water supply, education, housing and all the rest. This is relevant in all countries, but particularly important in poor countries. HIA requires health practitioners to look outward from the many concerns of the health sector itself. Successful HIA depends on understanding the agenda of other sectors, rather than expecting other sectors to understand the agenda of health. For example, good housing protects people from temperature extremes and helps reduce excess summer and winter deaths among the frail young and elderly.

The health sector may be more properly referred to as the medical sector. Its principal responsibility is to care for those who are clinically ill. The budget of the health sector is largely devoted to medicine. The average preventative and public health expenditure in Organisation for Economic Co-operation and Development (OECD) countries is about 3 per cent of total health expenditure (Butterfield et al, 2009). The comparative expenditure in England is about 4 per cent. This is further subdivided, in England, as prevention of non-communicable diseases (66 per cent); maternal and child health, family and counselling (17 per cent); school health services (3 per cent); prevention of communicable diseases (6 per cent); occupational health care (0.1 per cent); all other miscellaneous public health services (8 per cent). Although the budget is small, the preventative work has had important benefits. The last subcategory includes public health, environmental surveillance, public information on environmental conditions and HIA.

The root causes, or determinants, of illness and well-being may be physiological, social or environmental. The social and environmental determinants are affected by proposals. When these proposals cause negative health impacts they transfer a hidden cost from the sector responsible to the health sector, as sick people seek costly medical care. Equally, good proposals in other sectors can enhance health and reduce the costs to the health sector.

There are many elements of the health sector that are responsible for disease prevention and health promotion, rather than medical care. For example:

- Public health monitors and analyses the state of a nation's health; it proposes measures to improve health, advocates new policy and prevents pandemics.
- Environmental health has the responsibility for ensuring that agreed standards are maintained. It inspects existing facilities such as factories and restaurants and tests for hygiene and emissions.
- Health promotion conducts campaigns (such as to reduce smoking, improve diet and increase physical exercise), as well as advocating new policy.
- Health information measures and analyses trends in health status and health inequalities.
- Health and safety ensures safer living and working conditions.

Many of these activities have in common that they are reactive rather than proactive. The responsibility for ensuring that future proposals are safe and healthy has often rested elsewhere.

So the activities of non-health sectors have a crucial effect on public health. These activities often include the formulation of new proposals. Action is needed in order to ensure that these new proposals incorporate safeguards for promoting public health. Action has evolved at three levels and these can be referred to as policy, programme and project. At the policy level there has been increasing interest in healthy public policy. At the programme level there is increasing interest in including health in SEA. At the project level HIA is conducted alongside EIA in both the public and private domains.

1.3.1 Healthy public policy

The public health movement has built a case regarding the way that national and international policies in other sectors affect population health. Consensus is developing to ensure that all public policy has a tendency to promote and safeguard health and reduce health inequalities. This may be referred to as healthy public policy (WHO, 1986), health in all policies (HiAP) (Ståhl et al, 2006) and health in other sectors (WHO, 1986, 2005). Associated international initiatives include the Healthy Cities movement (Ashton, 1992; WHO, 1995; WHO Regional Office for Europe, 2005), the Ottawa Charter (WHO, 1986), the Commission on the Social Determinants of Health (CSDH, 2008) and the Performance Standards of the international banks, described below. For more details of international policy initiatives see Ritsatakis (2004).

In the UK, initiatives have included commissions on health inequalities such as the Black Report (DHSS, 1980), the Acheson Report (Acheson et al, 1998), the Royal Commission on Environmental Pollution (2007), public health priorities (Wanless et al, 2002) and the Marmot Review (Marmot, 2010).

Healthy public policy initiatives that support HIA are found in many other countries, as the following examples testify. In Canada, Quebec's Public Health Act requires laws and regulations to go through an HIA process (NCCHPP, 2002; Banken, 2004). In Thailand, the National Health Act includes HIA (Phoolcharoen et al, 2003; Government of Thailand, 2007). Section 11 of the Act gives people the right to request and participate in the HIA of a public policy, programme or project. Article 57 of the Thai Constitution establishes the right of a Thai citizen to receive information, explanation and justification for any national or local government project or activity that may affect health. Thailand has published rules and procedures for HIA (HIACU, 2010), strengthened the inclusion of health in environmental assessment (Nilprapunt, 2010) and established processes for promoting HIA throughout Thailand and, potentially, throughout Southeast Asia (Pengkam and Decha-umphai, 2009). Lao PDR has an HIA policy and Cambodia and Vietnam are formulating one by mentioning HIA in revisions of existing laws. There has been much work in Australia and New Zealand (WHO, 1988; Kickbusch et al, 2008; CHETRE, undated). In Lithuania, the Law on Public Health Care of 2002, amended 2007, requires public health authorities to conduct HIAs of proposed economic activities (Ingrida Zurlyt, pers. comm.). Roscam Abbing (2004) has described the development of HIA in the Netherlands. The Welsh Assembly government was established at devolution in 1998 and recognized the importance of all policies on health. The Wales Health Impact Support Unit was established in 2001 and is very active (WHIASU, undated). Other states and administrations are debating or implementing laws, including California and Geneva. Some other health policies that support HIA are discussed below under 1.5.2 External drivers.

1.4 THE WIDER LITERATURE

Over the last 20 years a plethora of guidelines on HIA have been published. In 2009, there were said to be over 30 existing HIA guidelines available on the Internet (Marla Orenstein, pers. comm.). (See, for example, Birley and Peralta, 1992; Birley et al, 1997; Scott-Samuel et al, 2001; EC HCPDG, 2004; WHIASU, 2004; IPIECA/OGP, 2005; Harris et al, 2007; ICMM, 2010.) Many of these have been targeted towards a specific audience. They are necessarily brief and cannot do the work of an introductory textbook. Some were written for specific sectors such as agriculture, energy, transport, housing or water management. Some were specific to a particular country, such as the public sector in Scotland or the federal government in Canada, or region, such as the European Community. Some referred to a single disease group, such as vector-borne diseases (Birley, 1991). Some were intended to integrate health into other forms of impact assessment, such as EIA (for example Birley et al, 1997, 1998). They were written for development banks, corporations, UN agencies, local authorities and national governments.

There are collections of edited papers and a wide and growing academic literature (Kemm et al, 2004; Wismar et al, 2007). Specialist journals that carry papers on HIA include *Impact Assessment and Project Appraisal* (IAIA, undated) and *Environmental Impact Assessment Review* (undated) among others. Papers are increasingly published in the more mainstream medical and health sciences journals. Chapters on HIA are included in many edited books in various disciplines including sociology as well as public health (for example, Scott-Samuel et al, 2006).

Much of this literature can be accessed online through the publishers or through specialist portals listed in Chapter 13 (and see HIA Gateway (APHO, undated)). Many of my own publications can be retrieved from www.birleyhia.co.uk.

1.5 DRIVERS: INTERNAL AND EXTERNAL

At the time of writing, the few countries with legislative requirements or regulations to carry out an HIA are listed above. Other countries are considering or preparing them. In addition there are international requirements (see External drivers below). In some cases, there are policy recommendations and these can be sufficient to persuade public authorities to undertake assessments. So why are many HIAs being carried out, worldwide? The drivers can be subdivided into internal and external groups.

1.5.1 Internal drivers

Internal drivers are derived from the mission, vision, aims and objectives of a particular institution. They are seen to be of benefit to the institution itself, assisting it to achieve its aims. For example, some transnational corporations (e.g. oil and gas, mining) have a 'business case' for carrying out impact assessments (Birley, 2005). The business case includes:

- reputation;
- licence to operate;
- staff satisfaction;
- risk management;
- differentiation;
- cost effectiveness;
- corporate social responsibility statements.

Reputation is regarded as an intangible asset and anecdote suggests that it may be valued at 10–20 per cent of annual income. Licence to operate applies both nationally and locally. At a national level, a corporation may have an agreement with the government to carry out its operations in a particular locality. At a later date, it may wish to ask that government for permission to carry out new operations in a new locality. It is more likely to receive permission if its operations are regarded as beneficial. At a local level, the corporation wishes to be on good terms with its neighbours. It does not wish them to sit down in the road in front of its gates and prevent its operation from continuing. Staff satisfaction is important because professional staff are often regarded as the most important asset of the corporation. The corporation should behave in a way which is acceptable to its staff or they may leave and take their skills elsewhere. Corporations can, of course, seek to maintain their reputation by using 'greenwash': in other words, by promoting themselves as behaving responsibly without actually doing very much.

However, it may be harder to maintain licence to operate and staff satisfaction by such means.

Corporations carry out lots of risk management. Managers are trained to identify risks of any kind to the success of the corporation, to analyse those risks and manage them. Damage to community health that can be attributed to the activities of a corporation represents an important risk, to both reputation and licence to operate. This may affect the perception of analysts regarding the financial risks associated with that corporation. This, in turn, may affect the interest rate at which it can borrow money. Interruptions in operations associated with a withdrawal of licence to operate can affect profitability and share price.

Differentiation refers to the need of a corporation to be distinguishable in the eyes of its customers from its competitors. For example, two companies may sell different brands of charcoal for barbecues. One company differentiates by certifying that their charcoal is obtained from local suppliers using sustainable sources.

So this 'business case' drives the corporation to take account of the unintended impacts of its operations. Although it has no altruistic elements, it can lead to much the same safeguards and mitigation measures as if it did. The activities look altruistic and enhance reputation. Internal drivers of this kind can be reversed by changes in the market or changes in leadership. An institution that commits to HIA may later abandon the process, or modify it until it has no value.

1.5.2 External drivers

External drivers are the actions that an institution must take because of treaties, laws, regulations or contractual conditions. There are currently a number of external drivers supporting HIA and more are expected.

1.5.2.1 Human rights

Human rights and HIA are both concerned with promoting and enhancing health and well-being (MacNaughton and Hunt, 2009). Human rights provide a set of values and an ethical and legal framework. Human rights are universal and inalienable, interdependent and indivisible, equal and non-discriminatory (OHCHR, 1948). They have been embodied in a range of treaties and conventions. Every state in the world has ratified at least one international treaty that includes the right to health (MacNaughton and Hunt, 2009).

The International Covenant on Economic, Social and Cultural Rights contains the most widely applicable provision. States that are party to this Covenant recognize the right of everyone to the enjoyment of the highest attainable standard of physical and mental health. This has been further elaborated to include the determinants of health including water supply and sanitation, housing, food and public participation. It is a right to control decisions that affect one's health and body. It is further elaborated as having four essential elements and six crucial concepts. One of these is progressive realization, meaning that states must take clear steps towards realizing the right of health for all and retrogressive steps are presumed not to be permissible. Impact assessments are tools for assuring progressive realization and avoiding retrogression in policy developments.

Summary of human rights requirement

Progressive realization of the right to health is presumed to mean that retrogressive steps are not permissible.

1.5.2.2 Treaty of Amsterdam and the European Commission

Health issues have been featured in European treaties since the formation of the European Union. The 1997 Treaty of Amsterdam revised an earlier EC Treaty (EC, 1997). Article 152 (replacing an earlier Article 129) has a wider scope than before. The Article states that a high level of human health protection shall be ensured in the definition and implementation of all Community policies and activities. The European Community can now adopt measures aimed at ensuring (rather than merely contributing to) a high level of human health protection. Among the areas of cooperation between member states, the new Article lists not only diseases and major health scourges but also, more generally, all causes of danger to human health, as well as the general objective of improving health. However, since most power remains in the hands of the member states, the Community's role is subsidiary and mainly involves supporting the efforts of the member states and helping them formulate and implement coordinated objectives and strategies. The Commission has published a guide to HIA (EC HCPDG, 2004).

The EC has also published a White Paper proposing a strategic approach to population health (CEC, 2007). This establishes a principle of health in all policies (HiAP). It recognizes that the population's health is not an issue for health policy alone. Other Community policies play a key role. The associated action is to strengthen integration of health concerns into all policies at Community, member state and regional levels, including the use of impact assessment (IA) and evaluation tools.

Evidence has emerged that corporations have influenced the Treaty so that new EU policies advance their own interests, including those that produce products damaging to health (Smith et al, 2010). The evidence suggests that the form of IA that has been adopted overemphasizes economic impacts, fails to adequately assess health impacts and favours corporate interests.

1.5.2.3 Environmental impact assessment in the US

The US National Environmental Policy Act established broad requirements 'to stimulate the health and welfare of man'. Therefore the policy requirement for EIAs meant that there was a policy requirement for HIA. However, in many jurisdictions including the US, health impacts have not been considered systematically (Steinemann, 2000). This is now changing (Dannenberg et al, 2008).

1.5.2.4 English legal system

In 1998, the UK enacted the Human Rights Act (HM Government, 1998) giving further legal effect in the UK to the fundamental rights and freedoms contained in the European Convention on Human Rights (Council of Europe, 1950).

According to the English legal system, any public authority, before approving a scheme or reaching a regulatory decision concerning an existing activity, which may impact on human rights, must consider the nature of the impact, its seriousness and whether it can be justified on public interest grounds. If it fails to do so, it may be held to have acted unlawfully under the Human Rights Act 1998 (Roger Seddon, pers. comm.).

Example of legal challenge

For example, eight applicants who lived in properties surrounding Heathrow Airport alleged that government policy on night flights at Heathrow Airport gave rise to a violation of their human rights and that they were denied an effective domestic remedy for this complaint (European Court of Human Rights, 2003). The European Court found the UK guilty of not providing adequate means to assess and mitigate potential health impacts. The Court declared 'a proper investigation and study should precede the relevant project'.

1.5.2.5 English planning system

At the time of writing, the English planning system was in a state of change. There was increasing recognition that planning was an important determinant of health, but legal requirements and obligations to give consideration to health impacts were limited (RTPI, 2009). Local authorities could formalize policies to require HIA for certain planning applications and some had done so. Barriers included lack of expertise by planners in health issues, lack of a statutory role by health authorities, lack of a sufficient evidence base and lack of dialogue between planners and health practitioners (Burns and Bond, 2008). See Chapter 11 for more details. The legal requirements for SEA provided new opportunity.

1.5.2.6 Strategic environmental assessment

Strategic environmental assessment (SEA) can be defined as follows: a systematic, objectives-led, evidence-based, proactive and participative decision-making support process for the formulation of sustainable policies, plans and programmes, leading to improved governance (Thérivel et al, 1992; Fischer, 2007).

Within Europe, health in SEA is supported by two key legal frameworks:

1   The Strategic Environmental Assessment Directive established a statutory requirement for the consideration of significant effects on human health in European Union member states and accession candidates (European Parliament and the Council of the European Union, 2001). The Directive is limited to 'certain plans and programmes' and tends to exclude many policies (Burns and Bond, 2008).
2   The United Nations Economic Commission for Europe (UNECE) Protocol on SEA was signed in 2003 by 35 European countries in Kiev and came into force in July 2010 (UNECE, 2003). It includes explicit references to human health and has a broad interpretation of health.

In addition, the Fourth Ministerial Conference on Environment and Health, Budapest, called for member states to take significant health effects into account in the assessment of strategic proposals (WHO Regional Office for Europe, 2004).

Together, these advocate closer links between environmental protection and health promotion and provide the health sector with an opportunity to take on a stewardship role in the environmental domain. A number of publications, meetings and research papers have sought to clarify how health is included in the SEA Directive (WHO, 2001; DCLG, 2005; Fischer, 2007; Nowacki et al, 2010). For example, the extent to which the social determinants of health can be included remains unclear.

Experience to date suggests that SEA practice is largely associated with spatial planning and transport (Fischer et al, 2010). For example, in England the SEA Directive's role in land-use planning is to examine the implications of regional and local plans and advise on modifications to make them more sustainable (Burns and Bond, 2008). Distributional and cumulative impacts (explained in Chapters 2 and 12) tend to receive insufficient attention; the Directive has a requirement to consider climate change but the health impacts of this are rarely mentioned. The SEA Directive became law in the UK in 2004. During 2007, the English Department of Health initiated a consultation on guidance on health in SEA (Williams and Fisher, 2007, 2008). A number of recommendations were proposed during the consultation and these included making health authorities statutory consultees for SEAs.

1.5.2.7 International Finance Corporation and the Equator Principles

A number of international development banks are situated throughout the world. They are financed by governments and lend to both governments and the private sector. Regional examples include the Asian Development Bank, African Development Bank and European Bank for Reconstruction and Development. Perhaps the most well known is the World Bank. The World Bank Group comprises a number of different entities. The International Finance Corporation (IFC) is the part of the World Bank that lends money to the private sector.

In 2006, the IFC published Performance Standards that it expects all its clients to follow (IFC, 2006). Performance Standard 4 is concerned with community health, safety and security. It requires the client to evaluate the risks and impacts to the health and safety of the affected community during the design, construction, operation and decommissioning of the project, and to establish preventative measures to address them in a manner commensurate with the identified risks and impacts. These measures are required to favour the prevention or avoidance of risks and impacts over minimization and reduction.

HIA is not mandated and borrowers can claim that health impacts have been evaluated by other means. HIA is one of the tools recommended by IFC that can be used to implement this Standard. The IFC has issued guidance on the contents of an HIA (IFC, 2009). The Standards are only a requirement in non-OECD countries because it is assumed that OECD countries have sufficient safeguards of their own.

The Equator Principles are a set of environmental and social benchmarks for managing environmental and social issues in development project finance globally (Equator Principles, 2006). They commit lending banks to refraining from financing projects that fail to follow the processes defined by the Principles. The banks chose to model the Equator Principles on the environmental standards of the World Bank and the social policies of IFC. Therefore, the banks follow Performance Standard 4 and are committed to the same health standard. Many financial institutions have adopted the Equator Principles, which have become the de facto standard for banks and investors on how to assess major development projects around the world.

The same Standards are promoted by institutions responsible for political risk insurance, such as the Overseas Private Investment Corporation (OPIC, undated) and the Multilateral Investment Guarantee Agency (MIGA, undated).

1.5.2.8 Health inequity as a driver

Another external driver that is promoting the use of HIA is the report of the International Commission on the Social Determinants of Health (CSDH, 2008). There is abundant evidence that different community groups within a single region experience substantially different levels of morbidity and mortality (for a summary see Wilkinson and Marmot, 2003).

Some writers consider inequality and inequity to be synonymous. Others make a distinction between facts and values. Health inequality could be considered a statement of fact: there are observable health differences between groups based on factors such as gender, race, socio-economic status and location. In some cases these are entirely appropriate, e.g. older people have a higher prevalence of poor eyesight. Health equity could be considered a statement of value: that all individuals should have a fair opportunity to attain their full potential. The CSDH proposes that health inequity is associated with policy failure. Governments may reasonably be expected to promote health equity, and tools are required to ensure that proposals do so. Tools are needed to assess the distributional impacts of proposals on the health of different community groups and to make justified recommendations to manage these impacts and to increase equality. The CSDH endorsed the use of HIA for this purpose. While the main focus of the CSDH is government policy, the need to assess distributional impacts applies to all proposals.

The CSDH report was influential in ensuring that the World Health Assembly in 2009 adopted a resolution to address health inequalities. This, in turn, gave WHO an increased mandate to promote HIA. In 2010, the Marmot Review gave further impetus to managing inequalities in England (Marmot, 2010). In 2010, equity impact assessment was a legal requirement in the UK (Local Government Improvement and Development, 2010).

1.5.2.9 National health objectives

Many countries establish national health objectives and strategies for tackling priority health issues, such as obesity. Increasingly, these refer to HIA as a tool for implementing strategy. They are often single disease focused. For example, the strategy for tackling childhood obesity in the US suggests the HIA of major developments and plans (Executive Office, 2010).

Table 1.1 *Examples of different contexts for HIA*

	Project	Programme	Policy
Public	Housing regeneration Water storage	Local transport strategy	EC agricultural policy Healthy public policy Health in all policies
Partnership	Incinerator Hydropower	Residential housing Land zonation	Promotion of joint ventures with NGOs, public sector and private corporations
Private	Extractive industry Land-use developments	Distribution networks	Safeguard policies of IFC Equator Principles

1.6 CONTEXTS FOR HIA

HIA takes place in a wide range of different contexts. Some of these can be captured in a map that has as its axes public and private, project and policy (see Table 1.1). Partnership – refers to public-private partnerships that are joint ventures between businesses, the public sector and, often, non-governmental organizations (NGOs). Programmes refer to collections of similar projects. These different dimensions are described below.

1.6.1 Public versus private

In the public sector, government departments may be expected to pursue their objectives within the broader context that benefits the community as a whole. In practice, the complexities of government often mean that departments are unable to work together for the common good. Furthermore, the common good may operate at the expense of a particular community. Indeed, one government department may act to the detriment of another department. For example, the European Union subsidized the production of dairy products in the livestock sector while, at the same time, the health sector was promoting a reduction in dairy fat consumption (Dahlgren et al, 1997). Arguably, the public sector is weak at regulating itself and governments often cannot work in a coherent way. In many countries, it may be difficult to justify the additional expenditure required to carry out a comprehensive impact assessment.

Example of intra-government conflict of objectives

A government may require a hydroelectric reservoir in order to supply the national grid. In order to achieve this common good communities may be involuntarily displaced from a river valley or exposed to health challenges.

In the private sector, businesses are expected to pursue their objectives for the benefit of their own shareholders rather than for the community as a whole. Yet, as we explored in Section 1.5 above, there are both internal and external drivers that may enable businesses to mitigate negative health impacts and enhance positive health impacts. There are project proponents such as developers, companies and corporations. The project proponents seek permission to implement their project from a range of decision-makers. The decision-makers can be government regulators, statutory consultees or financial lenders. The decision-makers may even be internal to their own organization. They are obliged to seek profits and control costs, but may be able to justify an HIA as a legitimate cost of doing business.

There is also the third sector, the non-governmental organizations. Like the public sector, these may be expected to pursue their objectives for the common good. However, their proposals may also have unintended impacts and require impact assessment. They may not have the resources to undertake such assessments.

At the project level, there is a tendency to form public–private partnerships in which government departments, private industries and specialist NGOs work together on community development projects in the vicinity of a private development.

1.6.2 Project and policy

The proposals subject to an HIA range in size from projects, through programmes to policies. In the public policy cell of Table 1.1, the intention for 'healthy public policy' is described above. This, in turn, has promoted a call for 'health in all policies'.

Many publications discuss the value of HIA as a tool for promoting healthy public policy (for example, Metcalfe and Higgins, 2009). These refer to the evidence of the huge burden of disease in society that would be preventable by changing policies in various sectors. For example, some data suggest that 80 per cent of cardiovascular diseases and type 2 diabetes and 40 per cent of cancers could be avoided by eliminating environmental risk factors (cited in Metcalfe and Higgins, 2009). HIA is respected as a tool that enables public decision-makers to make healthier policy choices.

In the private policy cell of Table 1.1, there are the safeguard policies of the lending banks and the mission statements of transnational corporations. For example, transnational corporations may seek to be 'good neighbours'. See the business case in Section 1.5.1 above. At the strategic level, there has been increasing emphasis on SEA.

One example of policy HIA is the discussion of common agricultural policies in the EC (Lock et al, 2003). This examined accession issues in Slovenia in relation to national food and agricultural policy. The need for better intersectoral working between health and agriculture was identified.

1.6.3 Sectors and sub-sectors

The contexts for HIA include all sectors and sub-sectors, countries and regions. In this text, the contextual diversity is illustrated by urban housing and planning, water resources and extractive industry. The diversity poses challenges to an HIA practitioner. One of the skills required of a practitioner is to understand quickly a new and unfamiliar context, the associated health determinants, and the culture and processes that govern its operations. As suggested before, the practitioner must understand the agenda of the sector and its proponents but not expect them to understand the agenda of the health sector.

No practitioner can expect to become familiar with all settings, so specialization is inevitable. However, the specialization required is different to that of academic fields. The practitioner must talk to key informants and review literature in order to understand something about all the health concerns and determinants associated with the context. From an academic perspective, the HIA practitioner is a generalist. For example, detailed knowledge of the health consequences of air pollution alone is insufficient for an HIA of a transport proposal; it omits, for example, the impact of transport on obesity.

1.7 IMPACT ASSESSMENT AND SOCIAL INVESTMENT

Corporations often use a part of their profits to benefit the community. They may seek to achieve this either through philanthropic donations or through community development, social investment and projects.

The distinction between philanthropy and social investment is an important one. Philanthropy frequently involves a single payment to support community well-being that is unconnected with the commercial project, not guaranteed to be sustainable and is a voluntary, charitable act. For example, a biscuit company may support an opera house in the nearest city. By contrast, a social investment is directly associated with the commercial project. For example, the company may decide to support vocational training courses in the nearby college. The courses are expected to provide a steady supply of potential employees for the company. Thus, the training courses are sustained by the company and the company is sustained by new recruits from the training courses. Or, social investment might provide general community development by local procurement of goods and services. This contributes to the company's business case by maintaining their licence to operate (Hastings, 2002; Shell UK, 2009). Social investment programmes sometimes include direct health components such as clean water supply, green spaces or mother and child clinics.

Social investments should be based on an analysis of community needs. These needs would probably be the same whether or not a proposal was being implemented in the vicinity of that community. The needs are not impacts of the proposal and are not strictly the responsibility of its proponent. Community health needs are an example. As HIAs document community health needs, they can provide some of the knowledge required to inform a social investment programme. The association between HIA and health needs assessment is discussed in Section 1.8.2 below.

A similar concept is found in UK town planning, where it is referred to as planning gain or Section 106 agreements. In this case it is not voluntary. A developer who wishes to build a substantial number of new homes is required to contribute to the development of the public realm, such as green spaces, schools and clinics. The additional cost is added to the price of the homes (Cullingworth and Nadin, 2006).

1.7.1 Strategic health management

Some corporations describe their approach to supporting the health needs of the community as strategic health management (OGP, 2000; Eni corporation, undated). The objective of a strategic health management plan is to introduce systematic, cooperative planning into each phase of the project life cycle, in order to ensure the health of the workforce and promote lasting improvements in the health of the host community. The corporation may believe that:

- industry cooperation on health is beneficial;
- industry can help host governments fulfil responsibilities for community health;
- early stakeholder involvement and consultation can achieve lasting improvements in community health.

Strategic health management is a partnership approach based on collaboration with government. From the perspective of the corporation, health projects should:

- address priority community health needs;
- be selected involving local, regional and national authorities and key stakeholders;
- promote social development of community groups geographically close to operational locations;
- have a self-sustainable future;
- have a detailed business plan;
- meet both local/national and international standards.

There are other corporations that state categorically that social investment is not their concern.

1.8 THREE KINDS OF HEALTH ASSESSMENT

There are usually three separate kinds of health assessment required for a large proposal. They are managed from different budgets, affect different communities and are analysed in separate reports. Table 1.2 is a summary. Health risk assessment (HRA) is concerned with the occupational health and safety of the future workforce associated with a proposal and, to some extent, with issues like explosions that could affect the peripheral community. Health needs assessment (HNA) considers the current health status of a population or community, independently of any development proposal. HIA, by contract, considers the future consequences of a proposal on the general community. There is often overlap (see Figure 1.3) and definitions vary between countries and institutions. They may share the same baseline studies and this is explained in more detail in Sections 1.8.1 and 1.8.2.

Table 1.2 *Summary of three different kinds of health assessment*

Types of health assessment	Community	Management plan
Health risk assessment	Future worker community and part of peripheral community	Occupational health and safety plan, hazard management plan
Health needs assessment	Current community	Social or health investment plan, health enhancement plan
Health impact assessment	Future community, often excluding occupational health and safety	Mitigation and enhancement plan

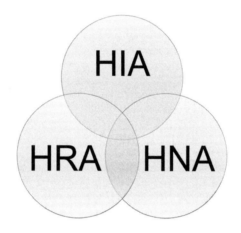

Figure 1.3 *Three kinds of health assessment*

1.8.1 Health risk assessment

Large organizations often have their own occupational health and safety specialists and a well-developed plan for managing and improving worker health and safety and ensuring safe operation of facilities. They undertake a health risk assessment (HRA) of proposals at the design stage. The assessment is often limited to 'inside the fence', in other words to localities where risks and facilities are directly under the control of management. It includes some engineering issues that extend beyond the fence, such as the risk of explosion of those facilities and the escape of hazardous chemicals. The HRA does not generally include issues like sexually transmitted infection, except sometimes for travelling employees such as truck drivers, air crews and managers on overseas assignments. The baseline health condition of the peripheral communities is relevant to HRA to the extent that it affects occupational risk. For example, if there is a high prevalence rate of malaria in the local community then the workforce may be at risk of malaria infection. For further information see Chapter 5.

While HRA is largely 'within the fence', HIA is largely 'outside the fence'. This is illustrated in Figure 1.4. As the health of the workforce is, for the most part, managed

Figure 1.4 *Illustration of the roles of HRA and HIA in an infrastructure project*

Copyright Tropix.co.uk

through HRA, it does not require detailed attention during the HIA. Exceptions occur where issues 'permeate the fence'. For example, there are a number of health issues associated with worker and community interactions. The HIA is broader than the HRA and should identify and manage any gaps.

In the public sector, HRA may have a slightly different meaning in some settings. For example, the health sector may use the term to describe assessments of health interventions. In other settings, occupational risks may be explicitly incorporated in HIAs.

1.8.2 Health needs assessment

Health needs assessment (HNA) is simply an investigation into the health status of the existing community, a rational prioritization of needs and a management plan for meeting those needs (Wright et al, 1998; NICE, 2005; Wright and Kyle, 2006). It has elements in common with HIA. The main difference is that an HNA looks at existing needs and an HIA looks at future needs. HNA focuses on the existing community, HIA focuses on the proposal.

For example, in a poor rural community an HNA might observe high levels of child mortality and morbidity from diarrhoea. The community may not have a clean and reliable water supply and the local primary care clinic may not be functioning because of a lack of medicines. The HNA might recommend changes to water supply and clinic management.

There is synergy between the HNA and the HIA. Both have a scope, baseline/profile, followed by a gap analysis and further data collection. The data are then analysed, prioritized and recommendations are made and incorporated into a management plan. The distinction lies in the analysis and recommendations. The needs assessment is strictly concerned with the existing community and its existing needs and priorities. The impact assessment is strictly concerned with the additional effects associated with the development proposal. This can include the introduction of communities who are not currently present, such as camp followers and construction workers.

The needs assessment provides an input to the social investment plan. For example, an industrial plan located in a rural area may include a rural development programme, which should provide a set of positive impacts to the community. The impact assessment should affect the design and operation of the industrial plan. This should mitigate any negative impacts to both the existing and future community and can also provide health gains.

Figure 1.5 illustrates some of the connections between an HIA and an HNA in the context of a rural project in a low-income economy. Both assessments require the preparation of a baseline or profile that summarizes the existing health, social and environmental conditions in the community. The HNA then focuses on existing conditions and may make recommendations for social or health investment. These would normally be implemented in partnership with a specialist, local NGO that has the appropriate connections and expertise, or an appropriate institution, such as the local health department. One of the recommendations may be to stimulate local businesses and to create new businesses in order to reduce poverty and promote health. For example, micro-finance initiatives and vocational training may be provided. At the same time, the impact assessment is concerned with the future conditions that will arise as a result of the project. One of the recommendations may be to procure local goods and services whenever possible in order to embed the project in the local community and promote health gains through poverty reduction. At this point, the two processes are linked: the social investment programme stimulates local businesses that in turn provide the goods and services that the impact assessment recommends should be procured locally.

This has budgetary implications for the proposal. Separate budgets are needed for establishing the baseline conditions, mitigating the health impacts of the proposal, and implementing the social investment programme.

1.9 QUANTIFICATION

Much of HIA uses qualitative methods and there is a debate about the value of quantitative methods. Different kinds of quantification may be used in HIA. The simple kind consists of surveys of community opinion. Quantitative methods can be used to decide on sample

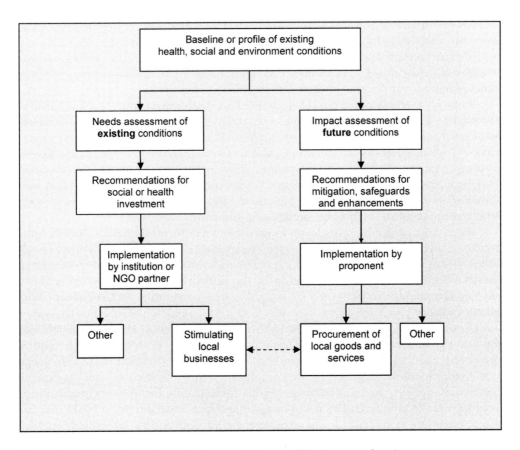

Figure 1.5 *Illustration of the links between HNA and HIA for a rural project in a low-income economy*

size and to count the frequency of responses. For example, representative samples of the target group are interviewed about their hopes and fears concerning the proposal. The results are presented as percentages with, or without, a standard error. For example, 10 per cent of a community may be asked what they think about an incinerator and 30 per cent of the sample may be concerned about air pollution. In addition, there are health statistics and special studies that portray baseline conditions in quantitative terms. This kind of quantification is descriptive but not predictive.

Predictive mathematical, statistical or dynamic models are greatly valued by some decision-makers because they appear to add objectivity and rigour, and enable different interventions to be compared. The numerical outputs can be used as inputs to further models, such as economic models. Modelling seems to be most appropriate in policy analysis, where there are very large populations and a statistical relationship between interventions and outcomes. Large resources of staff-months, public funding and academic interest may be needed in order to build and verify models. The opportunities for doing such modelling at project level are often limited. Models, and the pros and cons of quantification, are discussed further in Chapter 5.

Opponents of modelling may argue that a large number of simplifying assumptions are required and that a complex web of interactions underlies the association between a proposal and population health effect. Modelling may lend credibility to the assessment, but this may be misplaced because the figures produced are dependent on the assumptions made.

The literature review associated with an HIA should include published papers that report quantitative results. These will usually address a small area of the overall impact assessment, but do so in detail. They can be used to build an argument. For example, there is a large body of scientific research evidence on the association between particulate air pollution and population mortality rate. This can be used to estimate a reduction in mortality associated with a proposal that reduces air pollution. There is a danger that the small part of the assessment that has been quantified will be given more weight by decision-makers than the rest.

Much HIA analysis uses simple ranking in place of quantification. We want to assess whether a particular proposal will change a particular health outcome. The ranks might include: much better, better, worse, much worse or unchanged. We can also rank priorities: very high, low or insignificant. The rankings are justified by reasoned argument rather than numerical analysis. We might suggest that a recommendation would reduce the rank of an outcome from much worse to insignificant. The reasoned argument may be incorrect but providing it is made explicit, the reader is free to agree or disagree. In any case, as most proposals can be improved in order to safeguard human health it follows that many recommendations have some value, although they may not be cost effective.

1.10 ETHICS

HIA requires value judgement and this needs to be guided by a clear ethical framework. Several key ethical values have been proposed for the conduct of HIA (WHO European Centre for Health Policy, 1999; Scott-Samuel et al, 2001):

- democracy (participation in decision-making);
- equity;
- sustainable development;
- ethical use of evidence;
- adopting a comprehensive approach to health;
- respecting human rights.

In addition, the International Association for Impact Assessment has published a code of conduct for all impact assessors (IAIA, undated). Ethics is discussed in more detail in Chapter 12.

1.11 INTEGRATION AND FRAGMENTATION

In the case of projects, HIA is often undertaken at the same time as EIA and/or social impact assessment (SIA). The assessments may be made in parallel, or they may be integrated. Integrated assessments are sometimes referred to as environmental, social and health impact assessment (ESHIA) or environmental and social impact assessment (ESIA).

There is considerable overlap between health, social and environmental impact assessments. For example, poisoning associated with emissions is both a health and an environmental concern. Lack of employment, leading to mental stress, is both a health and a social concern. Water supply is a concern of all three. In addition, there are many similarities between health and social surveys. The overlap is represented by a Venn diagram in Figure 1.6. There are many components of the impact assessment that cannot be assigned logically to one of the three areas. How should a team decide which group addresses the overlapping issues? The decision must be made pragmatically, based on the skills, resources and timings available. However, this should not be construed as an opportunity for assessors with no health background to take responsibility for the HIA. There are many examples of impact assessment statements that contain paragraphs about health of dubious quality. Typical examples contain vague and general statements about 'health education' and curative medical services, or focus on toxicology to the exclusion of other health concerns.

The management of an integrated impact assessment is often controlled by an environmental consultancy company on behalf of a client. They must allocate resources to the three components and produce a balanced report. Collaboration should enable optimal sharing of resources. Some of the pitfalls are discussed in Chapter 4. The timing of the three assessments is important. For example, the three components should be undertaken in parallel and cross-referenced. The outputs of the EIA and SIA are often inputs to the HIA. So the HIA may have to be completed last (Birley, 2003).

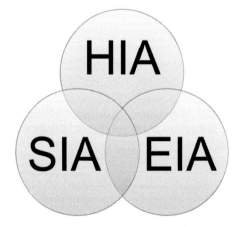

Figure 1.6 *HIA, EIA and SIA overlap*

HIA itself is increasingly subject to a process of fragmentation. The following are examples:

- health inequality impact assessment;
- health equity impact assessment;
- health systems impact assessment;
- environmental health impact assessment;
- mental well-being impact assessment.

There are arguments for and against such fragmentation. Each of these subtypes emphasizes a particular component of HIA and enables it to receive more attention. But do they achieve this at the expense of the whole? Does this detract from the overall aim of providing a holistic and balanced assessment? Is it a reflection of the institutional subdivisions that occur within the health sector and the competition for scarce resources by each division? How will such fragmentation be understood by observers from outside the health sector such as transport, education or agriculture? HIA is intended to be holistic and I do not support this fragmentation.

1.12 WHEN THINGS GO WRONG: BHOPAL

The assessment process in general provides an opportunity to safeguard communities and environments from the tragic consequences of mistakes. The chemical plant in Bhopal, India, formerly owned by the transnational corporation Union Carbide, provides a graphic example. In 1984, clouds of poisonous gas were released from a pressurized container to create one of the worst industrial incidents ever documented.

This is not strictly an example of HIA. Rather, it is an issue of health risk assessment. A buffer zone had been established around the plant at the time of construction. Over the years, this had been encroached upon by settlements. The release was the result of engineering failures that were entirely within the fence and under the control of the plant managers and their owners.

In June 2009, US congressmen sent a letter to the CEO of Dow Chemicals, the current owners of Union Carbide. While the incident occurred 25 years ago, the letter alleged that the company evaded responsibility before, during, immediately afterwards and still continued to do so (Bhopal Medical Appeal, 2009; Bhopal.net, 2009a).

It alleged that Union Carbide evaded responsibility before the incident as a result of inadequate technology, double standards and reckless cost cutting. It alleged that Union Carbide evaded responsibility during and in the aftermath of the incident by failing to admit responsibility; deliberate misinformation about medical issues; providing too little compensation too late; and evading trial in India.

It alleged that as a result of the incident:

- 500,000 people have been poisoned;
- 22,000 people have died;
- 15 people are still dying each month;
- the site remains heavily contaminated with chemicals that leach into the local water supplies and continue to poison the population.

It is further alleged that the company has spent more money on lawyers and public relations than it has spent on remediation and compensation.

The full story behind these allegations can be reviewed using the websites cited as well as the Bhopal website of Union Carbide itself (Union Carbide Corporation, 2008). One of the most disturbing allegations made concerns the complicity of the medical staff employed by Union Carbide (Bhopal.net, 2009b). The website asserts that Union Carbide's medical officer told questioners that methyl-isocyanate was only an irritant, like tear gas. It asserts that Union Carbide headquarters in the US continued to provide this advice while many people died.

What lessons can we learn from this example? They might include the following:

- the human consequences of (allegedly) placing profit before responsibility, allowing decisions to be made by lawyers rather than executive officers, taking poor risk management decisions and having double standards;
- the loss of reputation and licence to operate that can occur;
- the need for intent to implement recommendations made during assessments;
- the need to continue implementing the recommendations over the long term;
- the importance of openness and transparency;
- the need to have recovery procedures that function when accidents happen that give priority to protecting the community and only then to protect the institution responsible;
- the need for governments to have good regulation of industry and effective control over land zonation.

1.13 EXERCISES

Use the Internet to discover the views of the Asian and African Development Banks on the IFC Performance Standards.

Read the introductory sections of a few HIA guides retrieved from the Internet in order to get different perspectives on HIA.

1.14 REFERENCES

Acheson, D., D. Barker, J. Chambers, H. Graham, M. Marmot and M. Whitehead (1998) 'Independent inquiry into inequalities in health report', www.dh.gov.uk/en/publications andstatistics/publications/publicationspolicyandguidance/dh_4097582, accessed February 2011

Ashton, J. (ed.) (1992) *Healthy Cities*, Open University Press, Milton Keynes

APHO (Association of Public Health Observatories) (undated) The HIA Gateway, www.apho.org.uk/default.aspx?qn=p_hia, accessed July 2009

Banken, A. (2004) 'HIA of policy in Canada', in J. Kemm, J. Parry and S. Palmer (eds) *Health Impact Assessment, Concepts, Theory, Techniques and Applications*, Oxford University Press, Oxford

Bhopal Medical Appeal (2009) http://bhopal.org/index.php?id=23, accessed June 2009

Bhopal.net (2009a) 'International campaign for justice in Bhopal', www.bhopal.net, accessed July 2009

Bhopal.net (2009b) 'Union Carbide's medical response: Immediate aftermath', www.bhopal.net/bhopal.con/medical.html, accessed July 2009

Birley, M. H. (1991) 'Guidelines for forecasting the vector-borne disease implications of water resource development', World Health Organization, www.birleyhia.co.uk, accessed 2010

Birley, M. (2003) 'Health impact assessment, integration and critical appraisal', *Impact Assessment and Project Appraisal*, vol 21, no 4, pp313–321

Birley, M. (2005) 'Health impact assessment in multinationals: A case study of the Royal Dutch/Shell Group', *Environmental Impact Assessment Review*, vol 25, no 7–8, pp702–713, www.sciencedirect.com/science/article/B6V9G-4GVGT8V-1/2/01966b5af4f9ae9ecd390e4dd382a5a3

Birley, M. H. and G. L. Peralta (1992) *Guidelines for the Health Impact of Development Projects*, Asian Development Bank, Office of the Environment, Manila

Birley, M. H., M. Gomes and A. Davy (1997) 'Health aspects of environmental assessment', http://siteresources.worldbank.org/INTSAFEPOL/1142947-1116497775013/20507413/Update18HealthAspectsOfEAJuly1997.pdf, accessed October 2009

Birley, M. H., A. Boland, L. Davies, R. T. Edwards, H. Glanville, E. Ison, E. Millstone, D. Osborn, A. Scott-Samuel and J. Treweek (1998) *Health and Environmental Impact Assessment: An Integrated Approach*, Earthscan / British Medical Association, London

Burns, J. and A. Bond (2008) 'The consideration of health in land use planning: Barriers and opportunities', *Environmental Impact Assessment Review*, vol 28, no 2–3, pp184–197, www.sciencedirect.com/science/article/B6V9G-4P7FCR4-1/2/c3a98b071d3e41f8839505c3b5d7fe57

Butterfield, R., J. Henderson and R. Scott (2009) *Public Health and Prevention Expenditure in England*, Department of Health, London

CEC (Commission of the European Communities) (2007) 'White Paper, together for health: A strategic approach for the EU 2008–2013', EC, Brussels, http://ec.europa.eu/health-eu/doc/whitepaper_en.pdf, accessed 2010

CHETRE (Centre for Health Equity Training, Research and Evaluation) (undated) 'HIA Connect, building capacity to undertake health impact assessment', www.hiaconnect.edu.au/index.htm, accessed September 2009

Council of Europe (1950) 'European Convention on Human Rights', www.hri.org/docs/ECHR50.html, accessed July 2009

CSDH (Commission on Social Determinants of Health) (2008) 'Closing the gap in a generation: Health equity through action on the social determinants of health. Final report of the Commission on Social Determinants of Health', www.who.int/social_determinants/thecommission/finalreport/en/index.html, accessed July 2009

Cullingworth, B. and V. Nadin (2006) *Town and Country Planning in the UK*, 14th edition, Routledge, London

Dahlgren, G., P. Nordgren and M. Whitehead (eds) (1997) 'Health impact assessment of the EU common agricultural policy', Policy report from Swedish National Institute of Public Health

Dannenberg, A. L., R. Bhatia, B. L. Cole, S. K. Heaton, J. D. Feldman and C. D. Rutt (2008) 'Use of health impact assessment in the U.S: 27 case studies, 1999–2007', *American Journal of Preventive Medicine*, vol 34, no 3, pp241–256, www.sciencedirect.com/science/article/B6VHT-4RSS76V-C/2/094f349ebbf8eb230e003d37cf2548a2

DCLG (Department for Communities and Local Government) (2005) 'A practical guide to the strategic environmental assessment directive', www.communities.gov.uk/index.asp?id=1501988, accessed January 2007

DHSS (Department of Health and Social Security) (1980) 'Inequalities in health, report of a research working group', http://en.wikipedia.org/wiki/Black_Report, accessed September 2009

EC (European Commission) (1997) 'The Amsterdam treaty: A comprehensive guide', http://europa.eu/legislation_summaries/institutional_affairs/treaties/amsterdam_treaty/index_en.htm, accessed July 2009

EC HCPDG (European Commission Health and Consumer Protection Directorate General) (2004) 'European policy health impact assessment: A Guide', http://ec.europa.eu/health/ph_projects/2001/monitoring/fp_monitoring_2001_a6_frep_11_en.pdf, accessed July 2009

Eni corporation (undated) 'Guidelines on strategic health management', Eni E&P Division

Environmental Impact Assessment Review (undated), www.elsevier.com/wps/find/journaldescription. cws_home/505718/description#description, accessed January 2011

Equator Principles (2006) 'The Equator Principles', www.equator-principles.com, accessed October 2009

European Court of Human Rights (2003) 'Grand chamber judgement in the case of Hatton and others versus the United Kingdom', www.echr.coe.int/eng/Press/2003/july/JudgmentHatton GC.htm, accessed July 2009

European Parliament and the Council of the European Union (2001) 'Directive 2001/42/EC of the European Parliament and of the Council of 27 June 2001 on the assessment of the effects of certain plans and programmes on the environment', *Official Journal of the European Communities*, vol L197, pp30–37 , http://eur-lex.europa.eu/johtml.do?uri=oj:l:2001:197:som :en:html

Executive Office (2010) 'Solving the problem of childhood obesity within a generation, White House Task Force on Childhood Obesity, Report to the President', Executive Office of the President of the United States, Washington, DC, www.letsmove.gov/pdf/TaskForce_on_Childhood_Obesity_May2010_FullReport.pdf, accessed 2010

Fischer, T. (2007) *Theory and Practice of Strategic Environmental Assessment: Towards a More Systematic Approach*, Earthscan, London

Fischer, T. B., M. Matuzzi and J. Nowacki (2010) 'The consideration of health in strategic environmental assessment (SEA)', *Environmental Impact Assessment Review*, vol 30, no 3, pp200–210, www.sciencedirect.com/science/article/B6V9G-4XPXT0X-1/2/6c915b63f00e637144a4 192b836992c3

Government of Thailand (2007) 'National Health Act, BE 2550', National Health Commission Office, p44, www.nationalhealth.or.th, accessed February 2011

Harris, P., B. Harris-Roxas, E. Harris and L. Kemp (2007) *Health Impact Assessment: A Practical Guide*, Centre for Health Equity Training, Research and Evaluation (CHETRE), University of New South Wales, Sydney, www.health.nsw.gov.au

Harris, P. J., E. Harris, S. Thompson, B. Harris-Roxas and L. Kemp (2009) 'Human health and wellbeing in environmental impact assessment in New South Wales, Australia: Auditing health impacts within environmental assessments of major projects', *Environmental Impact Assessment Review*, vol 29, no 5, pp310–318, www.sciencedirect.com/science/article/B6V9G-4VTVJK2-2/2/1fd830648709abd9e78bac232eb4322e

Hastings, M. (2002) 'Neither Satan nor Santa: Shell, competitive advantage and stakeholders in the Peruvian Amazon', in T. de Bruijn and A. Tukker (eds) *Partnership and Leadership: Building Alliances for a Sustainable Future*, Kluwer, the Netherlands

HIACU (Health Impact Assessment Coordinating Unit) (2010) 'Thailand's rules and procedures for the health impact assessment of public policies', National Health Commission Office, Nonthaburi

HM Government (1998) 'Human rights', www.direct.gov.uk/en/Governmentcitizensandrights/Yourrightsandresponsibilities/DG_4002951, accessed July 2009

IAIA (International Association for Impact Assessment) (undated), www.iaia.org

ICMM (2010) *Good Practice Guidance on Health Impact Assessment*. International Council on Mining and Metals, London, www.icmm.com/document/792

IFC (International Finance Corporation) (2006) *Policy and Performance Standards on Social and Environmental Sustainability*, www.ifc.org/ifcext/enviro.nsf/Content/EnvSocStandards, accessed April 2008

IFC (2009) *Introduction to Health Impact Assessment*, IFC, Washington, www.ifc.org/ifcext/sustainability.nsf/AttachmentsByTitle/p_HealthImpactAssessment/$FILE/HealthImpact.pdf

IPIECA/OGP (2005) 'A guide to health impact assessments in the oil and gas industry', International Petroleum Industry Environmental Conservation Association, International Association of Oil and Gas Producers, London, www.ipieca.org

Kemm, J., J. Parry and S. Palmer (eds) (2004) *Health Impact Assessment: Concepts, Theory, Techniques and Applications*, Oxford University Press, Oxford

Kickbusch, I., W. McCann and T. Sherbon (2008) 'Adelaide revisited: From healthy public policy to health in all policies', *Health Promotion International*, vol 23 no 1, pp1–3, www.health.gov.au/internet/nhhrc/publishing.nsf/content/458/$file/458%20-%20h%20-%20sa%20health%20-%20kickbusch%20%20adelaide%20revisited.pdf

Local Government Improvement and Development (2010) *Introduction to Equality Impact Assessments*, www.idea.gov.uk/idk/core/page.do?pageId=8017174, accessed October 2010

Lock, K., M. Gabrijelcic-Blenkus, M. Martuzzi, P. Otorepec, P. Wallace, C. Dora, A. Robertson and J. M. Zakotnic (2003) 'Health impact assessment of agriculture and food policies: Lessons learnt from the Republic of Slovenia', *Bulletin of the World Health Organization*, vol 81, no 6, pp391–398, www.scielosp.org/scielo.php?script=sci_arttext&pid=S0042-96862003000600006&nrm=iso

MacNaughton, G. and P. Hunt (2009) 'Health impact assessment: The contribution of the right to the highest attainable standard of health', *Public Health*, vol 123, no 4, pp302–305, www.sciencedirect.com/science/article/b73h6-4w441fn-1/2/dd62787abded8202228f05ba8f948866

Marmot, M. (2010) *Fair Society, Healthy Lives: A Strategic Review of Health Inequalities in England post-2010*, Global Health Equity Group, UCL Research Department of Epidemiology and Public Health, www.ucl.ac.uk/gheg/marmotreview, accessed March 2010

McKeown, T. (1979) *The Role of Medicine: Dream, Mirage or Nemesis?* Blackwell, Oxford

Metcalfe, O. and C. Higgins (2009) 'Healthy public policy – is health impact assessment the cornerstone?', *Public Health*, vol 123, no 4, pp296–301, www.sciencedirect.com/science/article/B73H6-4VXJW0J-1/2/578b80443b0c537407737509e246925a

MIGA (Multilateral Investment Guarantee Agency) (undated) 'Insuring investments, ensuring opportunities', www.miga.org, accessed May 2010

NCCHPP (National Collaborating Centre for Healthy Public Policy) (2002) 'The Quebec Public Health Act's Section 54', www.ccnpps.ca/docs/Section54English042008.pdf, accessed October 2009

NICE (National Institute for Health and Clinical Excellence) (2005) 'Health needs assessment: A practical guide', www.nice.org.uk/aboutnice/whoweare/aboutthehda/hdapublications/health_needs_assessment_a_practical_guide.jsp, accessed November 2006

Nilprapunt, P. (2010) 'Notification of the Ministry of Natural Resources and Environment Thailand', Office of Natural Resources and Environmental Policy and Planning, Ministry of Natural Resources and Environment, Nonthaburi, www.thia.in.th

Nowacki, J., M. Martuzzi and T. B. Fischer (eds) (2010) 'Health and strategic environmental assessment, background information and report of the WHO Consultation Meeting, Rome 08./09. June 2009'. WHO Regional Office for Europe, Copenhagen, www.euro.who.int/en/what-we-do/health-topics/environmental-health/health-impact-assessment/publications/2010/health-and-strategic-environmental-assessment

OGP (2000) 'Strategic health management: Principles and guidelines for the oil and gas industry', International Association of Oil and Gas Producers, London, www.ogp.org.uk/pubs/307.pdf

OHCHR (Office of High Commissioner for Human Rights) (1948) 'United Nations: Human rights', www.ohchr.org, accessed July 2009

OPIC (Overseas Private Investment Corporation) (undated) 'Political risk insurance', www.opic. gov, accessed May 2010

OUP (Oxford University Press) (2010) 'Oxford dictionaries online', www.oxforddictionaries. com/page/askoxfordredirect, accessed August 2010

Pengkam, S. and S. Decha-umphai (eds) (2009) 'Chiang Mai declaration on health impact assessment for the development of healthy societies in the Asia-Pacific region'. Health Impact Assessment Coordinating Unit, Nonthaburi, www.nationalhealth.or.th

Phoolcharoen, W., Sukkumnoed, D. and Kessomboon, P. (2003) 'Development of health impact assessment in Thailand: Recent experiences and challenges', *Bulletin of the World Health Organization*, vol 81, no 6, pp 465–467, www.scielosp.org/pdf/bwho/v81n6/v81n6a20.pdf

Quigley, R., L. den Broeder, P. Furu, A. Bond, B. Cave and R. Bos (2006) 'Health impact assessment: International best practice principles', Special publication series No. 5, www.iaia. org, accessed September 2007

Ritsatakis, A. (2004) 'HIA at the international policy-making level', in J. Kemm, J. Parry and S. Palmer (eds) *Health Impact Assessment: Concepts, Theory, Techniques and Applications*, Oxford University Press, Oxford

Roscam Abbing, E. W. (2004) 'HIA and national policy in the Netherlands', in J. Kemm, J. Parry and S. Palmer (eds) *Health Impact Assessment: Concepts, Theory, Techniques and Applications*, Oxford University Press, Oxford

Royal Commission on Environmental Pollution (2007) 'Twenty sixth report: The urban environment', www.rcep.org.uk/reports/26-urban/26-urban.htm, accessed February 2011

RTPI (Royal Town Planning Institute) (2009) 'RTPI good practice note 5: Delivering healthy communities', www.rtpi.org.uk/item/1795/23/5/3, accessed April 2009

Scott-Samuel, A., M. Birley and K. Ardern (2001) 'The Merseyside Guidelines for health impact assessment', www.liv.ac.uk/ihia/IMPACT%20Reports/2001_merseyside_guidelines_31.pdf, accessed January 2007

Scott-Samuel, A., K. Ardern and M. H. Birley (2006) 'Assessing health impacts on a population', in D. Pencheon, C. Guest, D. Melzer and J. Grey (eds) *Oxford Handbook of Public Health Practice.* Oxford University Press, Oxford

Shell UK (2009) 'UK social investment', www.shell.co.uk/home/content/gbr/respsonsible_ energy/shell_in_the_society/social_investment, accessed July 2009

Smith, K. E., G. Fooks, J. Collin, H. Weishaar, S. Mandal and A. B. Gilmore (2010) '"Working the system" – British American Tobacco's influence on the European Union Treaty and its implications for policy: An analysis of internal tobacco industry documents', *PLoS Medicine*, vol 7, no 1, ppe1000202, doi:10.1371%2Fjournal.pmed.1000202

Ståhl, T., M. Wismar, E. Ollila, E. Lahtinen and K. Leppo (eds) (2006) 'Health in all policies: Prospects and potentials', Ministry of Social Affairs and Health, Finland, www.euro.who.int/ document/E89260.pdf

Steinemann, A. (2000) 'Rethinking human health impact assessment', *Environmental Impact Assessment Review*, vol 20, pp627–645, www.ingentaconnect.com/content/els/01959255/200 0/00000020/00000006/art0006; doi:10.1016/S0195-9255(00)00068-8

Thérivel, R., E. Wilson, S. Thomson, D. Heaney and D. Pritchard (1992) *Strategic Environmental Assessment*, Earthscan, London

UNDP (2008) Human development reports, http://hdr.undp.org/en/statistics/data, accessed 2008

UNECE (2003) 'Protocol on Strategic Environmental Assessment to the Convention on Environmental Impact Assessment in a Transboundary Context, Kiev', www.unece.org/env/ eia/sea_protocol.htm, accessed September 2009

UNECE (2010) Chapter A3 'Determining whether plans and programmes require SEA under the Protocol', www.unece.org/env/eia/sea_manual/chapterA3.html accessed August 2010

Union Carbide Corporation (2008) 'Bhopal information centre', www.bhopal.com, accessed July 2009

Wanless, D., M. Beck, J. Black, I. Blue, S. Brindle, C. Bucht, S. Dunn, M. Fairweather, Y. Ghazi-Tabatabai, D. Innes, L. Lewis, V. Patel and N. York (2002) *Securing Our Future Health: Taking a Long-Term View. Final Report*, www.hm-treasury.gov.uk./Consultations_and_Legislation/wanless/consult_wanless_final.cfm, accessed September 2007

WHIASU (2004) *Improving Health and Reducing Inequalities: A Practical Guide to Health Impact Assessment*, Welsh Health Impact Assessment Support Unit, Cardiff, www.wales.nhs.uk/sites3/Documents/522/improvinghealthenglish.pdf

WHIASU (undated) Wales Health Impact Assessment Support Unit, www.wales.nhs.uk/sites3/home.cfm?OrgID=522, accessed May 2010

WHO (World Health Organization) (1986) 'Ottawa Charter for Health Promotion', www.euro.who.int/AboutWHO/Policy/20010827_2, accessed September 2009

WHO (1988) 'Adelaide Recommendations on Healthy Public Policy', www.who.int/hpr/NPH/docs/adelaide_recommendations.pdf, accessed September 2009

WHO (1995) 'WHO healthy cities: A programme framework, a review of the operation and future development of the WHO healthy cities programme', World Health Organization, Geneva

WHO (2001) 'Health impact assessment in strategic environmental assessment', World Health Organization, Regional Office for Europe, Copenhagen, www.who.int/hia/network/en/HIA_as_part_of_SEA.pdf , accessed February 2011

WHO (2005) 'Health and the Millennium Development Goals', www.who.int/mdg/publications/en, accessed September 2009

WHO European Centre for Health Policy (1999) 'Health impact assessment, main concepts and suggested approach, Gothenburg consensus paper', www.who.dk/hs/echp/index.htm, accessed August 2000

WHO Regional Office for Europe (2004) 'Fourth Ministerial Conference on Environment and Health, Declaration, Budapest', www.euro.who.int/document/e83350.pdf, accessed September 2009

WHO Regional Office for Europe (2005) 'Healthy cities', www.euro.who.int/en/what-we-do/health-topics/environmental-health/urban-health/activities/healthy-cities, accessed February 2010

Wilkinson, R. and M. Marmot (eds) (2003) 'Social determinants of health: The solid facts', WHO Regional Office for Europe, Copenhagen, www.euro.who.int/__data/assets/pdf_file/0005/98438/e81384.pdf, accessed 2011

Williams, C. and P. Fisher (2007) 'Draft guidance on health in strategic environmental assessment: A consultation', Department of Health, www.dh.gov.uk/en/consultations/closedconsultations/dh_073261

Williams, C. and P. Fisher (2008) 'Draft guidance on health in strategic environmental assessment: Response to consultation', Department of Health, www.dh.gov.uk/en/Consultations/Responsestoconsultations/DH_083716

Wismar, M., J. Blau, K. Ernst and J. Figueras (eds) (2007) 'The effectiveness of health impact assessment', Brussels, European Observatory on Health Systems and Policies, Brussels, www.euro.who.int/__data/assets/pdf_file/0003/98283/E90794.pdf, accessed February 2011

Wright, J. and D. Kyle (2006) 'Assessing health needs', in D. Pencheon, C. Guest, D. Melzer and J. Grey (eds) *Oxford Handbook of Public Health Practice*, Oxford University Press, Oxford

Wright, J., R. Williams and J. R. Wilkinson (1998) 'Health needs assessment: Development and importance of health needs assessment', *British Medical Journal*, vol 316, pp1310–1313, www.bmj.com/cgi/content/full/316/7140/1310

CHAPTER 2

Health and its determinants

In this chapter you have an opportunity to study:

- The nature of health and health determinants.
- Classification systems that are useful in HIA.
- Health inequality.
- Types of health impacts.
- Disability adjusted life years.
- Health transitions associated with economic development.
- A mind-mapping exercise.

2.1 WHAT IS HEALTH?

Many people identify health with illness. The formal definition of health, however, is equally concerned with well-being. There are a number of different kinds of well-being that are included in the World Health Organization (WHO) definition of health (WHO, 1946). The following definition is based on that, with one modification: 'Health is not merely the absence of disease and infirmity, but a state of complete physical, mental, social, and spiritual well-being'. The original definition did not include spiritual well-being, although many people now do so (this is discussed further in Chapter 12).

Health is a resource for everyday life, not the objective of living. Health is a positive concept emphasizing social and personal resources, as well as physical capacities (WHO, 1986).

Health is a property of each individual but it is also useful to measure the average health of a defined population. The unit of measure for population health is a rate: the number of people with a particular condition divided by the total number of people in the population. This often has to be disaggregated by age, sex, location and socio-economic group, because the population is composed of different community groups. For example, the obesity rate may vary between children and adults.

When people are asked to list the characteristics of a healthy person or population, they can usually think of many. Table 2.1 contains examples of a mixture of different kinds of health characteristics. The following sections provide a classification system.

Table 2.1 *Some characteristics of a healthy person and population*

Individual health	Population health
Physical fitness and regular exercise	Long expectation of life
Healthy diet	Low levels of crime and disorder
Freedom from disease or disability	A pleasant physical environment to live and work in
Calm mental state	
Good social relationships	Absence of poverty
Satisfactory employment and income	Availability of food
Freedom from fear	Low prevalence rates of disease
Freedom from drug addiction, including nicotine, caffeine and alcohol	Low maternal and infant mortality rates
	Access to affordable health care for everyone
Stable and equal intimate relationships	Low risk of disease or injury
	Social insurance against illness and disability

There are a number of models of health, including the following.

2.1.1 Biomedical model of health

The biomedical model is concerned with the presence or absence of disease. It is a model of disease causation and is generally used in the context of the provision of health services. People who use this model would populate Table 2.1 with a list of diseases. It is familiar: we visit the doctor when we feel sick and expect medicines that will make us feel better.

2.1.2 Socio-environmental model of health

The socio-environmental model is concerned with the root economic, social, psychological and environmental causes of illness and well-being. It is reflected in the items in the table that are not directly associated with disease, such as the absence of poverty. The HIA method described in this book is based on the socio-environmental model of health. The items in Table 2.1 can be classified as a mixture of health determinants and health outcomes.

2.2 HEALTH DETERMINANTS AND OUTCOMES

HIA is a holistic and integrative activity that includes all the health concerns reasonably associated with a proposal. In order to progress, we need a way of classifying these (ontology). As yet, there is no universally agreed system for HIA. One method is to distinguish between health determinants and health outcomes. Figure 2.1 illustrates this. It is a simple model of HIA.

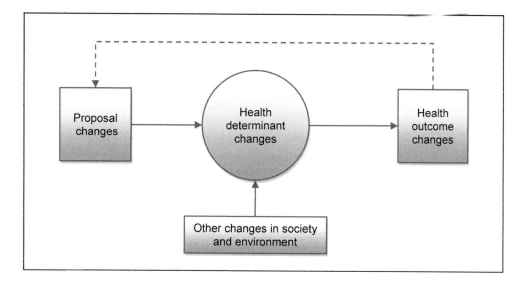

Figure 2.1 *Association between proposals, health determinants and health outcomes*

Table 2.2 *Examples of health determinants*

Category	Examples
Physiological	Age, sex, genetics
Behavioural	Exercise, smoking
Environmental	Pollution, economy, clean water supply, low crime rates
Institutional	Utilities supplying water, police, medical care

Health outcomes include medically defined states of disease and disability, as well as community-defined states of well-being. Examples include malaria, injury, asthma and fear.

Health determinants are the factors that cause these outcomes. For example, the presence of ozone in cities is a determinant of asthma. They are factors that are known or postulated to be causally related to states of health. Health determinants may also be referred to by many other names including risk factors or confounding factors. Health determinants are described in detail below and Table 2.2 provides examples.

Other changes are always taking place in the environment and society that are independent of the proposal. These also change health determinants and health outcomes. Consequently, the causal relationship between the proposal and health outcomes is often indistinct. To overcome this methodological challenge, HIA focuses on health determinants.

Feedback occurs as changes in health outcomes have an effect on the proposal itself either before or after implementation.

Example of feedback

An irrigation project was established on the Tana River in Kenya. Farmers were brought from the highlands, where there was little malaria, to the project, where malaria was common. The farmers became sick with malaria and were unable to farm, causing major project setbacks (Smith, 1984).

2.3 HEALTH OUTCOMES

There are many ways of categorizing health outcomes. For example, the International Classification of Diseases (ICD) is a system that is largely based on anatomical and organ groupings (WHO, 1977). The categories indicated in Table 2.3 have proved useful in HIA and are easily comprehensible to non-health specialists. Each category is fully explained below. In addition to biomedical conditions, the list includes well-being and psychosocial disorders.

Table 2.3 *Examples of the main categories of health outcomes that arise in a project*

Main categories of health outcomes	Examples
Communicable diseases (CD)	STI / HIV/AIDS, respiratory infections, malaria and diarrhoea
Non-communicable diseases (NCD)	Acute and chronic poisoning from hazardous chemicals and minerals, cardiovascular and dust-induced lung disease
Nutritional problems	Protein-energy and micronutrient deficiencies and excesses; food safety
Injuries	Associated with traffic, drowning, violence
Mental illness and psychosocial disorders	Suicide, depression, communal violence, drug abuse, stress, fear of disasters, compulsive gambling, compulsive sex
Well-being	Happiness, fulfilment, well-being, supportive communities and families

The top ten causes of death in the world are ranked in Table 2.4 (WHO, 2010c). Overall, causes of death are about 31 per cent communicable disease (CD), 60 per cent non-communicable disease (NCD) and 9 per cent injury. The data can be disaggregated by age, income, sex and region.

Statistics for morbidity (ill health) are different. In many localities, communicable diseases are a more common cause of morbidity and mortality than non-communicable diseases. See also Table 2.12 in this chapter. Death statistics are easier to record than

Table 2.4 *Leading global causes of death*

World deaths in millions	% of deaths (rounded)	Category
Coronary heart disease	12	NCD
Stroke and other cerebrovascular diseases	10	NCD
Lower respiratory infections	7	CD
Chronic obstructive pulmonary disease	5	NCD
Diarrhoeal diseases	4	CD
HIV/AIDS	4	CD
Tuberculosis	3	CD
Trachea, bronchus, lung cancers	2	NCD
Road traffic accidents	2	Injuries
Prematurity and low birth weight	2	NCD

illness statistics, although the cause of death can be mis-classified. For every death there are many cases of illness that cause suffering, pain, anxiety, distress, disability and financial burden.

2.3.1 Communicable diseases

The communicable or infectious diseases are different to the other health outcomes because they are self-perpetuating. Infectious agents, such as viruses, bacteria and protozoa, are transmitted from one person to another and multiply. Consequently, the number of people infected can grow exponentially and so can the risk. Influenza is an example. There are many modes of transmission of communicable diseases. These include air, water, food, vector (see below) and human contact (including sexually transmitted infections). Some are transmitted from human to human while others are transmitted from animals (zoonoses). Communicable diseases tend to be more common in poor countries and this is explained in Section 2.9. Vulnerability is increased by malnutrition, poverty, behaviour, poor environments and the presence of other disease.

2.3.1.1 Vector-borne diseases

Vector-borne diseases require an arthropod or other invertebrate host, such as a mosquito, for transmission. Transmission usually depends on the warm climates found in tropical and semi-tropical regions. In some cases, the pathogen comes from an animal or bird reservoir. Small changes in the physical environment associated with a proposal can affect the distribution and abundance of vectors and the animal reservoir. Control can include environmental management (Birley, 1991). Some examples are presented in Table 2.5. The aquatic snail associated with schistosomiasis is included for simplicity, although it is strictly an intermediate host and not a vector. Many are associated with water and these are described in more detail in chapter 9.

Table 2.5 *Examples of vector-borne diseases*

Vector-borne disease	Notes on environmental determinants
Malaria	Transmitted by anopheline mosquitoes and usually associated with rural habitats in warm climates. In each endemic country there are one or two principal mosquito species. Each species has specific breeding requirements in water which may be shaded or unshaded, stagnant or running, vegetated or unvegetated, fresh or saline. They generally do not breed in water that is contaminated with organic waste. Species have different blood-feeding habits: indoor/outdoor, animal/human preference.
Dengue	Transmitted by aedine mosquitoes and usually associated with urban habitats in warm climates. The breeding site is a container of rainwater. These vectors feed during daylight both indoors and outdoors and have a strong preference for human blood.
Japanese encephalitis	Transmitted by culicine mosquitoes. Pathogen reservoir in aquatic birds and pigs.
Schistosomiasis (bilharzia)	The intermediate hosts are aquatic snails associated with rural reservoirs, ponds, dam sites and irrigation systems in some tropical regions. They like shallow, vegetated margins and can be found underneath stones.

2.3.1.2 Water- and food-borne diseases

A large number of parasites, bacteria and viruses may be ingested with water or food. They are often classified according to their persistence in the environment. For example, *Ascaris* eggs persist for many months and may be ingested from soil or from uncooked plant surfaces. The eggs are excreted in human faeces and can pass through sewage treatment plants. Other pathogens are transmitted by direct contamination of food and water with faeces. These include the infectious hepatitis virus and the *Salmonella* bacterium. These diseases may be associated with food, agriculture, aquaculture, animal husbandry and organic waste disposal projects in warm climates. See Chapter 9 for more detail.

Acute diarrhoeal disease is a leading cause of infant and child mortality and morbidity in poor communities, especially in warm climates. There are many sources of infection. Neighbourhoods with clean water supplies and functioning toilets tend to have lower rates than those without. Contributing factors include hand-washing behaviour, early weaning and inappropriate treatment. Early weaning may be a function of the economic pressures on working mothers. Many of these determinants can be changed by proposals.

2.3.1.3 Sexually transmitted infections (STIs)

A number of diseases are transmitted by human sexual contact and these include HIV/AIDS, gonorrhoea, syphilis and chlamydiasis. Infection rates can be as high as 75 per cent among commercial sex workers in poor countries. High infection rates in the active working population can undermine economic development, health care, child survival and social structure. AIDS can aggravate other infections, such as tuberculosis. Large infrastructure construction projects in poor countries have a huge potential to contribute to STI epidemics. Indeed, all proposals that require temporary migration of men or women away from their families have this potential. Screening new employees for HIV infection is prohibited by international agreements. There is a need for prevention, care, and impact mitigation that targets vulnerable communities. Typical project components with high HIV risks include labour camps, labour migration, involuntary resettlement, tourism, marketing of agricultural produce and long-distance trucking. Countering the epidemic involves critical analysis of national responsiveness, identification of all the health determinants and cultural sensitivity. Many countries are in denial about the challenge and this makes matters worse.

HIV/AIDS in Africa

There are at least 28 million HIV infections in Africa south of the Sahara. These represent 70 per cent of the world total, and in some countries up to 25 per cent of adults in the age group 15–45 are infected. It is the primary cause of death amongst adults. About 12 million children are orphans. The economic impact is 0.5–1.2 per cent of GDP. The fight against the epidemic is one of Africa's top health priorities.

The root determinants of HIV risk include poverty and inequality. To these may be added the determinants associated with development projects:

- mobile;
- men;
- money;
- mixing.

HIV epidemics affect new infrastructure development in several ways, including:

- risk of infection of workers;
- reduced workforce availability;
- risk to reputation, if the development accelerates the epidemic.

One solution is to establish promotional programmes emphasizing abstinence and faithfulness, but these are not enough. Condoms must also be readily available and the knowledge of their correct use must be widespread. The recommendations for managing this issue include:

- abstinence;
- be faithful;
- condoms;
- denial – overcoming it.

2.3.1.4 Respiratory infections

Acute respiratory infections include colds, sinusitis, bronchitis, influenza and pneumonia. They are important causes of child mortality in poor communities. They include infections caused by other infectious diseases such as measles, whooping cough and varicella. Typical determinants include indoor and outdoor air pollution, overcrowding and poor ventilation.

Chronic respiratory infections include tuberculosis. Determinants include overcrowding, poor ventilation, poverty, alcoholism, drug abuse, homelessness, undernutrition and poor compliance with treatment. Successful medical treatment requires a complex and long-term drug regime.

Proposals can have a positive health impact on respiratory infections by reducing overcrowding and air pollution, and increasing wealth. Construction camps for large infrastructure projects should be designed and operated to international standards that prevent overcrowding. In poor countries, there are often large populations of camp followers that are attracted to construction camps and who live in unsanitary and overcrowded conditions where an increase in communicable disease transmission will occur.

2.3.2 Non-communicable diseases

Non-communicable diseases cannot be transmitted from one person or animal to another. They are not infectious. Consequently, they do not have the potential to multiply in the community and cause an epidemic. On the other hand, large numbers of community members may be exposed to environmental and social determinants that can cause high or low rates of morbidity or mortality. The amount of exposure varies, so that some diseases are apparent shortly after exposure, while others may not be apparent for many years. Some are associated with genetic or hereditary factors, gender and ethnicity. Some are associated with lifestyle factors such as improper diet, smoking or lack of exercise and these are referred to as the 'diseases of affluence'.

Health protective factors (such as emotional resilience), together with social, economic and environmental determinants (such as income, education, living and working conditions), determine differences in exposure and vulnerability. Many health determinants influence health opportunities, health-seeking and lifestyle behaviours as well as onset, expression and outcome of disease (WHO, 2006a). There are interactions between the diseases. For example, diabetes is a cause of cardiovascular disease.

Leading risk factors in Europe

In Europe, for example, about 60 per cent of the disease burden is accounted for by seven leading risk factors: high blood pressure (13 per cent); tobacco (12 per cent); alcohol (10 per cent); high blood cholesterol (9 per cent); overweight (8 per cent); low fruit and vegetable intake (4 per cent); and physical inactivity (4 per cent) (WHO, 2006a).

Examples of diseases influenced by genetics include diabetes, cardiovascular disease, many cancers, schizophrenia and Alzheimer's disease. Gene–environment interactions may play a major role. Examples of diseases associated with ethnicity include type 2 diabetes mellitus, which is up to six times more common in people of South Asian descent and up to three times more common among those of African and African-Caribbean origin (WHO, 2006a).

Gender has many influences on roles in society and consequently on exposure to risks, health-seeking behaviour, access to and control over resources, and decision-making needed to protect one's health. This leads to inequitable patterns of health risk, access to health services, use of health services and health outcomes.

Non-communicable diseases can result from the ingestion, inhalation or absorption of toxic chemicals, including pesticides, minerals and heavy metals. They may be associated with air-, water- or food-borne pollutants, poor occupational safety or poor domestic storage. The symptoms may be acute or chronic. For example, acute poisoning from pesticides can lead to acetyl cholinesterase inhibition. The symptoms of such inhibition include dizziness, weakness and coma, with severity depending on dose. Chronic neurological symptoms include blurred vision, dizziness, numbness or headache, and superficial or deep sensory loss (Amr et al, 1993).

Vulnerability to pollutants is increased by malnutrition, communicable disease and human behaviour. In addition, pollution can increase susceptibility to communicable disease. Non-communicable diseases sometimes require long latent periods and may be associated with many sub-acute exposures to toxic substances. The substances themselves may be associated with well-defined point sources such as chimneys. Control may depend on a general reduction in levels of use, emission and exposure. Chemical pollution can damage the resource base and destroy fisheries, make water unsuitable for irrigation, or damage crop vegetation and reduce agricultural productivity. This can lead to malnutrition due to food shortage.

Non-communicable diseases are often chronic conditions that cause dysfunction, or impairment in the quality of life, and they may develop over relatively long periods, at first without symptoms. After disease manifestations develop, there may be a protracted period of impaired health (Breslow and Cengage, 2002–2006). They also include injuries, which have an acute onset, but may be followed by prolonged convalescence and impaired function, as well as chronic mental diseases. In addition, these diseases cause pain, disability, loss of income, disruption of family stability and an impaired quality of life.

2.3.3 Nutritional disorders

Standard indices of nutritional status include weight for age, weight for height, height for age and body mass index. Body mass index is calculated as weight (kg) divided by height squared (m^2). The normal adult range is 18.5–25. Obesity is usually defined as a body mass index of greater than 30. There are a number of other indices, for example skin fold thickness.

Nutritional problems can be subdivided into under-nutrition and over-nutrition. Under-nutrition is associated with poverty, micronutrient deficiencies, food availability, food security, food safety, infection, workload, feeding practices and poorly regulated markets. Over-nutrition is associated with wealth, personal behaviour, junk food, food marketing practices and poorly regulated food industries, among other factors.

Undernourished people are more susceptible to communicable disease. The disease, in turn, may reduce their ability to assimilate whatever food is available. The effects of under-nutrition include less than average weight or height, blindness, cretinism, anaemia and poor skin conditions. About 20 per cent of children in developing regions are underweight (de Onis et al, 2004). In many regions the rate has been falling, but in the African region it is expected to increase. The main determinants are poverty, political instability and HIV/AIDS prevalence rates. There are more than 100 million underweight children in the world.

Proposals can have both positive and negative impacts on under-nutrition by changing food security and food availability. There may be unequal entitlements to food within the household that are affected in complex ways by development. See Section 2.8.2 for an example.

In wealthier communities and countries, the phenomenon of overweight and obesity is increasingly important (WHO, 2006b). A number of non-communicable diseases are associated with these conditions, including late onset diabetes, cardiovascular disease and various cancers. Determinants include:

- a global shift in diet towards increased intake of energy-dense foods that are high in fat and sugars but low in vitamins, minerals and other micronutrients;
- a trend towards decreased physical activity due to the increasingly sedentary nature of many forms of work, changing modes of transportation and increasing urbanization.

The HIAs of proposals that include food outlets and transport systems in wealthier communities are likely to include recommendations for managing obesity. Junk food outlets have permeated our societies and can be found in sports centres, hospitals, entertainment centres, schools and playgrounds.

Recent popular films document the association between the marketing practices of transnational food distributors and the obesity epidemic (Spurlock, 2004; Armstrong, 2005). There are many rigorous studies of obesity (Butland et al, 2007). In England, about 25 per cent of adults and 10 per cent of children are obese. By 2025, some 40 per cent of Britons are expected to be obese. In the US, one in three children are overweight or obese (Executive Office, 2010). Globally at the time of writing, some 1.6 billion adults are overweight and 400 million are obese.

Poor countries often face a double burden of disease associated with under-nutrition and over-nutrition. The two may exist in different communities such as the wealthier and poorer, the more urban and more rural (Birley and Lock, 1999). A nutritional transition can occur when the undernourished rural people adopt an urban diet with 'superior grains', more milled and polished grains, food higher in fat, more animal products and sugar, food prepared away from home and more processed food (Popkin, 1996).

2.3.4 Injuries

Injuries are subdivided into intentional and unintentional categories. Unintentional injuries include falls, motor vehicle traffic accidents, burns and scalds. Intentional injuries include homicides, suicides and violence, e.g. domestic violence.

Occupational injury is an important component but it will be excluded from the discussion as it lies outside the realm of HIA and is managed by other processes. However, low standards of occupational health and safety are commonplace in many countries and the results can spread out into the community. For example, an injured worker may no longer be able to support his or her extended family.

Globally, injury and violence are responsible for about 1 million child deaths (less than 18 years old) per year (Peden et al, 2008). The most common causes are road traffic injuries, drowning, fire-related burns, falls and poisoning. For every child killed, approximately 45 require hospitalization and 1300 are seen in an emergency department and discharged. Approximately 6 per cent of cases are attributed to homicide, 2 per cent to war and about 4 per cent to self-inflicted injuries. The most common types of unintentional injuries requiring admission to a health facility are intracranial, open wound, poisoning, fractures and burns. There are strong associations with poverty. For example, poor people may cook using open flames on unstable platforms at child height. Many injuries are avoidable through legislation, regulation and enforcement, product modification, environmental modification and provision of simple equipment.

Many surveys have found a high incidence of wife abuse (Heise, 1992). For example, 40 per cent of wives in one survey were 'beaten regularly'. Rates are high in all stages of economic development. Statistics are increasingly available about general violence against women and girls, including sexual violence (Watts and Zimmerman, 2002).

In the EU, injuries kill about 250,000 people per year and are the leading cause of death in children and young adults, consuming about 10 per cent of hospital resources (EUROSAFE, 2007). There are approximately 200 non-fatal injuries for each fatal injury and traffic injuries account for about 20 per cent of fatal and 6 per cent of non-fatal injuries. The most common setting is home and leisure. Male injury deaths are higher than female injury deaths in all age groups, rising sharply in young adults. Suicide and self-inflicted injuries account for 24 per cent, motor-vehicle traffic accidents for 21 per cent and accidental falls for 19 per cent. In young adults, motor-vehicle traffic accidents account for 51 per cent of fatal injuries.

The Injury Observatory of Britain and Ireland (IOBI) compile statistics for intentional and unintentional injury and death (IOBI, 2003). In 2002/2003 some 21,745 deaths were recorded. The highest rates were in Ireland followed by Scotland. Unintentional deaths comprise 65 per cent of all injury deaths. The unintentional injury deaths were attributed to falls (28 per cent), motor-vehicle traffic accidents (26 per cent), unintentional poisoning (7 per cent), fire/flame (3 per cent) and drowning (2 per cent). Intentional deaths were classified as follows as a percentage of all injury deaths: homicides (1 per cent), suicides (22 per cent) and undetermined (12 per cent).

There are many links between injury and psychosocial disorder. Proposals that affect the quality of life of disadvantaged groups may affect the incidence of violent behaviour.

2.3.5 Mental illness and psychosocial disorders

The term psychosocial refers to the psychological and social factors that influence mental health. Social influences such as peer pressure, parental support, cultural and religious background, socio-economic status and interpersonal relationships all help to shape personality and influence psychological make-up. Individuals with psychosocial disorders frequently have difficulty functioning in social situations and may have problems communicating effectively with others (Ford-Martin, 2002). Associated issues include substance abuse, psychosis (e.g. delusions), mood (e.g. depression), anxiety, sexual and gender identity, eating (e.g. anorexia), adjustment (excessive reaction to emotional event), personality (e.g. paranoia) and somatoform (physical symptoms that cannot be explained by a medical condition).

The determinants are many and complex and include biological, genetic, familial and social.

According to WHO, mental, neurological and behavioural disorders are common in all countries (WHO, 2010d). People with these disorders are often subject to social isolation, poor quality of life and increased mortality. It is estimated that 154 million people globally suffer from depression, 25 million people from schizophrenia, 91 million people are affected by alcohol-use disorders, 15 million by drug-use disorders, 50 million people suffer from epilepsy and 24 million from Alzheimer's and other dementias. Many people suffer neurological consequences of nutritional disorders and injuries. Mental illness affects and is affected by communicable and non-communicable disease. It is estimated that about 877,000 people die by suicide every year. Mental illness such as depression and anxiety is prevalent during pregnancy and maternity with consequences for infants and children (WHO, 2010b).

Expenditure on mental health is often a disproportionately small part of national health budgets, especially in low- and middle-income countries. Human rights abuses of patients are common (WHO, 2010e).

Proposals of all kinds produce change, stress and uncertainty and these are determinants of psychosocial disorders. For example, large influxes of migrant workers may lead to increased use of alcohol and other addictive drugs, undermining of local political systems and social values, or decreases in social capital.

2.3.6 Well-being

The definition of health used in this book subdivides well-being into physical, social, mental and spiritual categories. Incorporation of spiritual well-being is relatively new and is discussed further in Chapter 12. Well-being can be understood by considering its determinants and these are discussed in the next section.

Well-being is associated with quality of life and happiness. There is much debate about what constitutes a good life, which the ancient Greeks referred to as eudemonia. It has long been a subject of moral philosophy (Crisp, 2008).

The term quality of life is used in a comparative form in international development. The United Nations Development Programme has a Human Development Index that combines measures of life expectancy, education and standard of living at a national level.

In health care, quality of life often refers to an individual's emotional, social and physical well-being, including his or her ability to function in the ordinary tasks of living. It may be used in systems that require health care rationing or in the context of measuring chronic disability. Quality of life metrics have been developed such as quality adjusted life years (QALYs). These are used in economic cost-effectiveness analysis to allocate scarce resources.

There has been much research in recent years on happiness (Layard, 2005) and on positive psychology (Seligman, 2002). The new economics foundation (nef) promotes alternative measures of national economic progress based on happiness and well-being (nef, undated). It considers what policy-making and the economy would look like if the main aim were to promote well-being. It has developed an index that combines sustainable use of natural resources and happiness. International comparisons suggest that high levels of resource consumption are not a reliable indicator of well-being. The UK, for example, is ranked 74th out of 143 countries on nef's index. Research on happiness tends to conclude that it is derived mostly through connection and mindfulness. Connection with other people is derived from social relationships, cooperation, participation, engagement with the community, and assisting friends and strangers. Mindfulness, or taking notice, is derived from awareness of the present, the emotions, the good things in our lives and the world around us. HIA can benefit from this research by considering whether proposals encourage social connection, increase time for reflection, and improve the physical and social environment.

In England, evidence-based checklists and mental well-being impact assessment tools have been developed to assist with commissioning, delivering or developing proposals (National Mental Health Development Unit, 2010). This sub-field of HIA is very active.

2.4 DETERMINANTS OF HEALTH

Health determinants are factors that influence our state of health. The factors are personal, social, cultural, economic and environmental. They include our physical environment, income, employment, education, social support and housing. They act in unison and are synergistic. The italicized words in the box are examples of health determinants.

Examples of health determinants

We get malaria because our *immunity* is low, we do not have *access* to or are not *taking* the right drug, there is an infected *mosquito*, and the mosquito bites us. The infected mosquito is necessary but not sufficient.

   Asthmatic attacks do not only depend on the presence of a *pollutant* but also on the *age* and *immune* status of the individual and the medical *care* provided.

There are a number of different models and diagrams that are used to illustrate this concept. One of the most well-known is the lifeworld diagram of Dahlgren and Whitehead (1991), or its alternative representation (Barton et al, 2003). These diagrams envisage five successive spheres of influence starting with the individual and then ranging through social and community influences, living and working conditions, and general socio-economic cultural and environmental conditions.

In this book, three principal categories of health determinants are used:

1 individual/family determinants such as education, immune status and age;
2 physical and social environment determinants such as exposure to pollutants and employment opportunities;
3 institutional determinants such as medical care and clean water provision.

Within each category, there are a number of subcategories, illustrated in Table 2.6. Health determinants can also be subdivided into those that:

• can be managed, such as housing;
• cannot be altered, such as age;
• are negative, such as poverty;
• are positive, such as employment.

Table 2.6 *Checklist of some of the main health determinants affected by a proposal*

Principal categories	Subcategories	Examples of health determinants
Individual/ family	Physiological	Age, nutritional status, disability, sex, immunity, ethnicity, genetics
	Behaviour	Risk-taking behaviour, occupation, risk perception
	Socio-economic circumstances	Poverty, unemployment, education, social status
Environmental	Physical	Air, water and land, traffic, pollution, noise, dust, changes to natural environment, flaring, light, water use, land take, housing, crops and foods, vectors
	Social	Family structure, community structure, culture, crime, gender, inequality
	Economic	Loss of employment, investment
Institutional	Organization of health care	Primary care, specialist services, increased pressure on health care, access to health care, availability of drugs, quality of care
	Other institutions	Police, transport, public works, municipal authorities, local government, project sector ministry, local community organizations, NGOs, emergency services, boomtowns
	Policies	Regulations, jurisdictions, laws, goals, thresholds, priorities

They can also be represented as being about people on the one hand and the physical environment on the other. A balanced view gives equal attention to both.

The core of the assessment method is to judge how the health determinants are changed by the proposal and how these changes vary with each community, each stage of the proposal and each locality. The direction of change of the health determinants associated with a proposal may be positive, negative or both.

The checklist, and its hierarchical organization, was chosen so that both health gains and health risks can be assessed. There are other typologies in use among the HIA community. For example, some people prefer biological/behaviour to individual/ family. The distinction between sex and gender may require explanation: sex is a physiological property while gender is a social construct.

There are complex interactions between health determinants that we cannot generally capture in mathematical models. Many of the potential changes in health determinants associated with a proposal are not quantifiable, but are simply ranked as no change, increase or decrease. Health impacts occur when the individual, family and community group are exposed to health determinants associated with the proposal. The change may have a detrimental or enhancing influence on health. A common example is the exposure of individuals to chemical contaminants that are present in air, water or soil media.

The assessment proceeds by constructing a checklist of health determinants that may be relevant to the proposal – using literature reviews, key informant interview, mind-mapping and health baseline surveys. Not all health determinants will be relevant in every proposal.

2.4.1 Individual/family determinants

Many characteristics of the individual and family unit make them vulnerable. These include age, sex, educational level, training, immune status, poverty and place of origin. They are subdivided into physiological, behavioural and socio-economic circumstance factors. For example, poorly trained and overworked workforces who must endure long shifts for poor pay may be more vulnerable to infections. Studies of socio-economic inequality, described in Section 2.6, demonstrate the large differences in health status experienced by communities in the same locality.

2.4.2 Physical and social environment determinants

Proposals change the physical, social and economic environments, which in turn change the health status of affected communities. Discharge of airborne pollutants can occur in the industrial, domestic and occupational environments. Involuntary resettlement, economic displacement, financial insecurity, housing stress, rapid economic change, and poor design can produce dysfunctional communities with low social capital, few support networks, a sense of hopelessness, and fear of violence and robbery. For example, a highway built through a poor community reduces social support networks, often called severance, as well as increasing the risk of traffic injury. Economic changes such as employment opportunities are often health enhancing. However, loss of livelihood from expropriation of natural resources can damage health by creating poverty and uncertainty.

2.4.3 Institutional determinants

Institutional determinants are the capacity, capabilities and jurisdiction of health protection agencies. These include any agencies, organizations, service providers or ministries that have some responsibility for safeguarding human health. The range of institutions that have responsibilities for human health is very large. An example is the police force: it regulates traffic speed, prevents violent injuries, prevents crime and monitors commercial sex.

When analysing institutional determinants it is useful to distinguish:

- **capacity** – availability of staff and equipment;
- **capability** – knowledge and skill to use the resources available;
- **jurisdiction** – the area of responsibility of the agency.

Some institutions tend to make more positive contributions to health impacts than others. I suggest that they are likely to have some of the following characteristics: pro-choice, empowering, promoting social cohesion and community self-support, respecting and enhancing diversity, providing good models of ethical behaviour and responsive to new challenges. Other positive attributes of institutions can be accountability, transparency, openness and quality assurance. Key informant interviews and direct observations capture institutional attitudes and perspectives. For example, a public water supply utility that provides good services to existing customers is likely to provide good services to new communities.

2.4.3.1 Public policy

There are various components of public policy that may also affect the health determinants associated with a proposal. For example:

- There is usually a statutory requirement for EIA of major projects and this has a positive health impact by limiting pollution and safeguarding food chains.
- There may be a regional development plan.
- There may be implicit requirements to safeguard health in planning regulations.
- Government may be committed to reducing greenhouse gas emissions.
- There may be a commitment to healthy public policy or health in all policies.
- There may be policy guidance documents that justify health-promoting recommendations.
- There may be a transport policy promoting modal shift.

2.5 OTHER CATEGORIES

One of the key components of the HIA method is a clear and systematic process of classification and categorization. This ensures that all relevant issues are considered. Ideally, a single classification system would cover all eventualities, but this may not be practical. There are many possible classification systems and one way to choose between them is their broad applicability. The classification system should give appropriate balance to each subcategory. Another process of classification, referred to as environmental health areas, is discussed in Chapter 3.

2.5.1 Hazards and risks

In earlier work, my colleagues and I referred to 'health hazards' and 'health risks' rather than health determinants and some HIA practitioners may still prefer to do so. Health hazards are potential sources of harm while health risks represent the probability that a particular harm will affect a particular group of people in a particular setting with a particular severity. For example, the potential release of a dangerous chemical is a health hazard. The risk of the chemical escaping at sufficient concentrations to poison people might be estimated as one in a million. Some practitioners refer to a dangerous chemical as a hazard and reserve the word determinant to describe the various pathways by which the chemical comes into contact with vulnerable people.

Health hazards are a subset of health determinants. The following examples explain the difference between a hazard and a determinant:

● Traffic is a health hazard.
● Toxic chemicals are a health hazard.
● Poverty is a health determinant but would not normally be called a health hazard.

Because health hazards are a subset of health determinants, I now prefer to use the broader term health determinant. Further, changes in health determinants can have both a positive and a negative impact on a proposal. The term 'health risk' emphasizes negative impacts. HIA is equally concerned with positive impacts and these may be referred to as health gains or health opportunities.

Hazards/risk terminology is in common use among occupational health and safety practitioners and in health risk assessment. A bow-tie model is often used to describe the cause and consequences of unintended events – see Figure 2.2. The bow-tie model can be used to illustrate the HIA method if the proposal itself is identified as a primary hazard that contains component hazards. If these escape, they will trigger a 'top event'. In this case, the top event would be a change in determinants of health. The consequence is a change in the risk of adverse health outcomes. This, in turn, affects reputation, goodwill and licence to operate, in addition to the effect it has on the health of the community. Multiple barriers must be put in place to prevent the hazard from escaping and contingency plans must be made in case it does escape.

There are various other terminologies in use. For example, in epidemiology, health determinants are referred to as risk factors or confounders. All of this adds complexity and I think that the concept of health determinants provides the simplest solution.

2.6 HEALTH INEQUALITY

The definition of HIA, introduced in Chapter 1, referred to distributional impacts. One component of this is health inequality or inequity, also introduced in Chapter 1.

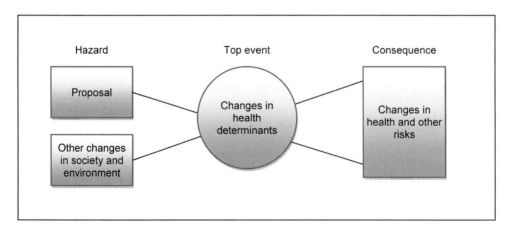

Figure 2.2 *Bow-tie model of HIA*

Health inequalities research is a tool for analysing the effect of poverty and associated socio-economic determinants on population health. It compares a measure of health outcome, such as life expectancy, with a measure of poverty, such as socio-economic rank. There is now a large body of evidence that demonstrates that poverty is a major determinant of health in all countries. This is increasingly referred to as the social gradient. A small number of examples of this evidence are provided below. It follows that any proposal that affects the poverty or socio-economic status of a community is likely to have a health impact.

The following are example of health inequalities in the UK (Acheson et al, 1998; Harding et al, 1999; Marmot and Wilkinson, 1999; Evans et al, 2001; Wilkinson and Marmot, 2003; CSDH, 2008; Marmot, 2010):

- The difference in life expectancy between the highest and lowest socio-economic groups is 4.7 years.
- Whitehall civil servants in the lowest grades have a four times higher mortality than the highest grades.
- Children in residential homes have 2.5 times the mortality of other children.
- About a third of households have no access to a car and these tend to be households with lower incomes who live closer to busy roads. Yet air pollution is more common close to roads and road junctions and in inner urban areas, which are often characterized by other indicators of disadvantage. The burden of air pollution tends to fall on people experiencing disadvantage, who do not enjoy the benefits of the private motorized transport that causes the pollution.
- The 'inverse care law' summarizes the observation that the least vulnerable people frequently receive the most medical care, and vice versa.

Table 2.7 Provides some additional numerical examples, subdivided by socio-economic groups (Acheson et al, 1998; Flynn and Knight, 1998).

Similar comparisons are available from countries like the US, as the following example illustrates using ethnic comparisons (Health Policy Institute of Ohio, 2004):

Table 2.7 *Health inequality in northwest England*

	Affluent	Less affluent (blue-collar)	Low	Lowest
Asthma rates	57	114	119	126
Standardized mortality	81	119	129	150

- The cancer incidence rate among African Americans is 10 per cent higher than among whites.
- The cancer death rate for African Americans is 35 per cent higher than the rate for whites.
- When hospitalized with acute myocardial infarction, Hispanics are less likely to receive aspirin and beta-blockers than whites.
- Nationally, African Americans, Hispanics, American Indians and Alaska Natives are about twice as likely as whites to have diabetes.
- The death rate is 27 per cent higher for African Americans with diabetes, and 40 per cent higher for Hispanics, than for whites.
- In Ohio, access to prenatal care among African American women is 15 per cent less than for whites.

These examples do not indicate whether the difference is biological, cultural or socio-economic in origin, or a mixture. More recent work in the US compares health status by county (Population Health Institute, 2010). Conditions in the US are also discussed in the CSDH report (2008).

Table 2.8 illustrates some health inequalities in other countries (World Bank, 1990, 1992, 1993; WRI, 1998; McGranahan et al, 1999; Wilkinson and Marmot, 2003).

Figure 2.3 illustrates how the average infant mortality rate per thousand in Iran varies with socio-economic quintile.

Table 2.9 illustrates how under-five child mortality varies with the level of the mother's education in Senegal.

The example in Table 2.10 is drawn from a study of mosquito bednet usage in the Gambia as a protection against malaria (Clarke et al, 2001). The community (sample size 618) were subdivided by socio-economic status based on their ownership of a radio, bed, livestock, concrete walls and metal roof. The rate of parasitaemia in children varied with socio-economic group, even though all these children lived in an environment where bites from malaria mosquitoes were commonplace.

Comparisons of the health and wealth of nations look like Figure 2.4 (World Bank, 1993). A typical national indictor for health is the expectation of life at birth. A typical indicator for wealth is average income per capita. About 50 per cent of the world population are to the left of the vertical dashed line, where small changes in average wealth produce large changes in average health. In the wealthier nations, the change is much more gradual.

There are many nations that do not lie close to the curve. Some nations have far worse average health than would be predicted by their wealth. Possible reasons include severe inequality within the country or economic reliance on abundant natural

Table 2.8 *Health inequality examples*

Share of income by poorest 20%	4% Nigeria 9% India
Life expectancy by region, Nigeria	40 years Borno State 58 years Bendel State
Safe drinking water, Nigeria	Urban 84% Rural 40%
Mexico	The difference in life expectancy between a rich and a poor community is 9 years
Indonesia	The poorest quintile shares 12% of the health budget while the richest quintile shares 29%
Russia, post-Soviet phase	The difference in life expectancy between white- and blue-collar workers was associated with accidents and violence. The difference in life expectancy between workers and non-workers was associated with chronic disease

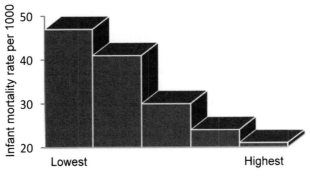

Figure 2.3 *Infant mortality rate and socio-economic status in Iran*

Source: Hosseinpoor et al, 2005

Table 2.9 *Inequality in Senegal*

Mother's education	Under-five mortality per 1000
None	225
Primary	140
Secondary	70

Source: WRI, 1998

Table 2.10 *Inequality and malaria in the Gambia*

Socio-economic status	Parasitaemia in children (%)
Poor	33
Poorer	42
Poorest	51

Source: Clarke et al, 2001, reproduced by permission of Elsevier

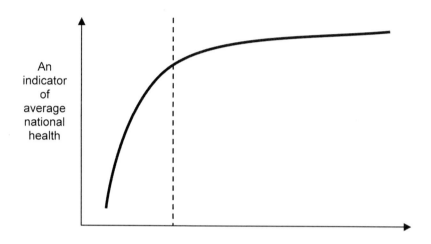

An indicator of average national wealth

Figure 2.4 *Association between national wealth and health*

Source: World Bank, 1993

resources, such as oil. Countries with higher than average national health also tend to have high levels of equality.

A similar graph can be constructed within one country. In England, for example, the x-axis could be geographical zones ranked by index of multiple deprivation (Marmot, 2010). The y-axis could be average expectation of life or years of disability-free life for the population in that zone. In England there is a difference of 17 years of disability-free life between communities in the least and most deprived areas. There is a marked social gradient in health and in environmental disadvantage.

This graph implies that proposals that affect the wealth of the community will have different health impacts in a poor country or region than in a wealthy country or region. In addition to the average effect, there will be differences between the socio-economic groups within the community. The poorer members of a poor community could receive substantial health benefits from a proposal – providing that they have access to its financial benefits. Proposals that increase equality are likely to contribute more to health. All of this emphasizes the importance, in HIA, of considering the differential impacts on community groups. The sub-field of equity-based HIA is very active.

Table 2.11 *Direct, indirect or cumulative health impacts*

Direct impacts	Changes in health determinants that immediately result from a proposal. Example: the traffic collisions and injuries that occur on a new section of road.
Indirect impacts	Changes in health determinants that are consequences of direct impacts. Example: as a result of perceived traffic injuries, people are less likely to cycle, physical fitness declines and circulatory diseases increase.
Cumulative impacts	The results of additive and aggregative actions producing impacts that accumulate incrementally or synergistically over time and space. Example: the air quality where two busy roads intersect may be poor and respiratory diseases may increase.

2.7 DIRECT AND CUMULATIVE IMPACTS

The health impacts of proposals can be direct, indirect or cumulative. The definitions in Table 2.11 are based on the IAIA glossary (IAIA, undated).

Cumulative impacts can result from individually minor but collectively significant actions taking place over a period of time. They can occur locally, nationally and globally. One example of a cumulative global impact is climate change. See Chapter 12 for more details. Ideally, all impact assessments should have a cumulative impact section.

2.8 EXAMPLES OF HEALTH IMPACTS

2.8.1 Transport

Figure 2.5 illustrates a health impact associated with transport in a low-income country. In the photograph, a badly damaged bus is visible beside a potholed road. The determinants of this event are many and complex. They are likely to include some of the following:

● The bus was poorly maintained, the headlights did not work and some of the tyres were bald.
● The driver was poorly trained and has never taken a driving test.
● The bus was being driven too fast and at night.
● The driver may have been on the road continuously for 24 hours. He may have been using some mild drug in order to stay awake.
● The bus was overcrowded and this included large numbers of women and children.
● There were no accident and emergency facilities in the neighbourhood.
● The road was poorly maintained and full of potholes.
● The oncoming vehicle was in a similar condition and being driven in a similar way.

Figure 2.5 *Poorly managed public transport*

Copyright Tropix.co.uk

- There was a shortage of spare parts in the country.
- The cost of transport had to be kept as low as possible in order to be affordable.
- The driver was required to maintain an unrealistic schedule.

Figure 2.6 depicts a heavily burdened woman carrying firewood and water along a rural road. She may also have a baby on her hip. This illustrates many determinants of health associated with a rural proposal that included a road. The proposal might use common land that had traditionally been used by the local community for gathering fuelwood, wild plants and water. These traditional activities would be undertaken by women. As a result of the proposal, the women have to walk longer distances in order to collect wood and water. The heavy burden causes physical injury. The extra distance travelled reduces the time they have to care for their families. The new road helps them to travel to distant resources, including medical facilities, public transport and markets. The road is designed for use by heavy, fast moving, motorized transport. There is no pavement or apron that can be used by pedestrians, cyclists or animals. The women have to share the road surface with fast moving vehicles.

Informal settlements will grow on the edges of the new road, especially where surface water pools. The pools will create foci for the transmission of vector-borne and

Figure 2.6 *Pedestrian use of roads*

Copyright Tropix.co.uk

waterborne diseases. The settlements will provide services such as food, alcohol and commercial sex for vehicle drivers.

2.8.2 Water

Figure 2.7 illustrates a cattle watering project in northern Benin. A dry riverbed has been dammed in order to store wet-season rainfall. Although the project is intended for cattle, it is also being used for a human drinking supply. The water is full of cattle faeces. Figure 2.8 indicates the alternative dry-season water supply in the same village and explains the popularity of the reservoir. Unfortunately, a small water snail called *Biomphalaria* has colonized the stones around the reservoir margin. The snail is the

Figure 2.7 *A cattle watering project in Benin*

Copyright Tropix.co.uk

intermediate host of a parasite called *Schistosoma mansoni*, which causes the disease schistosomiasis, or bilharzia, in humans. The parasite multiplies in the snail, is released into the water as a tiny swimming worm, and penetrates the skin of anybody who wades into the water. Consequently, the cattle watering project increases the prevalence rate of schistosomiasis.

As can be seen from Figure 2.8, the normal dry-season water supply consists of a shallow muddy pool that is some 2–3 metres below ground level. Collecting water from this pool is an arduous process. Many infrastructure development plans affect the level of groundwater. This community is living on the margins of existence and a small reduction of groundwater levels would be fatal for them. The consequence of drinking contaminated water is a high level of childhood mortality. The surviving adults are able to tolerate the contamination most of the time.

2.8.3 Resettlement

Figure 2.9 illustrates a resettlement village for a large reservoir project in Thailand. The resettlement site was taken from the protected forest reserve. The edge of the primary forest can be seen in the upper third of the picture. As compensation for loss of livelihood, the villagers were provided with a rubber plantation that can be seen in the

Figure 2.8 *The alternative water source*

Copyright Tropix.co.uk

middle third of the picture. They were also provided with financial compensation that enabled them to build elaborate housing. Unfortunately, in Southeast Asia a malaria mosquito breeds in small streams of clear water that emerge from the forest under light shade. The light shade of the rubber plantation increased the number of breeding sites and the village was located where malaria transmission was likely to be most intense.

Some of the other determinants of malaria include house design, impregnated bednet use, chemical control of breeding sites and availability of curative services.

2.8.4 Agriculture

Figure 2.10 illustrates a farmer in East Africa enjoying his traditional midday meal of maize and beans. The farmer and his family live on a rice irrigation system. They no longer grow maize and beans. In the traditional system, the women and children tended the land and grew the maize and beans, which they then harvested and stored and

Figure 2.9 *Resettlement village in Thailand*

Copyright Tropix.co.uk

cooked in the family home. In the commercial rice irrigation system, the women and children are working in the rice fields. The husband takes the rice crop to market and receives a cash payment. He then supplies cash to his wife who uses it to buy maize and beans. Or he may keep the cash in order to buy alcohol and cigarettes. Consequently, household entitlements to food may be reduced and, paradoxically, malnutrition may increase.

Figure 2.11 illustrates the wrong way to spray pesticides. The operator is wearing ordinary clothing that can easily absorb the spray. He has no face mask or gloves. This is an itinerant sprayer in Egypt who is poorly trained, poorly educated, and moves from farm to farm spraying every day. He mixes the concentrate using his bare hands. He falls outside the scope of health risk assessment because he is not part of a regulated workforce. The risk of poisoning by modern pesticides is far higher for the people who mix and use it than it is to the general public who will later consume the produce.

2.8.5 Air pollution

Figure 2.12 illustrates a cloud of pollution over the city of Damascus in Syria. There are many scientific studies of the causal relationship between air pollution and human health

Figure 2.10 *A Kikuyu farmer enjoys a lunch of maize and beans*

Copyright Tropix.co.uk

(Mindell, 2002). As a rule of thumb, in Europe, a 10µg/m² increase in PM$_{10}$ particulates in ambient air leads to about a 1 per cent increase in acute deaths. It also leads to a substantial rise in asthma medication, broncho-dilator use, chronic cardiopulmonary disease, emphysema, chronic bronchitis and cough. These studies have implications for the HIA of urban planning, industrial zoning, industrial emissions and transport in both wealthy and poor cities.

Figure 2.13 illustrates the dark, sooty, unventilated kitchens in which much of the world's food is cooked. There are now many studies of the association between such conditions and respiratory illness in women and children (WHO, 2010a). In developed economies, indoor air pollution is associated with damp, dust mites, smoking, and the solvents used in paints and glues (Wieslander et al, 1996). These studies have implications for the HIA of housing projects in both wealthy and poor communities.

Figure 2.11 *The wrong way to spray pesticides*

Copyright Tropix.co.uk

2.8.6 Construction and tourism

Construction brings mobile men with money, in their thousands. Away from their homes and their communities and working under stressful conditions for long periods, they seek solace in inappropriate sexual relationships. Women, who are excluded from education and paid work by culture or tradition, and lacking the basic means to support themselves and their children, find employment opportunities through selling their bodies. One way to avoid this issue is to shut men in construction camps and prevent them mixing with the local community. Consequences include men-to-men sex and smuggling of women. The result can be an epidemic of sexually transmitted infection, including HIV/AIDS.

Figure 2.14 is from the tourist information desk of a hotel in Southeast Asia. It offers visits to the local massage parlours or coach tours along the coast. It illustrates an association between tourism and business travel and STIs.

Figure 2.12 *Outdoor air pollution, Damascus*

Copyright Tropix.co.uk

2.9 DISABILITY ADJUSTED LIFE YEARS (DALYS)

It is difficult to compare the relative importance of different adverse health outcomes. For example, how do you compare a broken leg with a damaged lung? Some effects are fatal while others are disabling. Death occurs at different ages and disabilities last different lengths of time. To overcome this problem a common metric is used called the disability adjusted life year or DALY (Murray and Lopez, 1996). DALYs provide a tool for analysing baseline health conditions and may be used more often in future for analysing health impacts.

The DALY is a measure of time as it subtracts from expected duration of life the years lost due to disease-specific morbidity and mortality. The original analysis was based on 107 diagnostic entities. For each of these, estimates were made for age- and sex-specific incidents, duration, survival and disability. The number of years lost due to death was calculated by using the life expectancy at the age of death. For surviving cases, the time lost was based on the estimated duration of different diseases weighted with the degree of disability related to each disease. The weights for various levels of disability were estimated by debate.

Figure 2.13 *Indoor air pollution, Ghana*

Copyright Tropix.co.uk

There are a number of challenges. The concept is based on population averages and does not take adequate account of the distribution of health within the population. The concept has inherent value assumptions including discounting future health and weighting life years. There is a fundamental objection to reducing ill health and human suffering to a utility weight between zero and one. There is a dispute about who should make the judgement – the expert or the disabled community. The metric can be used both with and without these discounts and weightings. Despite these challenges, the DALY unit seems to provide a useful indicator of priorities when it is used with large populations.

Example of DALY use

Recent research in the UK has used the DALY method to investigate whether sustainable water management practices on new housing estates presented an acceptable risk (Fewtrell and Kay, 2008). A reference DALY risk threshold was assumed equivalent to a 10^{-6} risk of dying for a normally healthy adult.

Figure 2.14 *Massage parlours in Southeast Asia*

Copyright Tropix.co.uk

2.10 EPIDEMIOLOGICAL TRANSITION

Figure 2.15 indicates the percentage distribution of DALYs within each of the World Bank regions between three major health conditions (WHO, 2008). The stacked bars indicate the proportion of a nation's DALYs contributed by non-communicable conditions (medium grey), injuries (light grey) and a group of conditions that include communicable diseases, maternal and perinatal conditions and nutritional deficiencies (dark grey). This proportion is compared between regions with different levels of economic development. The major causes of ill health vary with economic development.

Table 2.12 provides the same data in more detail illustrating the relative contribution of maternal and perinatal conditions and intentional and unintentional injuries. In the poorest countries, childhood morbidity and mortality is largely attributed to infections. In the Middle East and Latin America, intentional injury is a significant issue. In some regions, maternal and perinatal conditions are a significant component.

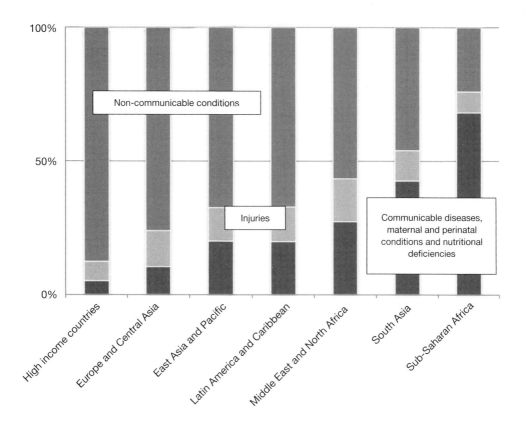

Figure 2.15 *Percentage DALYs between World Bank regions 2004*

The graph and table can assist with the prioritization of health impacts of proposals. For example, in high-income countries non-communicable conditions contribute the greatest percentage of the burden of disease experienced by the population: see Table 2.13. Where environmental pollution by chemical compounds is an important determinant of disease, reducing environmental pollution is a priority. EIA is one of the tools used to ensure that discharges associated with proposals are maintained within acceptable limits. On the other hand, in the poorest countries communicable diseases contribute the greatest percentage of the burden of disease. In these countries, priority lies with clean water supplies, sanitation, food safety, reducing overcrowding, and vector control. It may be inappropriate to prioritize the impact of non-communicable diseases over communicable diseases during an assessment of a proposal in a poor country, although this is frequently done.

During the 1990s, an estimate was made of the burden of disease in the European Union, using DALY units (Swedish National Institute of Public Health, 1997). The burden of disease was largely non-communicable. The most important were neuro-psychiatric and cardiovascular disorders, and all cancers. The study analysed some of the risk factors, or determinants of health, associated with these health outcomes. The results suggest that unemployment, tobacco and alcohol consumption, obesity and

Table 2.12 *Percentage DALYs between World Bank regions 2004 in more detail (rounded)*

	High-income countries	Europe and Central Asia	East Asia and Pacific	Latin America and Caribbean	Middle East and North Africa	South Asia	Sub-Saharan Africa
Infectious and parasitic diseases	2	4	9	8	8	18	41
Respiratory infections	1	2	3	4	5	8	11
Maternal and perinatal conditions	1	3	6	6	12	13	13
Nutritional deficiencies	0	1	2	2	3	3	3
Unintentional injuries	5	10	10	7	12	9	5
Intentional injuries	2	4	2	6	5	2	3
Non-communicable conditions	88	76	67	67	57	46	24

Table 2.13 *Summary of the different priorities in more and less developed economies*

Higher frequency in less developed economies	Higher frequency in more developed economies
Communicable diseases such as malaria, respiratory infections, diarrhoea, HIV/AIDS Protein-energy malnutrition Injuries	Non-communicable diseases such as heart, lung and circulation disorders and cancers Obesity Depressive illness

work-related diseases had become far more important determinants than chemical pollution. There were important differences between socio-economic groups.

The relative change in the importance of communicable and non-communicable diseases with economic development is referred to as the epidemiological or risk transition (Smith, 1991, 1997). It is an example of a number of such transitions, of which the demographic transition is probably the best known. During the course of economic development, 'traditional' risks (such as communicable diseases) decline and 'modern' risks (such as non-communicable diseases) increase. A similar phenomenon can sometimes be observed in low-income economies between rural and urban areas (Birley and Lock, 1999). In rural areas, traditional risks are more common. In urban areas modern risks are more common. Peri-urban areas may experience the worst of both worlds: crowding and poor sanitation associated with communicable diseases as

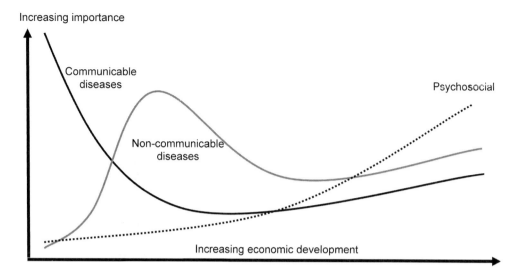

Figure 2.16 *A theoretical model of the change in health impact priorities with development*

well as chemical pollution of air and water. These are often the informal slums occupied by communities displaced from rural areas in search of employment in the cities. Some of that displacement is attributable to involuntary resettlement and may be a health impact of proposals. Introduction of large infrastructure proposals into rural areas can change the ratio of traditional and modern risks.

A number of transition scenarios are possible and it may be unwise to generalize, although Figure 2.16 is an attempt to do so. In this scenario, there is initially a decline in communicable diseases as economic development increases but an increase occurs later as a result of STIs. The decline in communicable diseases is followed by an increase in non-communicable conditions. Some of these are associated with environmental pollution and safety and they decline as pollution is controlled and safety is increased. However, other non-communicable diseases become more common as result of over-consumption of sugars and fats and reductions in physical activity. At the same time, a range of psychosocial conditions, such as drug addiction, are increasing.

2.11 HEALTH CARE

The quality and quantity of health and medical care available to the community is a determinant of health. There are huge differences between countries, between regions of a country and between rural and urban areas. This is a specialist subject and health care administrators, practitioners and health systems researchers should be consulted for detailed information. HIAs should always consider the impacts of a proposal on the delivery of existing and future health care services.

The following sections contrast conditions in low-income and high-income economies.

2.11.1 Low-income economies

In many low-income countries, the health administrative framework will start with the department or ministry of health at national level. The structure will be repeated at regional and perhaps local level. Government officers will be responsible for all the public medical services including primary care clinics, secondary referral clinics and tertiary referral hospitals. They will also be responsible for public pharmacies. They will not be responsible for the private sector. This includes traditional healers, traditional birth attendants, traditional herbalists, private clinics and pharmacies.

The ministry of health in a poor country may have a budget of about US$20 per person per year. By comparison, the health budget in a developed country may represent US$5000–10,000 per person per year.

In poor countries, the available budget is often not distributed equitably. It tends to be clustered in tertiary units, in capital cities, where the rich and powerful live. The amount that reaches poor rural areas, where a proposal may be located, can be tiny.

Example of primary care limitations in some low-income countries

Primary care units in rural areas may have no running water or latrines, an intermittent drug supply, and be staffed by nurses who are paid irregularly. They will have very little functional equipment, including transport. These are the clinics that may have to cope with a large influx of migrants who are attracted by a proposal. When the proposal is operational, the staff in these clinics are likely to ask the owners of the proposal for emergency assistance. It is best to understand these health needs in advance and design for them. The proposal should include a social investment budget.

2.11.2 Developed economies

In the UK, for example, health care services are universally available. However, a proposal that substantially increases local community size or density will require a corresponding increase in service provision.

In London, the Healthy Urban Development Unit (HUDU) has developed a model for estimating the additional provision (HUDU, 2007). The model enabled the NHS in London to secure over £10 million for additional health facilities over two years. The model takes account of baseline demographics and health data at local level, the household profiles and the population gain, among other factors. It calculates:

- amount of hospital beds or floor space required for that population in terms of acute elective, acute non-elective, intermediate care, mental health and primary care;
- the capital cost of providing the required space;

- the revenue costs of running the necessary services before mainstream NHS funding takes account of the new population.

The impacts of national or international policies on health systems and services are even more substantial and have given rise to a call for special health systems impact assessments. An online tool has been developed for the assessment of EU policies on health systems (Health-EU, 2007).

2.12 HEALTH INDICATORS

Many indicators are available to measure the state of health determinants. Good indicators are relatively easy to measure, are not intrusive and have relevance to a large number of health outcomes. For example, Table 2.14 provides some examples of environmental health indicators (WHO, 1999).

The US Centers for Disease Control has a simple set of comparative indicators used to measure health-related quality of life (Centers for Disease Control and Prevention, 2000). Communities are surveyed and asked to provide subjective answers to four questions (see Table 2.15). Despite their apparent simplicity and subjectivity, the statistics generated from these questions have good comparative and predictive value.

There are many indicators for assessing the quality of medical care. Table 2.16 provides examples.

Table 2.14 *Examples of environmental health indicators*

Type	Measure	Form of data
Emissions	Emission rates	Tonnes per year
Water supply	% community with access	Number of litres per person per day
Outdoor and indoor air quality	Suspended particulates, SOx, NOx	mg/m³ per hour
Soil oontaminaιils	Heavy metals	mg/kg
Housing	Crowding	Average people per room

2.13 VULNERABILITY AND RESILIENCE

Everyone has the capacity to cope with small changes in their lives. Some changes may overwhelm our personal coping capacity. These are the changes to which we are vulnerable and for which we require external support. The impacts of proposals include large changes of this kind. They may create vulnerabilities in affected communities. For example, when children are walking to school and must cross busy roads they are vulnerable to changes in traffic density. When house owners are living by rivers they

Table 2.15 *Core healthy days measures*

- Would you say that in **general** your **health** is excellent, very good, good, fair, or poor?
- Now thinking about your **physical health**, which includes physical illness and injury, for how many days during the past 30 days was your physical health not good?
- Now thinking about your **mental health**, which includes stress, depression and problems with emotions, for how many days during the past 30 days was your mental health not good?
- During the past 30 days, for about how many days did poor physical or mental health keep you from doing your **usual activities**, such as self-care, work or recreation?

Table 2.16 *Examples of medical service indicators*

Variable	Measure
Hospitals	Beds per population; average in-patient stay by disease category (long is not always better)
Staffing	Doctors, nurses, nursing assistants and community volunteers per 10,000 population, distribution of staff between primary, secondary and tertiary care facilities
Location	Maps of primary, secondary and tertiary care locations, access
Stocks	Drug availability, storage conditions, cold chains
Equipment	Functioning, maintenance, sterilization
Standards	Referral practices, hygiene, waste disposal
Financial	Cost recovery, charges, insurance, salaries
Record-keeping	Evidence that statistics are collected and used in decision-making

are vulnerable to changes in flooding regimes. Vulnerabilities can arise from location, poverty, education, livelihood, gender, age and many other determinants.

Vulnerability can be defined as the set of factors associated with an individual or group that increases their probability of experiencing a reduction in well-being associated with change. Vulnerable communities are 'brittle': when they are changed they are likely to break. Resilience is the opposite of vulnerability. It can be defined as the set of factors associated with an individual or group that increases the probability that their well-being remains unaffected by change. For both vulnerable and resilient communities, the change may be associated with a proposal or other factors in the environment.

Most of the determinants of health affect the vulnerability and resilience of communities. For example:

- The individual is vulnerable because of his or her immunity, behaviour, occupation, education and poverty.
- The physical environment affects vulnerability, for example when a community is close to a hazard.
- The social environment provides the support networks and coping mechanisms that enable the community to be more resilient.

Resilience and brittleness

A useful analogy for comparing brittleness and resilience is the difference between a brick and a stick. When you bend a brick it breaks, but when you bend a stick it springs back.

Factors external to the community that affect their vulnerability include:

- the economic environment that provides the community with the financial opportunities used for coping;
- the services provided to the community by government or other agencies that are not under community control;
- the climate;
- policies, programmes and projects of all kinds.

One of the objectives of an HIA should be to make recommendations that safeguard communities from being overwhelmed by the changes associated with a proposal. In other words, the recommendations should reduce vulnerability and increase resilience. For example, micro-finance initiatives may assist the community to increase their livelihoods and in so doing increase their resilience.

There is a close relationship between vulnerability and inequity. Some community groups will be more vulnerable as a result of their poverty, location and other determinants. Some community groups will be more resilient as a result of their education, livelihood, capital and other determinants.

Vulnerability and resilience are also properties of systems. Social, climate and ecological systems have feedback loops that tend to maintain a steady state. External pressures displace the system from the steady state. A resilient system can be displaced relatively far and will still return. When the system is displaced too far, it will be permanently displaced into a new state. For example:

- The world's fisheries were able to continue functioning effectively even when large quantities of fish were removed for human consumption. Eventually, the rate of fishing exceeded the resilience of the system and the fisheries collapsed.
- The climate has remained in a steady state for millions of years, with natural variation and oscillation. Now, the release of vast quantities of geological carbon may be overwhelming the feedback mechanisms and the climate may flip into a different steady state that will be much more hostile for human life.
- The complex webs of health determinants that are the cause of obesity have been mapped (Butland et al, 2007). Change that overwhelms coping strategies may transform a lean, active person with a healthy diet into an overweight, sedentary person who consumes junk food and alcohol.

2.14 AN EXERCISE

The following exercise can be done individually but is more rewarding when it is done in a group.

2.14.1 Mind-mapping

Imagine that you are the consultant to the investment group that is funding a large processing facility on a coastal corner of a country called San Serriffe (see Figure 2.17). You have been asked to advise on the change in health that the proposal may cause. The site is a bay where there is considerable fishing activity.

The objective of this exercise is to construct a list of health concerns associated with the proposed project. It is a mind-mapping exercise and there are no right or wrong answers. You are not asked to categorize or prioritize the health concerns you identify, or to develop recommendations.

2.14.1.1 Description of the community

The proposed project site is currently a fishing village of approximately 2000 inhabitants who will have to move. The nearest town is of 50,000 inhabitants and is located 10 kilometres north. Several other small human settlements of no more than 200 inhabitants surround the site. In the village there is a religious shrine. The villagers are mostly fishing folk, but some work in town and there is some metal craftwork. There is local agriculture based on livestock and vegetables. Village water supply is poor and there are few functional water closets. There is a road from the town to the village. It continues to a small coastal settlement. The land is low lying with a high water table, wooded with coastal mangroves. There are both private and public medical facilities in town. Much transportation is by boat.

The community includes two major ethnic groups. Many of them are Christians and some Muslims. In addition, many follow a traditional, pagan, religion, which holds all nature to be sacred. Trees, shrubs, ponds and mounds are believed to be imbued with Spirit. Whatever their religion, there is a strong tradition that men are superior to women and in charge of the household. The role of men is clearly defined to include fishing, politics, mending the house, the boat and the fishing equipment, and owning the family finances and capital. The role of women is clearly defined to include gathering firewood, wild foods and medicine; collecting water; growing and cooking food; looking after the family and trading. Functional literacy is more common among males than females.

2.14.1.2 Description of the development proposal

The entire bay will become a processing complex (see Figure 2.18):

- The processing plant will receive raw materials through pipelines.
- There will be a jetty for loading ships with processed products.
- The road from town will be upgraded.

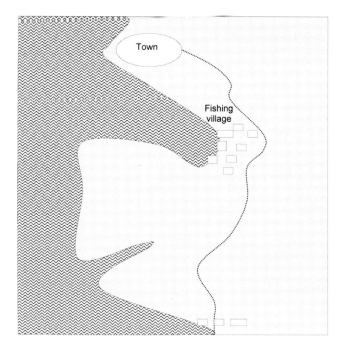

Figure 2.17 *San Serriffe before the project*

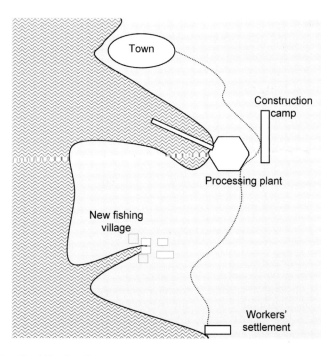

Figure 2.18 *San Serriffe after the project*

- Some dredging and land reclamation is required.
- Fishing will be forbidden in the bay.
- Villagers will be relocated to a new fishing village.
- A construction camp of approximately 4000 workers will be established throughout the construction period of two years.
- A modern settlement will be constructed on the coast for workers. Some mangrove trees will be cleared.
- Local employment opportunities will be provided.

2.14.2 Organizing the health concerns

The objective of this exercise is to identify health determinants and health outcomes using the categories discussed in the chapter.

Take another look at your list of health concerns from the mind-mapping. Now categorize them as follows:

- Health determinants.
- Health outcomes.
- Concerns about the HIA process itself.
- Concerns about the method of assessment.

Examples of process concerns follow. These are discussed in a later chapter:

- When should we do the assessment?
- Who should do it?
- What happens to the assessment when it is completed?
- What controls are in place to assure quality?

Examples of method concerns follow. These are also discussed in a later chapter:

- Who could be affected?
- What baseline data are available?
- How far into the future should we consider?
- How far away from the project site will impacts occur?
- What should we do to safeguard health?

2.15 REFERENCES

Acheson, D., D. Barker, J. Chambers, H. Graham, M. Marmot and M. Whitehead (1998) 'Independent inquiry into inequalities in health, www.dh.gov.uk/en/publicationsandstatistics/publications/publicationspolicyandguidance/dh_4097582, accessed February 2011

Amr, M. M., M. M. Salem and M. S. El-Beshlawy (1993) 'Neurological effects of pesticides', Biological monitoring conference, Kyoto University, Kyoto, Japan

Armstrong, F. (2005) 'McLibel', Spanner Films Ltd, UK: 85 minutes, www.spannerfilms.net/films/mclibel, accessed February 2011

Barton, H., M. Grant and R. Guise (2003) *Shaping Neighbourhoods: A Guide for Health, Sustainability and Vitality*, Spon Press, London and New York

Birley, M. H. (1991) 'Guidelines for forecasting the vector-borne disease implications of water resource development', World Health Organization, www.birleyhia.co.uk, accessed 2010

Birley, M. H. and K. Lock (1999) *The Health Impacts of Peri-urban Natural Resource Development*, Liverpool School of Tropical Medicine, Liverpool, www.birleyhia.co.uk/Publications/periurbanhia.pdf

Breslow, L. and G. Cengage (eds) (2002–2006) 'Noncommunicable Disease Control, Encyclopedia of Public Health', eNotes.com, www.enotes.com/public-health-encyclopedia/non communicable-disease-control

Butland, B., S. Jebb, P. Kopelman, K. McPherson, S. Thomas, J. Mardell and V. Parry (2007) *Tackling Obesities: Future Choices*, www.foresight.gov.uk, accessed September 2009

Centers for Disease Control and Prevention (2000) *Measuring Healthy Days: Population Assessment of Health-Related Quality of Life*, US Department of Health and Human Services, Atlanta, www.cdc.gov/hrqol/pdfs/mhd.pdf

Clarke, S. E., C. Bogh, R. C. Brown, M. Pinder, G. I. L. Walraven and S. W. Lindsay (2001) 'Do untreated bednets protect against malaria? ', *Transactions of the Royal Society for Tropical Medicine and Hygiene*, vol 95, pp457–462

Crisp, R. (2008) 'Well-being', http://plato.stanford.edu/entries/well-being/, accessed February 2010

CSDH (Commission on Social Determinants of Health) (2008) 'Closing the gap in a generation: Health equity through action on the social determinants of health. Final Report of the Commission on Social Determinants of Health', www.who.int/social_determinants/thecommission/finalreport/en/index.html, accessed July 2009

Dahlgren, G. and M. Whitehead (1991) 'Policies and strategies to promote social equity in health', Institute for Future Studies, Stockholm

de Onis, M., M. Blossner, E. Borghi, E. A. Frongillo and R. Morris (2004) 'Estimates of global prevalence of childhood underweight in 1990 and 2015', *Journal of the American Medical Association*, vol 291, no 21, pp2600–2606, http://jama.ama-assn.org/cgi/content/abstract/291/21/2600

EUROSAFE (2007) 'Injuries in the European Union, Summary 2003–2005', Kuratorium für Verkehrssicherheit (Austrian Road Safety Board), Vienna, www.injuryobservatory.net/Injuries_euro_union.html

Evans, T., M. Whitehead, F. Diderichsen, A. Bhuiya and M. Wirth (eds) (2001) *Challenging Inequalities in Health – From Ethics to Action*, Oxford University Press, Oxford, New York

Executive Office (2010) 'Solving the problem of childhood obesity within a generation, White House Task Force on Childhood Obesity, Report to the President', Executive Office of the President of the United States, Washington, www.letsmove.gov/pdf/TaskForce_on_Childhood_Obesity_May2010_FullReport.pdf, accessed 2010

Fewtrell, L. and D. Kay (eds) (2008) *Health Impact Assessment for Sustainable Water Management*, IWA Publishing, London

Flynn, P. and D. Knight (1998) ' Inequalities in health in the North West', NHS Executive, North West, Warrington

Ford-Martin, P. A. (2002) *Gale Encyclopedia of Medicine*, The Gale Group, Inc., Detroit, www.healthline.com/galecontent/psychosocial-disorders

Harding, S., J. Brown, M. Rosato and L. Hattersley (1999) 'Socio-economic differentials in health: Illustrations from the Office for National Statistics longitudinal study', *Health Statistics Quarterly*, Spring, pp5–15

Health-EU (2007) 'Health Systems Impact Assessment Tool', http://ec.europa.eu/health/index_en.htm, accessed February 2010

Health Policy Institute of Ohio (2004) *Understanding Health Disparities*, www.healthpolicyohio.org, accessed September 2009

Heise, L. (1992) 'Violence against women: The hidden health burden', *World Health Statistics Quarterly*, vol 46, no 1, pp78–85

Hosseinpoor, A., K. Mohammad, R. Majdzadeh, M. Naghavi, F. Abolhassani, A. Sousa, N. Speybroeck, H. Jamshidi and J. Vega (2005) 'Socioeconomic inequality in infant mortality in Iran and across its provinces', *Bulletin of the World Health Organization*, vol 83, pp837–844, www.who.int/bulletin/volumes/83/11/837.pdf

HUDU (Healthy Urban Development Unit) (2007) HUDU Model, www.healthyurbandevelopment.nhs.uk/pages/hudu_model/hudu_model.html, accessed February 2010

IAIA (undated) International Association for Impact Assessment, www.iaia.org

IOBI (2003) Injury Observatory for Britain and Ireland, www.injuryobservatory.net/index.html, accessed January 2010

Layard, R. (2005) *Happiness: Lessons from a new science*, Penguin, London

Marmot, M. (2010) *Fair Society, Healthy Lives: A Strategic Review of Health Inequalities in England Post-2010*, Global Health Equity Group, UCL Research Department of Epidemiology and Public Health, www.ucl.ac.uk/gheg/marmotreview, accessed March 2010, now at www.marmotreview.org

Marmot, M. and R. Wilkinson (eds) (1999) *Social Determinants of Health*, Oxford University Press, Oxford

McGranahan, G., C. Hunt, M. Kjellén, S. Lewin and C. Stephens (1999) 'Environmental change and human health in Africa, the Caribbean and the Pacific', The Stockholm Environment Institute, Stockholm, http://sei-international.org/

Mindell, J. S. (2002) 'Quantification of health impacts of air pollution reduction in Kensington & Chelsea and Westminster', Imperial College of Science, Technology and Medicine, University of London, London

Murray, C. and A. Lopez (eds) (1996) *The Global Burden of Disease – A Comprehensive Assessment of Mortality and Disability from Diseases, Injuries and Risk Factors in 1990 and Projected to 2020*, Harvard University Press, Boston

National Mental Health Development Unit (2010) Mental Wellbeing Checklist and Mental Wellbeing Impact Assessment Toolkit, www.nmhdu.org.uk

nef (new economics foundation) (undated) 'Well-being', www.neweconomics.org/programmes/well-being, accessed February 2010

Peden, M., K. Oyegbite, J. Ozanne-Smith, A. A. Hyder, C. Branche, A. F. Rahman, F. Rivara and K. Bartolomeos (eds) (2008) *World Report on Child Injury Prevention*. World Health Organization, Geneva, www.injuryobservatory.net/documents/WorldReport_childinjuryprevention.pdf

Popkin, B. (1996) 'Understanding the nutrition transition', *Urbanisation and Health Newsletter*, vol 30, September, pp3–19

Population Health Institute (2010) 'County health rankings, mobilizing action toward community health', www.countyhealthrankings.org, accessed November 2010

Seligman, M. E. P. (2002) *Authentic Happiness: Using the New Positive Psychology to Realize Your Potential for Lasting Fulfillment*, Free Press, New York, www.ppc.sas.upenn.edu

Smith, D. H. (1984) 'Bura Irrigation Settlement Project – Mid-term evaluation, health sector report', Liverpool School of Tropical Medicine, Liverpool

Smith, K. R. (1991) 'Managing the risk transition', *Toxicology and Industrial Health*, vol 7, no 5–6, pp319–327

Smith, K. R. (1997) 'Development, health and the environmental risk transition', in G. S. Shahi, B. S. Levy, A. Binger, T. Kjellstrom and R. Lawrence (eds) *International Perspectives on Environment, Development and Health: Towards a Sustainable World*, Springer, New York

Spurlock, M. (2004) 'Supersize Me', US: 100 minutes, http://en.wikipedia.org/wiki/Super_Size_Me

Swedish National Institute of Public Health (1997) 'Determinants of the burden of disease in the European Union', European Commission, Directorate General 5, Stockholm

Watts, C. and C. Zimmerman (2002) 'Violence against women: Global scope and magnitude', *The Lancet*, vol 359, no 9313, pp1232–1237

WHO (World Health Organization) (1946) 'Preamble to the Constitution of the World Health Organization as adopted by the International Health Conference, New York, 19 June–22 July 1946; signed on 22 July 1946 by the representatives of 61 States (Official Records of the World Health Organization, no 2, p100) and entered into force on 7 April 1948', www.who.int/suggestions/faq/en/index.html, accessed July 2009

WHO (1977) *Manual of the International Statistical Classification of Diseases, Injuries, and Causes of Death*. World Health Organization, Geneva

WHO (1986) 'Ottawa Charter for Health Promotion', www.euro.who.int/en/who-we-are/policy-documents/ottawa-charter-for-health-promotion,-1986, accessed September 2009

WHO (1999) 'Environmental health indicators: Framework and methodologies', WHO, Geneva, http://whqlibdoc.who.int/hq/1999/WHO_SDE_OEH_99.10.pdf

WHO (2006a) 'Gaining health, the European strategy for the prevention and control of noncommunicable diseases', www.euro.who.int/en/what-we-do/health-topics/environmental-health/urban-health/publications/2006/gaining-health.-the-european-strategy-for-the-prevention-and-control-of-noncommunicable-diseases

WHO (2006b) 'Obesity and overweight', www.who.int/mediacentre/factsheets/fs311/en/index.html, accessed September 2009

WHO (2008) 'The global burden of disease: 2004 update', www.who.int/healthinfo/global_burden_disease/GBD_report_2004update_full.pdf, accessed September 2009

WHO (2010a) 'Indoor air pollution', www.who.int/indoorair/en, accessed April 2010

WHO (2010b) *Maternal Mental Health & Child Health and Development*, www.who.int/mental_health/prevention/suicide/MaternalMH/en/index.html, accessed January 2010

WHO (2010c) Media Centre fact sheets, www.who.int/mediacentre/factsheets/fs310/en/index.html, accessed January 2010

WHO (2010d) 'Mental health', www.who.int/mental_health/en, accessed January 2010

WHO (2010e) *Mental Health, Human Rights and Legislation: WHO's Framework*, www.who.int/mental_health/policy/fact_sheet_mnh_hr_leg_2105.pdf, accessed January 2010

Wieslander, G., D. Norbäck, E. Björnsson, C. Janson and G. Boman (1996) 'Asthma and the indoor environment: The significance of emission of formaldehyde and volatile organic compounds from newly painted indoor surfaces', *International Archives of Occupational and Environmental Health*, vol 69, no 2, pp115–124, www.metapress.com/content/6y4q8y2yv4akrqc9/

Wilkinson, R. and M. Marmot (eds) (2003) 'Social determinants of health: The solid facts'. WHO Regional Office for Europe, Copenhagen, www.euro.who.int

World Bank (1990) *World Development Report 1990: Poverty*, Oxford University Press, New York, http://econ.worldbank.org/external/default/main?pagePK=64105259&theSitePK=469372&piPK=64165421&menuPK=64166093&entityID=000178830_98101903345649

World Bank (1992) *World Development Report 1992: Development and the Environment*, Oxford University Press, New York, www-wds.worldbank.org/external/default/main?pagePK=64193027&piPK=64187937&theSitePK=523679&menuPK=64187510&searchMenuPK=64187283&siteName=WDS&entityID=000178830_9810191106175

World Bank (1993) *World Development Report 1993: Investing in Health*, Oxford University Press, New York, www-wds.worldbank.org/external/default/WDSContentServer/WDSP/IB/1993/06/01/000009265_3970716142319/Rendered/PDF/multi0page.pdf

WRI (World Resources Institute) (1998) *World Resources 1998–99, A Guide to the Global Environment – Environment for Change and Human Health*, World Resources Institute, United Nations Environment Programme, United Nations Development Programme and World Bank with Oxford University Press, New York and Oxford, www.wri.org/publication/world-resources-1998-99-environmental-change-and-human-health

CHAPTER 3

History of HIA

This chapter will cover:

- Record of personal experience.
- Multiple origins of HIA.
- Evolution of HIA in the context of international development and water resource development.
- Other trends.

3.1 INTRODUCTION

A complete scholarly analysis of the origins of HIA is beyond the scope of this book. In this chapter, my purpose is to record the history of HIA as I personally experienced it. I also mention many of the people I have encountered on the way, with apologies that I cannot name all. The main trend in which I participated was associated with international development projects.

There is no obvious historical origin for HIA, simply an aggregation of trends in different regions, sectors, projects and policies. Some of those trends go back to at least the 19th century with the inquiries that took place concerning the health of populations. The hygienic improvements in water supply and sanitation in the more developed nations, and the siting of settlements in the less developed nations, involved some kind of systematic assessment of the effect of infrastructure on health.

One of the origins of HIA lies in the report published in 1842 entitled *An Inquiry into the Sanitary Condition of the Labouring Population of Great Britain* (Hamlin, 1998; Hennock, 2000). Then, as now, an intense debate was under way on the principal cause of morbidity: was it the physical environment or the socio-economic conditions? People were migrating in large numbers from the countryside to the towns, where new factories had been established. An unfettered free market ensured that labour was sold at the lowest possible price. Adults and children were working more than ten hours per day in appalling conditions and then returning, exhausted and starving, to squalid and overcrowded housing without waste disposal or clean water. Epidemics of cholera, typhoid, typhus and other communicable diseases were common in Europe. The state recognized a duty of support to the sick and the destitute, but not to the extent of enforcing a living wage. Reformers among a uniting medical profession

Table 3.1 *Lalonde's health fields*

Environment	All matters related to health external to the human body and over which the individual has little or no control. This includes both the physical and social environment.
Human biology	All aspects of health, physical and mental, developed within the human body as a result of organic make-up.
Lifestyle	The aggregation of personal decisions, over which the individual has control. Self-imposed risks created by unhealthy lifestyle choices can be said to contribute to, or cause, illness or death.
Health care organization	The quantity, quality, arrangement, nature and relationships of people and resources in the provision of health care.

recognized poverty and overwork as key health determinants, but this idea was politically unacceptable. Others, including Chadwick, focused on the filthy physical environment and this was politically popular, as it posed a risk to the wealthier class.

Local boards of health were established in the first Public Health Act of 1848. Sanitary authorities were established in 1893 and these evolved into local authorities with public health functions. The origins of town and country planning in the UK are often ascribed to public health (Cullingworth and Nadin, 2006). The first UK Town Planning Act was introduced in 1909. New towns and houses were designed to ensure standards of airspace, light, ventilation, sanitation, fire control and stability. The planning process was conceived as an opportunity for 'securing many of the necessary elements for a healthy condition of life'.

Another origin of HIA was the evolution of healthy public policy and health promotion movements. Some aspects of healthy public policy are discussed in the introductory chapter of this book. One of the health promotion landmarks is the Lalonde Report, Canada (Lalonde, 1974). This referred to four 'health fields', which could now be considered health determinant categories: environment, human biology, lifestyle and health care organization (see Table 3.1).

3.2 INTERNATIONAL DEVELOPMENT

My personal experience grew from the sector and paradigm in which I worked: academic research into medical pest control and water resource development in tropical environments. Water is a scarce natural resource that must be collected and channelled in order to increase agricultural production, generate hydroelectricity, provide for domestic and industrial usage, and for many other purposes. In warm climates, water can be a breeding site for vectors and pathogens. I was privileged to work at the Liverpool School of Tropical Medicine and to have the time and resources to reflect on these issues.

By the 1980s, pest control was dominated by the use of chemicals and there was an active discourse about integrated methods that would use environmental approaches and more selective use of pesticides. This discourse drew on the earlier experiences of the sanitary movement, before modern pesticides were invented (Snellen, 1987). At that earlier time, vector-borne diseases had sometimes been successfully controlled using environmental methods alone. During the colonial occupation of India, there were repeated expressions of concern that engineering construction should be designed to prevent malaria (Mulligan and Afridi, 1938).

A complete analysis of the origins of HIA in the context of international development is also beyond this book. However, some specific reference must be made to earlier literature in order to strengthen the argument. My own database contains the following early references and these shaped my thinking during the 1980s. This small sample serves to illustrate the rich discourse on which HIA was founded in the international setting:

- *Malaria and Agriculture in Bengal* (Bentley, 1925).
- 'The prevention of malaria incidental to engineering construction' (Mulligan and Afridi, 1938).
- 'The medical aspects of the Kariba hydro-electric scheme' (Webster, 1960).
- 'Health implications of the Kafue river basin development' (Hinman, 1965).
- 'Effects of irrigation on mosquito populations and mosquito-borne diseases in man' (Surtees, 1970).
- 'Health aspects of man-made lakes' (Brown and Deom, 1973).
- 'Water, engineers, development and diseases in the tropics' (McJunkin, 1975).
- 'Bura Irrigation Settlement Project' (Smith, 1978).
- *Health Impact Guidelines for the Design of Development Projects in the Sahel* (Family Health Care Inc./USAID, 1979).
- 'Schistosomiasis and development' (Prescott, 1979).
- 'Health and nutritional problems in the Nam Pong water resource development scheme' (Sornmani et al, 1981).
- 'Environmental health impact assessment of irrigated agricultural development projects' (WHO, 1983).
- *The Environment, Public Health and Human Ecology: Considerations for Economic Development* (World Bank, 1983).
- 'The social and environmental effects of large dams' (Goldsmith and Hildyard, 1084).

This literature demonstrates that changes in the incidence rate of tropical diseases in poor communities as a consequence of the construction of dams, irrigation systems and other infrastructure were being observed and debated. By the 1970s, there were some proactive and commissioned assessments by development agencies in the UK, US, Thailand, Australia and no doubt elsewhere.

During the 1980s, there was also a debate about the effect of international development policies on the well-being and livelihoods of poor people in poor countries. This was a critique of the prevailing model, which was thought to benefit the rich and powerful at the expense of the poor and marginalized. The policies of the World Bank were subject to multiple criticisms (Hancock, 1989; Cooper Weil et al, 1990; Caufield, 1997).

These debates led to the establishment of the WHO/FAO/UNEP Panel of Experts on Environmental Management for Vector Control (PEEM) in 1981. The aim of PEEM was to create an institutional framework for effective interagency and intersectoral collaboration and to strengthen collaboration among the participating agencies. It did so by bringing together various organizations and institutions involved in health, water and land development, and the protection of the environment. The objective was to promote the extended use of environmental management measures for vector control within development projects. It was felt, at the time, that water resource developments were being designed, constructed and operated by engineers without regard to the health consequences (see also Chapter 1 of Fewtrell and Kay, 2008).

In 1984, PEEM commissioned guidelines on forecasting the vector-borne disease implications of water resource development. I was offered the commission and set about systematizing an assessment process. My colleagues and I were unaware of parallel developments in the broader field of public health. One of my first steps was to attend a course on EIA at the University of Aberdeen. The course was supported by the European office of WHO, thanks to the foresight of Eric Girault who was regional environmental health adviser and concerned about the health and safety aspects of EIA. A collection of papers from the Aberdeen course was later published as an edited book (Turnbull, 1992).

PEEM's principal idea was derived from existing concepts of malaria control that classified risk factors as vulnerability, receptivity and vigilance. Communities had certain characteristics that made them more or less vulnerable to malaria. This enabled us to subdivide communities into groups with differential vulnerability, recognizing inequality and drawing on the work of social anthropologists. The biophysical environment had certain characteristics that made it more or less receptive to the malaria vector. Vigilance described the capacity and capability required of a health service to deal with an increased health hazard. We also borrowed from occupational safety literature by distinguishing between hazards and risks (WHO, 1987). We field tested our ideas in Zambia, Thailand, Malaysia and Pakistan. PEEM experimented with numerical systems for prioritizing impacts and finally settled for a simple ranking approach. The project was ably led by Robert Bos from the PEEM Secretariat, WHO Geneva. The guidelines were first published in 1989 (Birley, 1991). The risk factors included many, but not all, of the social, environmental and institutional determinants that are recognized today (see Chapter 9 for more details). Other colleagues who influenced my thinking are gratefully acknowledged at the beginning of this book.

PEEM guidance addressed the method of assessment but not the procedural context. We did not describe the management structure in which the assessment might be required (see Chapter 4 for more on HIA management). Looking back, I am struck by how technical it was. For me, it was a tentative step away from quantitative, predictive, computer modelling. At the same time, the biomedical scientific community to which I belonged considered it hopelessly general because it did not supply all the technical details that might be required for each disease. For example, in Africa the important malaria vector is a member of a species complex that can only be identified using genetic or molecular tools. Precise knowledge of the species is needed in order to formulate management strategies (see Chapter 9 for more details).

During preparation of this systematic approach we experimented with two computer-based methods of assessment: expert systems and hypertext (Birley, 1990). The expert system approach provides information to a computer and enables the computer to make decisions. By contrast, in the hypertext approach the computer provides information to a person and the person makes decisions. A hypertext version of PEEM guidance was published. The hypertext concept evolved into the hyperlinks that are now so commonplace on the Internet.

From 1990 to 1995 I led the Liverpool Health Impact Programme at the Liverpool School of Tropical Medicine, funded by ODA/DFID, the foreign development department of the UK government. The concepts that had evolved in a disease- and sector-specific context were generalized to other sectors, at the project level. One of our tasks was to write *Guidelines for the Health Impact Assessment of Development Projects* for the Asian Development Bank Office of the Environment (Birley and Peralta, 1992). This was written in collaboration with Gene Peralta, University of the Philippines, and is archived as a set of Word Perfect files on my website. In this guidance we referred to hazards and risks and to vulnerable communities, environmental factors and the capability of health protection agencies.

The field of discourse broadened from water resource development to include a wide range of development sectors, but still focused on economically less developed countries. The Liverpool team undertook a systematic review of the available literature linking development projects in particular sectors with known health outcomes (Birley and Peralta, 1992; Birley, 1995). The institutional role of the health sector represented by 'vigilance' was broadened. We recognized that there are many agencies responsible for protecting human health including the police and the fire brigade. We separated the procedure of impact assessment from the method. We used categories of communicable disease, non-communicable disease, nutrition and injury as well as project stages of location, planning and design, construction and operation. We added mental disorder to our list of categories and this was considered contentious because it diverged from the biomedical model. The book was published in hardback and hypertext in 1995. I have often referred to that publication as a coffee table HIA book because its message could be understood by flicking through the pages at random. At that time there was very little interest in the topic of HIA and our messages had to be simple, vivid and promotional. Our audience were primarily non-health specialists who did not believe that their favourite development projects could impact on health.

Between 1989 and 2003 I was part of a collaborative project, supported by the Danish development agency, with Robert Bos from WHO, Peter Furu from Danish Bilharziasis Laboratory and Charles Engel from University of London. The project was entitled Health Opportunities in Development (Birley et al, 1995, 1996; Bos et al, 2003). One objective of this project was to develop training material for HIA that would be accessible to a wide range of professionals who were not health specialists. Another objective was to run HIA capacity-building workshops in many countries. The focus was water resource development in less developed economies. One specific aim was to bring together government officers from the irrigation and energy sectors with the health sector in order to explore intersectoral working. We hypothesized that in order to safeguard health, the health sector needed to understand the agenda, priorities and ways of working of those other sectors.

The training material was field tested in Zimbabwe, Ghana, Tanzania, India and Honduras and then published as a manual. We developed problem-based learning materials designed for small group work. The course evolved and was adapted for many different countries, but always maintained the same format. For example, it formed the basis of training courses run in Liverpool, and later by Royal Dutch Shell in the Netherlands.

Courses were also held in the Middle East in collaboration with my colleague Amir Hassan (Hassan et al, 2005) and in WHO workshops in Southeast Asia. HIA was taught in many courses in the Liverpool School of Tropical Medicine between 1986 and 2002.

In 1995, I joined the International Association of Impact Assessment (IAIA) and first attended the annual conference, in Durban. At that conference I raised the possibility of forming a health section and including HIA alongside EIA and SIA, which then formed the main sectors of the Association. The IAIA produced an edited book on environmental and social impact assessment and this included a chapter on HIA (Birley and Peralta, 1995). In 1997, the health section of the IAIA was formed at the New Orleans conference and co-chaired by Roy Kwiatkowski and myself. At that time IAIA was still dominated by EIA and most of the literature referred to the biophysical environment. During the decade that followed, I helped the IAIA to broaden its focus to include all forms of impact assessment equally. Many long-standing members of IAIA were supportive throughout, including Rita Hamm. In addition, Robert Bos and I brokered a memorandum of understanding between WHO and IAIA that established an institutional interest in HIA. The health section was later chaired by Lea den Broeder, Supakij Nuntavorakarn, Ben Cave and Francesca Viliani (to date).

In 1995, Gene Peralta and I formulated a problem tree to explain why health risks frequently seem to increase as a result of development (Birley and Peralta, 1995). The main factors that we identified were lack of training materials, lack of institutional requirements, lack of technical skills and a lack of economic analysis. To these could be added a lack of academic interest that made publication of research papers and acquisition of research grants surprisingly difficult. More broadly, we identified a lack of political interest.

From 1997 onwards, political interest in the UK started to grow as government policy changed with the departure of the more right wing Conservative government, moulded by Margaret Thatcher, and the arrival of the more left wing New Labour, under Tony Blair. The new government was interested in healthy public policy and health inequality. This meant that new policies at all levels needed to be scrutinized for their health impacts. HIA provided an appropriate instrument. Alex Scott-Samuel, University of Liverpool, and I devised a plan to introduce HIA into the UK public health arena and we were supported by Margaret Whitehead, then professor of Public Health at the University of Liverpool. This led to the merging of concepts from our different disciplines. Greater emphasis was given to inequality and the social determinants of health, reflecting the priorities of UK public health policy in contrast to the priorities of less developed countries, as discussed in Chapter 2.

We obtained a government grant to build capacity for HIA in the UK. The grant enabled us to launch the first of an annual series of HIA conferences. These conferences have now been held in Liverpool, Birmingham, Cardiff, Dublin and Rotterdam. The conference history and proceedings are listed in 'Sources of further information' at the end of this book. We founded the International Health Impact Assessment Consortium (IMPACT), a collaborative venture between the Department of Public Health and the School of Tropical Medicine at the University of Liverpool. Other early collaborators included Mike Eastwood, specialist in environmental health, Kate Ardern, who had participated in the HIA of the Manchester airport second runway (Will et al, 1994), Debbie Abrahams and Andy Pennington. The directors of public health in the Metropolitan Borough of Liverpool, including Ruth Hussey, provided opportunities for pilot testing assessment methods and procedures. We jointly developed and ran the IMPACT HIA training course, which was the first course of its kind in the country (Birley et al, 1999). We produced the 'Merseyside Guidelines' on HIA and this became core reading for students and practitioners (Scott-Samuel et al, 2001).

Some other commissions during the 1990s were also significant. At that time, as now, we were debating whether HIA should be a stand-alone topic or whether health should be included more constructively within the existing procedures of EIA. The World Bank had defined EIA to include natural environment (air, water and land); human health and safety; social aspects (involuntary resettlement, indigenous peoples and physical cultural resources); and trans-boundary and global environmental aspects (World Bank, 1999). In 1996, the World Bank commissioned me to update its EIA sourcebook on the health aspects of EIA (Birley et al, 1997). The British Medical Association commissioned Alex Scott-Samuel, other colleagues and myself to write a book on integrating health and EIA (Birley et al, 1998). The Department for International Development (DFID) commissioned a book on the health impacts of peri-urban natural resource development, written with Karen Lock (Birley and Lock, 1999). In addition, further work took place in the Middle East, especially Bahrain, with Amir Hassan and Balsam Ahmed, which led to the production of regional guidelines (Hassan et al, 2005). Around that time, workshops were taking place in the Netherlands, organized by Lea den Broeder, and Germany, organized by Rainer Fehr. HIA was an idea whose time had come.

Although global interest in HIA was growing, unfortunately divisive influences were affecting UK university research structures. Only publication in certain high-profile journals was regarded as valid. The profile of these journals is measured by an impact factor that was based on the number of times they were cited and this was affected by fashions in science. HIA, in general, has never lent itself to that kind of publication. It is essentially holistic, intersectoral and interdisciplinary, rather than reductionist. HIA breached a biomedical model that was then increasingly oriented towards molecular biology. HIA did not supply grant income commensurate with laboratory research. Throughout the journey, there had been challenges of professional isolation and opposition. The challenges increased when I was asked by my department to stop working on HIA. This was, perhaps, an example of paradigm shift in the sense of Kuhn (1962). The two relevant paradigms could be identified as attempting to resolve the global burden of disease. There were those who seek resolution by discovering new molecules and distributing new drugs and those who seek resolution by reducing poverty, increasing equity and ensuring that proposals are beneficial. Arguably, both approaches are needed.

More recent policy developments have been summarized in the introductory chapter of this book. By 2001, some transnational corporations were adding HIA to their project planning processes and in 2002 I was asked by Royal Dutch Shell to be their senior health adviser on HIA. By 2006, the International Finance Corporation (IFC) had issued new standards that implicitly required the HIA of large projects. This has empowered practitioners and academics alike. However, at the time of writing, a career that is focused on HIA is still problematic, as is publication in high-profile journals.

3.3 SOME OTHER NATIONAL TRENDS

Roy Kwiatkowski (2004) has described the evolution of HIA in federal Canada in the context of projects and EIA. A taskforce was established in 1995 to address deficiencies in how health was assessed in EIAs. This process seems to have occurred in parallel with the development of HIA in healthy public policy described by Reiner Banken (2004) as well as other initiatives in Quebec (National Collaborating Centre for Healthy Public Policy, 2002) and Ottawa (Canadian Public Health Association, 1997). There was a separate initiative in British Columbia (Frankish et al, 1996; Institute of Health Promotion Research, 1999). The Health Canada (2004) handbook on HIA was published.

In Australia, early interest in HIA can be dated at least to concerns associated with large dam development and arboviruses (Ackerman et al, 1973; Stanley and Alpers, 1975). In 1993 I attended a WHO workshop in Kuala Lumpur on environmental health impact assessment where I met Dennis Calvert, an author of the *Australian National Framework for Health Impact Assessment in Environmental Impact Assessment* (Ewan et al, 1992; NHMRC, 1994). One of the authors of the first New Zealand guide was also present (Public Health Commission of New Zealand, 1995). In 1995, the Townsville conference on water resources, health, environment and development in Townsville, organized by Brian Kay (a PEEM member), commemorated the earlier work from the 1970s and reported new work (Kay, 1999). One of the papers described the policy context in Australia and New Zealand in 1995.

In Australia, government interest waned with a change in politics until 2000, when a new initiative was proposed by the Environmental Health Council (2001). In this context, HIA was seen as an addition to EIA and so the focus was projects and environmental issues (Wright, 2004). At the same time, there was growing interest from the Centre for Primary Health Care and Equity in the University of New South Wales through the work of Ben Harris-Roxas, Elizabeth Harris and Patrick Harris amongst others (Harris et al, 2007, 2009; CHETRE, undated). In New Zealand, government interest diminished until a change of politics in 2000 when the policy context was again discussed (Signal et al, 2006). Colleagues such as Rob Quigley and Richard Morgan continued to promote it. See the CHETRE website for a more detailed discussion of HIA in Australia and New Zealand.

In Thailand, there have been two waves of interest in HIA. The first was led by Professor Sansori Sornmani, a fellow member of PEEM (Sornmani et al, 1981; Butraporn et al, 1986). This wave was based in the Department of Tropical Medicine at Mahidol University. Various studies were commissioned by the Electricity Generating Authority of Thailand, which was engaged in the construction of large dams. For example, I was a guest speaker at the 30th SEAMEO-TROPMED Seminar 'The impact of water resources development on the health of the communities and preventative measures for adverse effects', held in Surat Thani, 1988. The second wave was broader and more associated with healthy public policy. It has been comprehensively described elsewhere (Phoolcharoen et al, 2003) and remains strong.

The devolved countries of Wales and Scotland have both developed strong advocacy and practice of HIA through the work of Eva Elliott, Liz Green and colleagues in Wales and Margaret Douglas, Martin Higgins and colleagues in Scotland (Health Scotland, undated; WHIASU, undated). In Ireland, there is a joint venture between the Republic and Northern Ireland, managed by Owen Metcalfe and colleagues (Institute of Public Health in Ireland, undated). Ireland and Wales have hosted national conferences. Important contributions have also been made by colleagues in the Netherlands, Germany, Sweden and Finland. I have had the privilege of many discussions with Rainer Fehr in Germany and Lea den Broeder in the Netherlands.

Interest has been growing in the US thanks to the work of Andy Dannenberg, Rajiv Bhatia, Aaron Wernham and colleagues. In 2009, two philanthropic organizations – the Robert Wood Johnson Foundation and the Pew Charitable Trusts – provided considerable funds for the advancement of HIA in the US (Health Impact Project, 2009). There have also been international contributions to HIA. This includes the earlier work by Bill Jobin (a PEEM member) and colleagues on water resource development, Robert Goodland at the World Bank, the support by Larry Canter and colleagues, and the later work by Gary Krieger and colleagues.

Additional work in the UK has included the establishment of an HIA listserver by IMPACT, a Wiki community website by Salim Vohra and the HIA Gateway, managed by John Kemm and colleagues.

3.4 HIA IN WHO

Within WHO, there have been several different threads that support HIA. These have been combined through the able work of Carlos Dora. One was the health policy thread, formerly managed by Anna Ritsitakis, which led to the 'Gothenburg consensus' workshop (WHO European Centre for Health Policy, 1999). This workshop brought together public health policy practitioners and produced a standard definition of HIA. There was also the work of PEEM in the water and sanitation division, which focused on water-related diseases associated with projects in less economically developed countries. The Healthy Cities movement has promoted HIA, with the able assistance of Erica Ison (see Chapter 11). In addition, there is an environmental health impact assessment (EHIA) thread promoted by environmental health divisions in headquarters (e.g. Corvalán and Kjellstrom, 1995) and regional offices (e.g. PEPAS, 1991; CEHA, undated). The determinants of health, as commonly used in HIA, include both the physical and social environment. On the other hand, the term EHIA must have a different meaning to the term HIA, otherwise there would be no need to adopt it. Presumably it is used to emphasize the biophysical determinants as opposed to the social determinants. The WHO website does not provide clarification on this point.

WHO has a normative function and it is possible that this could be used, in the future, to promote a preferred categorization (WHO, unpublished). However, WHO has its own institutional structure with an associated division of responsibilities and objectives. This leads to the disaggregation of HIA into the sub-areas listed above.

3.5 ENVIRONMENTAL HEALTH IMPACT ASSESSMENT

In the historical development of HIA some tension has developed between the terms environmental health impact assessment and health impact assessment. This seems in part to be the result of different definitions of the word environment. Strictly, environment refers to everything that is not self. In many contexts, the term environment is used to emphasize, sometimes exclusively, the biophysical. For example, one dictionary of environmental health and safety defines environmental health as the body of knowledge concerned with the prevention of disease through the control of biological, chemical or physical agents in air, water and food and the control of environmental factors that may have an impact on the well-being of people (Lisella, 1994). This definition emphasizes biophysical factors but does not clarify the meaning of environmental factors. By contrast, Lalonde (1974) defined environment as all matters related to health external to the human body and over which the individual has little or no control. This includes the physical, social and economic environment.

One solution to this conundrum is to place biophysical and social determinants in different parts of the web of causality. For example, social determinants can give rise to biophysical determinants, and vice versa, which in turn give rise to states of health (WHO, 1997). In the driving forces, pressures, states, exposures, effects, actions (DPSEEA) model, described in Chapter 5, driving forces such as economic development, poverty and other social factors create pressures on production, consumption and waste (WHO, 2005). The pressures give rise to states that include pollution levels. Exposure to these states produces effects on well-being, morbidity and mortality.

The term EHIA can also be a reflection of the proposal to integrate health within EIA (Birley, 2002). There has been a debate about integration in a number of countries, including the Philippines – *National Framework and Guidelines for Environmental Health Impact Assessment* (Philippine Environmental Health Services, 1997); Canada – the handbook on HIA (Health Canada, 2004); Australia – the national framework and other work (Ewan et al, 1992; NHMRC, 1994; Environmental Health Council, 2001; Harris et al, 2009); and, no doubt, elsewhere. The term EHIA is also used in environmental epidemiology (Hurley and Vohra, 2010).

3.5.1 Environmental health areas

The International Finance Corporation (IFC) issued an introduction to HIA to support its Performance Standards (2009). As the IFC is an influential body, the HIA methods it promotes are important. In the document, the authors chose a different approach to categorizing health impacts, in place of outcomes and determinants. They referred to their categories as 'environmental health areas' (EHAs). These are listed below:

- vector-related diseases;
- respiratory and housing issues;
- veterinary medicine and zoonotic issues;
- sexually transmitted infections;
- soil- and water-sanitation-related diseases;

- food- and nutrition-related issues;
- accidents and injuries;
- exposure to potentially hazardous materials;
- social determinants of health;
- cultural health practices;
- health services infrastructure and capacity;
- non-communicable diseases (NCDs) such as hypertension, diabetes, stroke, cardio-vascular disorders, cancer and mental health issues.

EHAs are a mixture of health outcomes, such as vector-borne diseases, and health determinants, such as exposures. Particular importance is given to the communicable diseases, which appear in perhaps five separate categories. The global importance of communicable diseases varies with region according to the epidemiological transition. In Africa, for which the EHA was designed, they are priority areas. At the other end of the epidemiological transition they are not priority areas and this suggests that the EHA do not have universal relevance. In addition, the general concept of well-being is not explicitly represented, nor is the concept of health inequalities that is discussed in Chapter 2.

The main determinants of health included in the EHA are housing issues, social determinants of health, cultural health practices, health services infrastructure and capacity, and exposure to potentially hazardous materials. When this classification of health determinants is compared with the system used in this book, a number of gaps are apparent. For example, the institutional determinants of health include health services infrastructure and capacity. But they also include a range of other infrastructures and capacities such as water supply, sanitation, power, transport, law, markets, education and police.

Environmental health is defined in the IFC document as the body of knowledge concerned with the prevention of disease through control of biological, chemical or physical agents in the air, water and food, and the control of environmental factors that may have an impact on the well-being of people. Based on this definition, social determinants and cultural health practices must be environmental factors. The list seems to exclude consideration of a range of health determinants. In the classification system that I prefer, the biophysical environment and the social environment are given equal weight. The task of assigning priorities becomes part of the assessment process for a particular proposal. It is not predetermined.

The basis for the EHA approach was work carried out at the World Bank during the 1990s (Listorti and Doumani, 2001). This work focused on infrastructure development in sub-Saharan Africa and demonstrated that a significant proportion of the ill health could be attributed to a relatively small number of environmental health issues in which intervention was practical. This in turn was based on a gradual consensus built within the World Bank regarding the need to incorporate health impacts within existing EIA practice (Birley et al, 1997; Mercier, 2003). It was also influenced by the division of responsibilities between the formal departments in the World Bank, such as housing and transport. The HIA work was sponsored by the environmental department, not the health department, and responded to its perceived needs. As a consensus grew within Europe and elsewhere with regard to HIA and the determinants of health, terminology changed. The World Bank and its partners developed a divergent consensus. A synthesis is needed.

Listorti and Doumani's achievement must not be underestimated. It was made during a period when there were many counter-arguments prevailing within the World Bank. The health department was arguing for a substantial expansion in loans to the health sector and supporting this with evidence that investment in the sector led to substantial national economic benefits. Health in other sectors was not receiving sufficient attention, despite initiatives to promote it (Cooper Weil et al, 1990). Since then, interest in policy-level HIA and health inequality has increased.

The EHA approach has one large advantage over the methods preferred in this book: it is a closed system. It prescribes in advance for the client what the HIA will contain. This is very reassuring for the client. By contrast, the preferred method is open-ended and the health impacts identified during the HIA may contain anything. The disadvantage of EHA is that one size cannot fit all. I argue that the list of what is to be included should vary between regions and sectors.

The EHA approach has influenced other guidelines including those written for the oil and gas sector (IPIECA/OGP, 2005). It promoted a debate about which approach to adopt in the mining sector, during the preparation of draft HIA guidelines by the International Council on Mining and Metals (personal observation, 2009).

In conclusion, this chapter has mapped some of the threads in the history of HIA and provided a personal account of one of them. There are many others accounts yet to be written. New strands continue to appear, including mental well-being impact assessment, policy HIA, health systems impact assessment and health equity assessment.

3.6 REFERENCES

Ackerman, W., G. White and E. Worthington (eds) (1973) *Man-made Lakes, Their Problems and Environmental Effects.* American Geophysical Union, Washington, DC

Banken, A. (2004) 'HIA of policy in Canada', in J. Kemm, J. Parry and S. Palmer (eds) *Health Impact Assessment, Concepts, Theory, Techniques and Applications,* Oxford University Press, Oxford

Bentley, C. A. (1925) *Malaria and Agriculture in Bengal: How to Reduce Malaria in Bengal by Irrigation,* Government of Bengal Public Health Department, Calcutta

Birley, M. H. (1990) 'Assessing the environmental health impact: An expert system approach', *Waterlines,* vol 9, no 2, pp12–16

Birley, M. H. (1991) 'Guidelines for forecasting the vector-borne disease implications of water resource development', World Health Organization, www.birleyhia.co.uk, accessed 2010

Birley, M. H. (1995) *The Health Impact Assessment of Development Projects,* HMSO, London

Birley, M. H. (2002) 'A review of trends in health impact assessment and the nature of the evidence used', *Journal of Environmental Management and Health,* vol 13, no 1, pp21–39

Birley, M. H. and K. Lock (1999) *The Health Impacts of Peri-urban Natural Resource Development,* Liverpool School of Tropical Medicine, Liverpool, www.birleyhia.co.uk/publications/periurbanhia.pdf

Birley, M. H. and G. L. Peralta (1992) *Guidelines for the Health Impact Assessment of Development Projects,* Asian Development Bank

Birley, M. H. and G. L. Peralta (1995) 'The health impact assessment of development projects', in F. Vanclay and D. A. Bronstein (eds) *Environmental and Social Impact Assessment,* Wiley, New York

Birley, M. H., R. Bos, C. E. Engel and P. Furu (1995) 'Assessing health opportunities: A course on multisectoral planning', *World Health Forum,* vol 16, pp420–422

Birley, M. H., R. Bos, C. E. Engel and P. Furu (1996) 'A multi-sectoral task-based course: Health opportunities in water resources development', *Education for Health: Change in Training and Practice*, vol 9, no 1, pp71–83

Birley, M. H., M. Gomes and A. Davy (1997) 'Health aspects of environmental assessment', http://siteresources.worldbank.org/intsafepol/1142947-116497775013/20507413/update 18healthaspectsofeajuly1997.pdf, accessed October 2009

Birley, M. H., A. Boland, L. Davies, R. T. Edwards, H. Glanville, E. Ison, E. Millstone, D. Osborn, A. Scott-Samuel and J. Treweek (1998) *Health and Environmental Impact Assessment: An Integrated Approach*, Earthscan / British Medical Association, London

Birley, M. H., A. Scott-Samuel, K. Ardern and M. Eastwood (1999) 'Health impact assessment training course, course report', Merseyside HIA training consortium, Liverpool Public Health Observatory, University of Liverpool, Liverpool

Bos, R., M. Birley, P. Furu and C. Engel (2003) *Health Opportunities in Development: A Course Manual on Developing Intersectoral Decision-making Skills in Support of Health Impact Assessment*, World Health Organization, Geneva

Brown, A. W. A. and J. O. Deom (1973) 'Health aspects of man-made lakes', in W. C. Ackerman, G. F. White and E. B. Worthington (eds) *Man-made Lakes, Their Problems and Environmental Effects*, American Geophysical Union, Washington, DC

Butraporn, P., S. Sornmani and T. Hungsapruek (1986) 'Social behavioural housing factors and their interactive effects associated with malaria in East Thailand', *Southeast Asian Journal of Tropical Medicine and Public Health*, vol 17, no 3, pp386–392

Canadian Public Health Association (1997) 'Health impacts of social and economic conditions: Implications for public policy', Canadian Public Health Association, Ottawa

Caufield, C. (1997) *Masters of Illusion: The World Bank and the Poverty of Nations*, Macmillan, London

CEHA (undated) Centre for Environmental Health Activities, World Health Organization, Eastern Mediterranean Regional Office, www.emro.who.int/ceha, accessed 2011

CHETRE (Centre for Health Equity Training, Research and Evaluation) (undated) 'HIA Connect, building capacity to undertake health impact assessment', www.hiaconnect.edu.au/index.htm, accessed September 2009

Cooper Weil, D. E. C., A. P. Alicbusan, J. F. Wilson, M. R. Reich and D. J. Bradley (1990) *The Impact of Development Policies on Health: A Review of the Literature*, World Health Organization, Geneva

Corvalán, C. and T. Kjellstrom (1995) 'Health and environment analysis for decision making', *World Health Statistics Quarterly*, vol 48, no 1, pp71–77

Cullingworth, B. and V. Nadin (2006) *Town and Country Planning in the UK*, 14th edition, Routledge, London

Environmental Health Council (2001) *Health Impact Assessment Guidelines*, Department for Health and Aged Care, Commonwealth of Australia, Canberra, www.dhs.vic.gov.au/nphp/enhealth/council/pubs/pdf/hia_guidelines.pdf

Ewan, C., A. Young, E. Bryant and D. Calvert (1992) *Australian National Framework for Health Impact Assessment in Environmental Impact Assessment*, University of Wollongong

Family Health Care Inc./USAID (1979) *Health Impact Guidelines for the Design of Development Projects in the Sahel: Volume 1: Sector-Specific Reviews and Methodology*, United States Agency for International Development, Washington, DC

Fewtrell, L. and D. Kay (eds) (2008) *Health Impact Assessment for Sustainable Water Management*, IWA Publishing, London

Frankish, C., L. Green, P. Ratner, T. Chomik and C. Larsen (1996) 'Health impact assessment as a tool for population health promotion and public policy', Health Promotion Development Division of Health Canada, Vancouver

Goldsmith, E. and N. Hildyard (eds) (1984) 'The social and environmental effects of large dams – overview', Wadebridge Ecological Centre, Wadebridge

Hamlin, C. (1998) *Public Health and Social Justice in the Age of Chadwick, 1800–1854*, Cambridge University Press, Cambridge

Hancock, G. (1989) *Lords of Poverty*, Macmillan, London

Harris, P., B. Harris-Roxas, E. Harris and L. Kemp (2007) *Health Impact Assessment: A Practical Guide*, Centre for Health Equity Training, Research and Evaluation (CHETRE), University of New South Wales, Sydney, www.health.nsw.gov.au

Harris, P. J., E. Harris, S. Thompson, B. Harris-Roxas and L. Kemp (2009) 'Human health and wellbeing in environmental impact assessment in New South Wales, Australia: Auditing health impacts within environmental assessments of major projects', *Environmental Impact Assessment Review*, vol 29, no 5, pp310–318, www.sciencedirect.com/science/article/b6v9g-4vtvjk2-2/2/1fd830648709abd9e78bac232eb4322e

Hassan, A. A., M. H. Birley, E. Giroult, R. Zghondi, M. Z. Ali Khan and R. Bos (2005) *Environmental Health Impact Assessment of Development Projects, A Practical Guide for the WHO Eastern Mediterranean Region*, World Health Organization, Regional Centre for Environmental Health Activities (CEHA), Jordan

Health Canada (2004) *Canadian Handbook on Health Impact Assessment*, www.hc-sc.gc.ca, accessed November 2009

Health Impact Project (2009) 'Health Impact Project, advancing smarter policies for healthier communities', www.healthimpactproject.org, accessed May 2010

Health Scotland (undated) 'Scottish Health Impact Assessment (HIA) Network', www.health scotland.com/resources/networks/shian.aspx, accessed May 2010

Hennock, P. (2000) 'The urban sanitary movement in England and Germany, 1838–1914, a comparison', *Continuity and Change*, vol 15, no 2, pp269–296

Hinman, E. H. (1965) 'Health implications of the Kafue river basin development, with special reference to bilharziasis', World Health Organization

Hurley, F. and S. Vohra (2010) 'Health impact assessment', in J. G. Ayres, R. M. Harrison, G. L. Nichols and R. L. Maynard (eds) *Environmental Medicine*, Hodder Arnold, London

IFC (International Finance Corporation) (2009) *Introduction to Health Impact Assessment*, IFC, Washington, DC, www.ifc.org/ifcext/sustainability.nsf/attachmentsbytitle/p_healthimpact assessment/$file/healthimpact.pdf

Institute of Health Promotion Research (1999) 'Canadian conference on shared responsibility and health impact assessment: Advancing the population health agenda', *Canadian Journal of Public Health*, vol 90, no S1, ppS1–S75

Institute of Public Health in Ireland (undated) 'Health impact assessment', www.publichealth. ie/hia, accessed May 2010

IPIECA/OGP (2005) 'A guide to health impact assessments in the oil and gas industry', International Petroleum Industry Environmental Conservation Association, International Association of Oil and Gas Producers, London, www.ipieca.org

Kay, B. H. (ed.) (1999) *Water Resources: Health, Environment and Development*. Spon, London

Kuhn, T. S. (1962) *The Structure of Scientific Revolutions*, University of Chicago Press, Chicago

Kwiatkowski, R. (2004) 'Impact assessment in Canada: An evolutionary process', in J. Kemm, J. Parry and S. Palmer (eds) *Health Impact Assessment, Concepts, Theory, Techniques and Applications*, Oxford University Press, Oxford

Lalonde, M. (1974) 'A new perspective on the health of Canadians', www.phac-aspc.gc.ca/ph-sp/pdf/perspect-eng.pdf, accessed September 2009

Lisella, F. S. (1994) *The VNR Dictionary of Environmental Health and Safety*, Van Nostrand Reinhold, New York

Listorti, J. A. and F. Doumani (2001) *Environmental Health: Bridging the Gaps*, World Bank, Washington, DC, www.worldbank.org/afr/environmentalhealth

McJunkin, F. E. (1975) 'Water, engineers, development and diseases in the tropics', USAID, Washington, DC

Mercier, J. (2003) 'Health impact assessment in international development assistance: The World Bank experience', *Bulletin of the World Health Organization,* vol 81, pp461–462, www.who.int/bulletin/volumes/81/6/mercier.pdf

Mulligan, H. W. and M. K. Afridi (1938) 'The prevention of malaria incidental to engineering construction', *Health Bulletin,* vol 25, pp1–52

National Collaborating Centre for Healthy Public Policy (2002) 'The Quebec Public Health Act's Section 54', www.ccnpps.ca/docs/Section54English042008.pdf, accessed October 2009

NHMRC (National Health and Medical Research Council) (1994) 'National framework for environmental and health impact assessment', www.hiaconnect.edu.au/files/NHMRC_EHIA_Framework.pdf, accessed September 2009

PEPAS (1991) 'Summary of activities', World Health Organization, Pacific Regional Centre for the Promotion of Environmental Planning and Applied Studies

Philippine Environmental Health Services (1997) *National Framework and Guidelines for Environmental Health Impact Assessment,* Department of Health, Manila

Phoolcharoen, W., D. Sukkumnoed and P. Kessomboon (2003) 'Development of health impact assessment in Thailand: Recent experiences and challenges', *Bulletin of the World Health Organization,* vol 81, no 6, pp465–467, www.scielosp.org/pdf/bwho/v81n6/v81n6a20.pdf

Prescott, N. M. (1979) 'Schistosomiasis and development', *World Development,* vol 7, no 1, pp1–14

Public Health Commission of New Zealand (1995) *A Guide to Health Impact Assessment,* Public Health Commission, Rangapu Hauora Tumatanui, Wellington, New Zealand

Scott-Samuel, A., M. Birley and K. Ardern (2001) 'The Merseyside Guidelines for health impact assessment', www.liv.ac.uk/ihia/IMPACT%20Reports/2001_merseyside_guidelines_31.pdf, accessed January 2007

Signal, L., B. Langford, R. Quigley and M. Ward (2006) 'Strengthening health, wellbeing and equity: Embedding policy-level HIA in New Zealand', *Social Policy Journal of New Zealand,* no 29, pp17–30

Smith, D. H. (1978) 'Bura Irrigation Settlement Project – Project planning report – Public health annexe', publisher unknown

Snellen, W. B. (1987) 'Malaria control by engineering measures: Pre-World War II examples from Indonesia', *ILRI Annual Report 1987,* pp8–21

Sornmani, S., F. P. Schelp and C. Harinasuta (1981) 'Health and nutritional problems in the Nam Pong water resource development scheme', *Southeast Asian Journal of Tropical Medicine and Public Health,* vol 12, no 3, pp402–405

Stanley, N. and M. Alpers (1975) *Man-made Lakes and Human Health,* Academic Press, London

Surtees, G. (1970) 'Effects of irrigation on mosquito populations and mosquito-borne diseases in man, with particular reference to ricefield extension', *International Journal of Environmental Studies,* vol 1, pp35–42

Turnbull, R. G. H. (ed.) (1992) *Environmental and Health Impact Assessment of Development Projects: A Handbook for Practitioners,* Elsevier, London and New York

Webster, M. H. (1960) 'The medical aspects of the Kariba hydro-electric scheme', *Central African Journal of Medicine,* vol 6, supplement, pp1–36

WHIASU (undated) Wales Health Impact Assessment Support Unit, www.wales.nhs.uk/sites3/home.cfm?OrgID=522, accessed May 2010

WHO (World Health Organization) (1983) 'Environmental health impact assessment of irrigated agricultural development projects', WHO/EURO, Copenhagen

WHO (1987) *Health and Safety Component of Environmental Impact Assessment,* World Health Organization

WHO (1997) *Health and Environment in Sustainable Development: Five Years after the Earth Summit,* World Health Organization, Geneva

WHO (2005) 'The DPSEEA model of health-environment interlinks', www.euro.who.int/EH indicators/Indicators/20030527_2, accessed November 2009

WHO (unpublished) 'Health impact assessment in development lending – a reference guide', WHO, Geneva

WHO European Centre for Health Policy (1999) 'Health impact assessment, main concepts and suggested approach, Gothenburg consensus paper', www.who.dk/hs/echp/index.htm, accessed 21 August 2000

Will, S., K. Ardern, M. Spencely and S. Watkins (1994) 'A prospective health impact assessment of the proposed development of a second runway at Manchester International Airport', Manchester and Stockport Health Commissions, Stockport

World Bank (1983) *The Environment, Public Health and Human Ecology: Considerations for Economic Development*, World Bank, Washington, DC

World Bank (1999) 'OP 4.01 – Environmental assessment', http://web.worldbank.org/wbsite/external/projects/extpolicies/extopmanual/0,,contentmdk:20064724~menupk:4564185~pagepk:64709096~pipk:64709108~thesitepk:502184,00.html, accessed November 2009

Wright, J. S. F. (2004) 'HIA in Australia', in J. Kemm, J. Parry and S. Palmer (eds) *Health Impact Assessment: Concepts, Theory, Techniques and Applications*, Oxford University Press, Oxford

CHAPTER 4

HIA management

This chapter will help you:

- Learn how to manage the HIA process.
- Understand screening and scoping.
- Find out how to procure competent HIA consultants.
- Check what to look out for.
- Appreciate budget, timing and integration.

4.1 PROCEDURE

The HIA procedure, or process, consists of actions that have to be taken before, during and after the assessment: see Table 4.1. These actions require two separate groups of people with different skills and experience: the managers and the assessors. The managers initiate the assessment, recruit the assessors and use the results. The assessors apply methods, collect information, analyse, prioritize, develop recommendations and report.

Some clarity about terms is needed at this point. The proposal is put forward by a proponent. The proponent is usually an institution rather than an individual, It contains managers who are decision-makers. The managers are commissioners of impact assessments. The terms proponent, manager, decision-maker and commissioner are all used in this book synonymously.

An alternative view of these stages is provided in Figure 4.1. The items on the left describe the procedure and the items on the right describe the method. These are idealized views and in practice much variation of content and terminology can be expected.

4.2 SCREENING

The first step in this procedure – screening – has received a large amount of attention and some controversy. The term derives from the process for sorting objects of different sizes. Imagine a number of proposals and suppose that there are insufficient resources

Table 4.1 *Overview of HIA procedure*

Phase	Objective	Actions include
Before	Prepare for the assessment	Screening Scoping Procurement Stakeholder engagement Prepare terms of reference
During	Do the assessment	Baseline and literature review Analysis Prioritization Recommendations
After	Check and use the assessment	Appraisal/evaluation/review Negotiation Implementation Monitoring

to conduct an HIA of all of them. The proposals have to be sorted into those requiring an assessment and those that do not. What criteria should be used?

Two different approaches to this question have evolved. The first approach is based on the assumption that the decision-makers have no knowledge or understanding of public health. They are managers who are employed to get a proposal designed and implemented. They need to ask themselves: 'could this proposal have any health impacts?' These decision-makers need very simple rules. This is sometimes called pre-screening. Typical rules might be as follows:

- If an EIA is required then an HIA may be required as well.
- If you don't know, ask the health team.
- If there is an external requirement for an HIA then do one.

The second approach is based on the assumption that the decision-makers, or their advisers, include health specialists. They will receive a brief summary description of a proposal and they will advise on whether or not an HIA should be initiated. They have a very limited amount of time and information with which to make and justify their decision.

The proposal managers have a natural desire to avoid the need for an HIA. They think it involves delays, costs and controversies. They have a target and they don't like subsidiary activities that create obstacles and which they may not understand. If they are provided with a checklist with which to make the decision they are likely to tick all the 'no' boxes. The health team know this. On the other hand, the health team has a natural desire to ensure that there is an HIA. It provides them with resources and safeguards human health while also increasing their status. The proposal managers know this.

The tensions between proponents and health impact requirements need careful management. In some transnational corporations this is achieved successfully by tight prescription of the screening process. The following example is from such a process

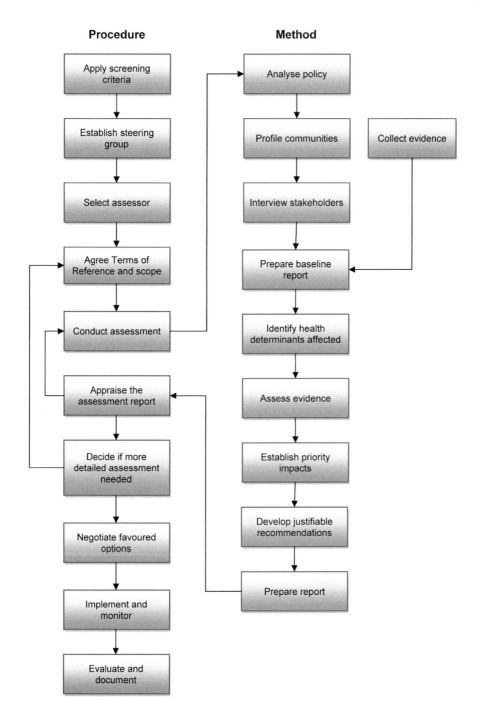

Figure 4.1 *HIA flow chart of procedure and method*

Source: based on Scott-Samuel et al, 2001, 2006; EC Health and Consumer Protection Directorate General, 2004

and refers to a competency framework that will be described in more detail in Section 4.7 (see Table 4.7):

- The screening process must be facilitated by a health professional with knowledge-level competence.
- All other participants must have awareness-level competence.
- The manager must ensure that these people are present during the screening process.
- The health professional is accountable for quality assurance.
- The screening team must include the following people: someone with a good understanding of the proposal; someone with a good understanding of the communities affected; someone with a good understanding of the likely health impacts; someone with a good understanding of the local health profile.
- The names and competencies of the screening team members must be included in the screening report.
- The conclusion must be documented in the proposal risk register and signed off by a person with skilled-level competence.

Similar tensions are likely to occur in the public sector and in financial institutions. For example, financial institutions contain project managers who are seeking to lend large sums of money to proponents. They are assisted by advisers who are seeking to ensure that safeguard policies are implemented. In the public sector, there are planning authorities who are balancing conflicting interests and there are statutory consultees who are promoting sectoral interests.

The screening criteria that trigger an HIA may be external to the proposal. They could be based on national legislation, local planning requirements, lending bank criteria or some other factor that is outside the control of the proposal managers. In EIA, typical requirements are based on the size, cost and location of the project. There is often a list of the kind of projects that must have an EIA. In HIA, there are no general criteria of a similar nature (at the time of writing).

The requirement for an EIA could provide a useful simple criterion to trigger an HIA. However, some proposals do not impact on any human communities other than the workforce. For example, a drilling rig might be sited in the deep ocean. In these cases, an HIA may not be required. But this conclusion will always require justification and consideration of the consequences of catastrophic failure. There are also instances where an HIA is desirable even though no EIA is required because the proposal has no infrastructure components. For example, a change in employment practice could have health impacts.

Experience of similar proposals of a similar size and in similar locations may provide a useful screening trigger. The best predictor of the health impacts of a new proposal is likely to be the experience of the outcomes of a similar proposal that has already been implemented. Generally speaking, impacts will be repeated unless new action is taken. Experience can be used systematically when there are a large number of small projects. For example, an oil company may wish to build many filling stations or an agricultural department may wish to build many small irrigation projects. It would be impractical to carry out an individual HIA on each small project. Instead, an HIA is carried out on one or more pilot projects and extrapolated to the rest.

Screening is usually a desk exercise but in some cases a brief site visit may be included and this is always very informative.

The outcome of screening should allocate a proposal to one of three categories:

A. The proposal could have health impacts and more HIA work is needed.
B. The proposal will have health impacts but these are well understood and manageable and no further HIA work is needed.
C. The proposal will have negligible health impacts and no further HIA work is needed.

A screening report should be produced which justifies the category selected.

A similar process is used in EIA. For example, the environmental and social screening process of the Equator Principles states that environmental screening of each proposed project is undertaken to determine the appropriate extent and type of EIA (Equator Principles, 2006). Proposed projects are assigned to one of three categories, depending on the type, location, sensitivity and scale of the project, and the nature and magnitude of its potential environmental and social impacts:

- **Category A**: A proposed project is classified as Category A if it is likely to have significant adverse environmental impacts that are sensitive, diverse or unprecedented. A potential impact is considered 'sensitive' if it may be irreversible (e.g. lead to loss of a major natural habitat) or affect vulnerable groups or ethnic minorities, involve involuntary displacement or resettlement, or affect significant cultural heritage sites. These impacts may affect an area broader than the sites or facilities subject to physical works. A full EIA is normally required. This examines the potential negative and positive environmental impacts, compares them with those of feasible alternatives (including, the 'without project' situation), and recommends any measures needed to prevent, minimize, mitigate or compensate for adverse impacts and improve environmental performance.
- **Category B**: A proposed project is classified as Category B if its potential environmental impacts on human populations or environmentally important areas are less adverse than those of Category A projects. These impacts are site specific; few if any of them are irreversible; and in most cases mitigating measures can be designed more readily. The scope of the assessment is narrower than that of Category A. It also examines the project's potential negative and positive environmental impacts and recommends mitigation measures.
- **Category C**: A proposed project is classified as Category C if it is likely to have minimal or no adverse environmental impacts. Beyond screening, no further EIA action is required.

The language used in this classification system suggests that health impacts are included, but this is usually restricted to issues like pollution. The point of the example is to illustrate how a detailed definition of the categories can be made.

4.3 AN ITERATIVE PROCESS

One approach to overcoming the resistance expressed by proposal managers at the screening stage is to recognize that screening does not have to be an all-or-nothing decision. Assessments can be carried out iteratively. At each step of the iteration the manager can ask: 'Do we have enough information to make a decision?'. If the answer is no, further resources of time and money can be allocated in order to obtain more information. The screening process itself represents the first iteration.

The resource input increases at each cycle of the iteration. The initial screening is the least resource-intensive part of the process, requiring a few staff-hours at most. This might lead to a rapid assessment process requiring one member of staff, deskwork, brief field visit, advice and a week or two of activity. The process called 'scoping' (see below) is used at each stage to decide on what is to be included in the assessment. Terminology varies. In some cases, the term 'high-level assessment' may be used while in others it may be 'preliminary HIA', 'rapid HIA', or 'scoping report'. Eventually, this may lead to the decision to undertake a full impact assessment requiring many staff, field surveys, key informant interviews and months of work. The size and complexity of the proposal will influence the decisions. Small proposals may only require a rapid assessment. Figure 4.2 summarizes screening and iteration.

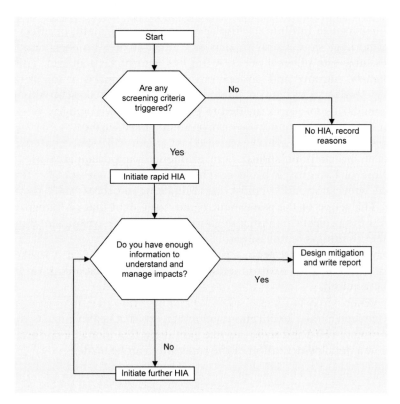

Figure 4.2 *An iterative approach to screening and HIA*

4.3.1 Screening criteria

Many of the available guidelines on HIA include lists of screening criteria. There is no unique list. The criteria need to be adapted to the sector and region of the world. Table 4.2 is illustrative. In the wrong hands, such checklists are dangerous because the 'no' answer can be given to all criteria.

4.4 RESOURCES

Table 4.3 indicates the time and resources that may be required to do the assessment during each iteration. Conducting an HIA may require both internal and external resources. The managers provide the internal resources in the form of procurement, communication and oversight. The assessors undertake the technical components of the assessment and are often external and contracted.

These time resources are indicative and depend on many factors, such as the size and complexity of the proposal.

4.5 BIAS

If a proponent pays for an HIA, there is a risk of bias. The proponent naturally wishes to have their proposal approved at minimum cost and therefore the proponent has an interest in minimizing the negative and maximizing the positive impacts identified. The consultant who prepares the HIA report wishes to remain on good terms with his or her clients and therefore has a conflict of interest. The solution depends in part on the professionalism of the assessor and in part on other checks and balances. For example, the IAIA has a set of principles that members agree to abide by.

A number of procedures have been devised for managing this bias. Examples include:

- The proponent pays for the HIA but the work is commissioned by an independent government agency such as the local authority or the local health department.
- The scope of the HIA is determined through widespread consultation that includes both professional and community stakeholders.
- A steering group is formed to oversee the process. This should have wide representation, including the communities affected by the proposal.
- The HIA report is sometimes challenged in a commission of inquiry.

Commissioning the HIA through a government department or agency does not strictly remove the bias. The public sector is just as subject to bias as the private sector, because one department may seek to achieve its objectives at the expense of other departments. There may be political commitment to a proposal that a public service manager cannot oppose. In addition, knowledge of HIA may be limited or non-existent in government departments.

Table 4.2 *Illustrative screening criteria appropriate for a large infrastructure project in a low-income economy*

Criteria (requiring yes/no answers)	Examples
Is an EIA or SIA required? Are any people affected? Then an HIA is required.	
Is the project, or construction camp, site located within 2 kilometres of homes or communal structures such as residences, commercial, schools, hospitals, or meeting halls? Would the project site reduce the buffer zone between the plant and the neighbouring community?	People living close to the project site or sometimes in between camps and works sites might be exposed to health risks, noise, injury from vehicular accidents, dust, air emissions or social issues associated with the presence of project workers and camp followers.
Is the project planned for a new a site, or does it add a new processing unit to the site that is dissimilar from existing processes?	Addition of a similar process to a site that has already been assessed is unlikely to create additional impacts. However, there may be cumulative impacts.
Has there been a previous similar project or project phase that caused adverse health impacts?	The best evidence for future impacts is the record of what has happened on similar projects in similar locations. Assume that we can repeat the same mistakes.
Could the project introduce any new chemical emissions or other hazards like noise and light to which the local communities could be exposed either routinely or in emergency?	Health risk assessment at the design level may already indicate hazardous exposures that could get into the communities' air, water and soil.
Will there be a substantial demand on public infrastructures to meet project needs, e.g. health care, domestic water supply, sewerage or transportation systems?	Public utilities may already be stretched and unable to accommodate additional burden from project workers and their camp followers.
Is there any potential that the project may generate high levels of public health concerns, expectations or perceptions?	Community fears and concerns about risk of explosions or imported diseases. High expectations of jobs and income with indirect potential health benefit.
Will this project involve the employment of temporary workers including temporary housing arrangements?	Temporary immigration of large groups of male workers without their families is a cause of much disruption and can lead to increased incidence of sexually transmitted infections, sexual harassment, violence and drug abuse. Temporary housing places strains on water supply, waste disposal, police and other services.
Do the prevailing local or international regulations, or funding agency, require HIA?	If there are funding agencies involved in the project they may have strict HIA requirements. Some countries also have guidelines they expect a company to follow.
Is any involuntary resettlement required?	Involuntary resettlement is a major challenge. World Bank guidelines require that the settlers should be at least as well off as they were before. This includes their houses, assets, communities and livelihoods.
Could the project size and scope have a significant effect on the local economy, food supply or social structure?	Projects can affect the social determinants of health including poverty and inequality.

Table 4.3 *Indicative time resources for different levels of health impact assessment*

	Level	Managers	Assessors	Note
1	Screening	2–8 hours	none	Must have sufficient competence
2	Rapid, high-level, scoping or preliminary assessment	1–3 staff-days	2 staff-weeks	Desk exercise, brief field visits, consultation with key informants and stakeholders
3	Moderate assessment	10 staff-days	2–3 staff-months	Collection and analysis of reports, widespread key informant interviews
4	Full assessment	10–20 staff-days	3–12 staff-months	Several specialist staff, social surveys, may include modelling and sampling of physical environment

There are those who believe that bias should be given a top priority because no HIA is robust until it is resolved. However, bias is simply one of many important and unresolved issues in HIA, and by no means exclusive to HIA.

4.5.1 Form a steering group

One of the most practical methods of managing the charge of bias is to commission the HIA through a steering group. Steering group members should include the proponent, government officers and representatives of the affected communities. Ideally, the steering group would control the budget, manage procurement of the consultant and take receipt of the HIA report. This will not remove the charge of bias completely, but may do much to reduce it. For example, ensuring true representation is often difficult. Steering groups are also very valuable for gaining access to relevant information, obtaining stakeholder buy-in and helping to determine the scope of the assessment.

4.6 SCOPING

Scoping provides an opportunity to decide what is to be included and what is to be excluded from the assessment. Scoping sets the boundaries for the assessment in space and time and helps identify the stakeholders. It builds upon the information available during screening and provides a first opportunity for stakeholder engagement. It helps ensure that the HIA is properly scaled, planned and executed. Poor scoping creates bias (see Section 4.6.5). The inputs and outputs of scoping can be summarized as follows:

- Inputs to scoping:
  - screening report;
  - proposal information;
  - stakeholder opinion.

Table 4.4 *The technical content and the processes of scoping*

Technical content	Processes
Geographical boundaries	Stakeholder engagement
Time boundaries	Scoping workshop
Stakeholders	Data mapping
Baseline	Terms of reference
Health concerns	Scoping report
Relevant laws and regulations	Overlap with environmental and social assessment
	Roles of experts

- Outputs from scoping:
  - scoping report
  - arrangement for additional work.

The key components of the scoping stage can be subdivided into the technical content and the processes that enable that content to occur. These are summarized in Table 4.4.

A detailed set of scoping questions has been published for the mining sector (ICMM, 2010).

4.6.1 Stakeholders

The stakeholders consist of many different community groups. Care should be taken to identify them and to make explicit which ones have been or will be included in data collection and analysis. They may be distinguished by age, gender, ethnicity, economic rank, locality, occupation, influence and more besides.

4.6.2 Geographical boundaries

The scoping process should identify the geographical boundaries of the assessment. This is sometimes called the area of influence. It has local, regional and sometimes global components. In the case of an infrastructure proposal, the most obvious local area of influence includes the site itself and the periphery. See, for example, Figure 4.3. A pragmatic decision will have to be made about how many kilometres from the site should be considered as part of the zone of influence. Typical criteria might include: downstream or downwind areas where emissions are detectable; human communities that are within daily travel distance; and linear features such as roads and shipping routes.

Regional components could include major towns and cities from which migrant workers will relocate or goods and services will be procured. Large proposals can influence the economy of an entire region. There may be cumulative impacts at local, regional and global levels and these are discussed in Chapter 12.

Including the potential consequences of catastrophic failure in the HIA may suggest other criteria. For example, major oil spills can destroy fisheries and the livelihoods that depend upon them.

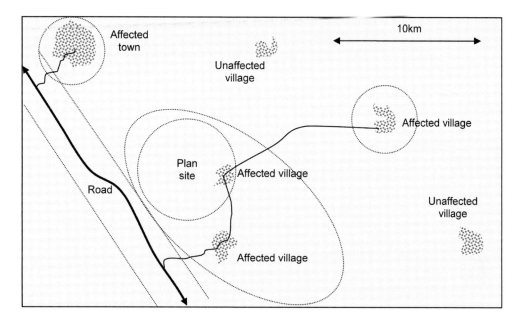

Figure 4.3 *Zones of influence of a proposal on town, villages and road*

4.6.3 Time boundaries

All proposals have four main time phases or stages. The terms used vary according to whether there is an infrastructure component or not: design, construction (or roll-out), operation and decommissioning (or deactivation). The health impacts are likely to vary with each phase.

These phases can be illustrated by reference to a manufacturing project. During the design phase the community may become aware of the proposal and apprehensive about how it will affect their lives. They may also have unrealistic expectations of the benefits it will bring them. Design may last several years. During the construction phase there may be an in-migration of construction workers and camp followers who affect the well-being of the local community and are, in turn, affected. Construction may last two years. During the operational phase there are likely to be emissions, traffic movements, flows of materials and risks of catastrophes. The workforce is probably much smaller and more permanent than during the construction and this phase may last 50 years. Decommissioning may involve removal of old plant and there may be risks of soil or water contamination from hazardous materials that have accumulated on site.

Priority health impacts will vary between these phases. During construction the priorities may include acute conditions such as injury, communicable diseases and loss of well-being (e.g. anxiety, frustration). During operation they may include more chronic conditions such as non-communicable diseases, injuries, nutritional issues and mental health issues (e.g. depression, learned helplessness).

The decommissioning phase is usually far in the future and relatively little can be said about the health priorities. However, an HIA can recommend that records are made and archived of all materials used and major operations on site. A further HIA should be undertaken for decommissioning and should have access to those records.

In the case of policies, review dates may be set in advance and these provide a specific time for undertaking a new HIA.

4.6.4 Data mapping

Scoping also provides an opportunity to identify relevant documents and other data. These may include recent policy statements, sources of health data and descriptions of the proposal. It also includes the contact details for possible key informants such as health officials, community representatives, community liaison officers and local universities. The literature is identified and categorized. References are sifted for relevance, and suitability is determined from the abstract or executive summary. Key informants are contacted and these can identify additional documents and informants. The interdependence of evidence is examined. The objective is to map the tasks ahead. The content of the data and sources of information are discussed in Chapter 6.

Information gathering of this kind is an important component of HIA and it can be frustrating. As a rule-of-thumb, it will require at least twice as much time as it should. Many key informants, including the proponent, will respond very slowly to requests for information or meetings. Frequent reminders will be needed. It is helpful to maintain a database indicating who was contacted for what purpose and on which date. The dates of all follow-ups should be recorded, together with the date on which the information was received and the item was closed. A written record should be made of all interviews and stored in the database. The database not only provides an efficient method of managing the data mapping, it also provides an audit trail to demonstrate how time was used and where delays occurred.

4.6.5 Scoping pitfalls

Many proponents have a vested interest in minimizing the scope of work because they fear that the assessment will interfere with the delivery and cost of their proposal. Consequently, they may try to exclude issues that clearly should be included. This is illustrated in the boxed example, where the individual risks to health associated with tobacco and foods with high sugar/fat content are well known and probably much higher than the risks associated with petroleum products. Acknowledgement of those risks would mean stopping the sale of the associated products. This would significantly damage the profitability of the enterprise. The managers proposed that the HIA would be conducted in two phases. The first phase would exclude the shop sales and the second phase would include them. They also argued that purchase of the non-petroleum products was a personal lifestyle choice that was the customer's responsibility and not the oil company's responsibility. In effect, they were prejudging the HIA. I opposed this view and was excluded from the assessment of the first phase. I was not invited to assist with the second phase, if there was one.

An even more extreme example of exclusion was observed in a different oil company. Here, the criteria were decided by corporate lawyers based on the need to

---

Example from the oil industry

The client was planning to build a large number of new petrol filling stations and wanted a general HIA that would cover all of them. The scope included the sale of petroleum products on the forecourt and the movement of vehicles. The scope excluded the sale of non-petroleum products in the service station shop. These products were tobacco, junk food, junk drinks and confectionery and may account for about half of net sales.

---

minimize the risk of litigation. It was well understood in the corporation that oil projects in poor countries are associated with commercial sex and that there is a risk of HIV transmission (IPIECA/OGP, 2005). They managed the litigation risk by prohibiting any mention of HIV in any written communication whatsoever. This was a world of 'double-think' where reality could not be acknowledged.

A partial solution to these scoping issues is to enable the affected stakeholders, or the steering group, to propose the scope of an HIA, rather than the proponents. In some cases, the HIA consultant can refer to national or international regulations and standards to ensure that the scope is broadened appropriately. In other cases, there is a real dilemma and new solutions are needed. Should the assessor stick to principles or do what the client asks?

4.7 COMMISSIONING AN HIA

The commissioners of an HIA may use internal staff resources or they may contract out to a specialist consultant. As HIA is unlikely to be a core component of the commissioner's staff expertise, the second option is often used.

The process may happen in several stages. First, there is an invitation to tender that describes the setting of the proposal and the main aims and objectives of the assessment. The proponent should obtain health advice when writing the tender document, but many do not. The HIA contractor will bid on the basis of this document and the bid will include a statement of competence, a list of staff and their CVs, a statement of understanding of the brief and a budget. Competence is discussed in Section 4.7.3. The client will make a shortlist of consultants and then choose the preferred bid. The choice will be made partly on the technical content of the bid and partly on the budget. In many cases, shortlisted candidates will be invited to an interview with a selection panel. Ideally, the client will be able to distinguish a good and bad bid. Commissioners will be under pressure to accept the cheapest bid, but this will not necessarily be the best. See Section 4.7.1 for an example.

There are many variations on this basic process. For example, the invitation to tender may be for an integrated environmental, social and health impact assessment (ESHIA), or for an SEA. The lead bidder is then usually an environmental consultancy company. The company may use internal staff for the social and health components, or may subcontract to specialists.

4.7.1 Example of a poor procurement process

This is an example of a poor ESHIA procurement process from the oil and gas sector (Birley, 2007). Five environmental consultancies were invited to bid (A–E). All budget information was removed from the technical evaluation. The environmental, social and health components of the bid were evaluated separately. The HIA component was scored from 5 (very good) to 1 (very bad). The scores were based on the number of staff, their experience of health and HIA, and their technical understanding of the brief. Company B received a high score because of the number and skills of the staff listed. Company C received the lowest score because the number of staff was fewest and the staff had limited health background.

The bid was won by company C because they had a strong environmental team and because they proposed the lowest budget. Their ability to complete the HIA component of the bid was poor. The consequences were that the assessment ran over time and budget, a lot of rewriting was necessary, the result was generic and the recommendations were limited.

4.7.2 Figureheads

There is a disturbing tendency for large consultancy companies to name figureheads on their bids in order to win contracts (Birley, 2007). The figureheads are internationally renowned professors or people of similar standing. These figureheads then have no further involvement in the contract after it has been won. Instead, their junior research staff do the work. A variant of this practice is known as 'bait and switch'. Figureheads are placed on the proposal but when the contract is won neither they nor their team have any further involvement. Instead, the lead contractor chooses its own internal staff, or recruits temporary junior staff to do the work. Practices of this kind do not lead to good quality and are likely to bring HIA into disrepute. Clearly, service procurers need to be on the lookout for this and instigate some monitoring to ensure that figureheads actually contribute.

4.7.3 Procurement and competence

At the time of writing, there is a global shortage of suitably qualified HIA consultants. There are probably less than 200 with suitable competencies, plus 1000 with some familiarity with HIA. The skills/competencies required of HIA practitioners are poorly defined so that procurement officers do not know who to select and do not set adequate competence criteria. Consequently, lead contractors do not know who to propose for the HIA component. There is no quality assurance. Those procuring will accept any standard – so there is no incentive to improve quality. The market is unclear and supply and demand is skewed.

Table 4.5 illustrates the strengths and weaknesses in HIA of all the main actors required for an international proposal such as a major infrastructure project (Birley, 2007). As a consequence of new performance standards, financial institutions are promoting HIA (IFC, 2006). However, those institutions employ generalist staff and

Table 4.5 *Strengths and weaknesses in HIA of international proposals*

	Strength in HIA	Weakness in HIA
International financial institutions	Condition of borrowing	Appraisal and monitoring competence
National government	Legislative and regulatory system	No legislation or regulation
Proponent	Management system, requirement for HIA	Procurement, competence, consistency
Lead contractor	Management system, logistics	Knowledge and experience of HIA
HIA subcontractor, international	Competence	Knowledge of country, experience of project sector
HIA subcontractor, national	Health background, country experience	Knowledge and experience of HIA and project sector

may not have the competence to appraise the assessment reports that are produced or to determine who is competent to undertake HIA. National governments regulate large infrastructure projects in their countries and have a suite of legislation. However, very few countries have legislated for HIA and even if they do, their officers are unlikely to have experience and competence. The proponent may be a transnational corporation with a strong internal management system. However, the proponents themselves may have little experience of HIA and be unable to procure competent services. Transnational corporations may also have difficulty ensuring that management is consistent in every country in which they operate. The lead consultant is likely to be an environmental consultancy that is well known to the proponent. They would have a strong management system and be able to send international staff to remote locations. On the other hand, they may have limited knowledge and experience of HIA and need to subcontract to an international HIA consultant. That consultant may be competent to undertake an HIA, but have no knowledge of the country and language where the project is planned. He or she may also be unfamiliar with the sector and its associated health impacts, for example, extractive industry, urban planning or water management.

A national HIA subcontractor will be required who is familiar with the country, language, local institutions, local health issues and health information systems. However, that person may have no experience or training in HIA and no experience of the sector.

For such an international proposal to receive a successful HIA, all the actors need to develop competence in HIA to an appropriate level. In order to achieve this, there must be international agreement regarding a competency framework for HIA. Table 4.6 indicates what such a competency framework might look like. The framework illustrated is based on one used in the oil and gas industry (Birley, 2005).

At the time of writing, there is no general agreement in the HIA community about such a framework and no governing body that can certify courses, promote standards or accredit individuals. There is no agreed system that procurement agencies at international or national level can use to judge competence. There are no university courses to undergraduate or master's level in HIA, although there are single lectures, part-modules and short courses. A recent initiative by the IAIA has sought to develop a

Table 4.6 *Competence framework for HIA*

Levels	How obtained	Some uses
Awareness	Attends introductory HIA course	Knows what needs to be managed
Knowledge	Attends HIA course	Can contribute to an HIA as a team member
Skilled	Experience of HIA Has attended both courses Health background	Can lead an HIA
Expert	Substantial experience of HIA International reputation Health background	Can improve HIA methods and procedures

professional standard for all impact assessment (IA) skills. At the time of writing this is in draft (IAIA, pers. comm.). It distinguishes three levels of competence: IA practitioner, senior IA practitioner and lead IA practitioner. Competencies include training, experience, understanding of methods, management, understanding of sustainable development, administrative systems, professional development and mentoring. A similar framework is being considered for IA administrators.

In the HIA competency framework outlined in Table 4.6, those seeking to procure services would require the first level of competence: awareness. They would be unable to do an HIA themselves, but would understand what needs to be managed. This understanding could be obtained by attending a short training course of a half to one day duration. The second level, knowledge, would be based on a longer training course, see below. Higher levels would be based on experience and background.

4.7.4 Content of a knowledge-level course

Managers may wish to know what a typical HIA knowledge-level course would contain. The following course outline is based on a five-day training programme that has been run successfully throughout the world for many years. The course is based on a slightly longer one that was created for use in the water resource sector (Bos et al, 2003). It has evolved into a number of different forms and been adopted by different institutions. A regular course has been run at IMPACT, in the University of Liverpool. The course is normally provided to a group of 12–24 participants who work in teams of six for the majority of the time. Participants have been drawn from government departments, local authorities, transnational corporations, UN agencies and NGOs. There have been hundreds of participants, but many of these may have had no subsequent opportunity to practice their skills and may have been reassigned to different duties. Over time, their knowledge will fade and a refresher course would be advisable.

The aim of the knowledge-level course is to introduce the participants to HIA. The course consists of five tasks:

- task 1 – the context for HIA;
- task 2 – the methods used;

Table 4.7 *Summary of task aims and objectives for knowledge-level course*

Task	Aim	Objectives
	The aim of this task is to:	By the end of this task you will have had an opportunity to:
1	Put HIA into context in terms of what it is and how it works. Provide you with the background you need to make the most out of the rest of the course.	Develop a working definition of health and HIA. Consider examples of health impacts. Gain a contextual base to the work you will undertake on the course.
2	Explore methodological questions by undertaking a rapid HIA of a relevant case study.	Consider what might be involved in conducting an HIA. Arrange the problem into logical components. Focus on the methodological components.
3	Explore procedural questions associated with HIA.	Decide when an HIA is needed. Examine what should happen before and after the actual assessment. Consider how to assure the quality of an HIA.
4	Explore different approaches to prioritization taking account of perceptions of risk, resource constraints and interests of project proponents.	Explore different approaches to setting priorities and their limitations. Consider the perception of the health risks and gains of each stakeholder community.
5	Consider what constitutes a 'good' HIA study. Learn how to critically appraise aspects of the procedures and methodology of HIA.	Revise what you learned in the earlier tasks. Critically appraise aspects of an HIA. Check elements of an HIA. Consider the issues that contribute to successful HIAs. Appreciate the extent to which the various constraints may influence the outcomes of an HIA.

- task 3 – the procedures used;
- task 4 – how to determine priorities;
- task 5 – critical appraisal/review of a completed assessment.

Each task has aims and objectives and these are summarized in Table 4.7. As the course has been adopted and adapted, these aims and objectives have evolved to fit new settings.

There have been attempts to adapt this HIA course for an e-learning environment. But these have not always been successful. One reason for this is that professional staff have been asked to complete the e-course on top of their existing duties. By contrast, when attending a physical course they are away from their desks for the whole period.

The course has an explicit aim of transferring knowledge. But there is also an implicit (or sometimes explicit) aim of building intersectoral work skills. The participatory nature of the course means that attendees have the opportunity to learn from each other, acquire informal networks of contacts, learn the jargon, agendas and assumptions of colleagues in other disciplines, and improve communication skills. Much of this depends on face-to-face interaction and ensuring attendance at the whole course (e.g. by holding it in a venue distant from homes and workplaces).

4.7.5 Necessary and desirable skills

Ideally, HIAs would be led by experienced and trained practitioners with the skill level of competence. As there is a shortage of such people, a lower level of expertise may have to be accepted. There is an unfortunate tendency to assume that any medically qualified person is competent to undertake an HIA. This is a mistake. On the other hand, education in one of the many health-related disciplines is likely to ensure a basic understanding of the issues.

Table 4.8 lists the necessary and desirable qualities of a lead HIA consultant and the evidence required to demonstrate it. Consultants who do not meet the minimum requirements can be offered training at the initiation stage, but this has financial implications. Suitable people for training may have a background in public and/or environmental health and a demonstrable interest in environmental and social, as well as health, issues. Contractors should be informed in the invitation to bid that they must propose only suitably qualified consultants. It is desirable that lead HIA consultants are skilled as teachers so that they can transfer their knowledge to their team.

4.7.6 Specialist or generic skills?

Many, but not all, of the skills required to do HIA are generic. They are part of the normal skills required of a public health consultant. Generic skills include the following (Kemm, 2007):

- project management;
- negotiation;
- team working;
- community engagement;
- research;
- robust common sense;
- ability to pull together disparate elements;
- capacity to persuade different people to cooperate.

It has been argued that HIA does not require rare skills and a new profession. Many people are capable of conducting HIAs and their confidence can be promoted. In this view, the focus should be on the quality of the product rather than qualifications of the assessor. This perspective has merit for individuals with strong public health backgrounds. However, HIA may require the exercise of skills that are not routinely used.

Table 4.8 *Person specification for a technically competent, lead HIA consultant*

	Necessary	Desirable	Evidence
Education	Degree in a health-related subject	Postgraduate degree in environmental or public health	Degree certificates
Training	Training in environmental or public health	Training in HIA Training in project management	Certificates of attendance, name of teacher, syllabus
Experience	Consultancy Impact assessment Participating, leading or managing an HIA	Leading an HIA on relevant project	List of HIA projects completed
Interests	Membership of professional bodies active in impact assessment such as IAIA Has attended HIA conferences	Knowledgeable about social, health, international development, inequality or environment issues	Activities and affiliations listed in CV List of HIA conferences and workshops attended
Skills	Skills in at least one of fieldwork, analysis or management Can review and summarize reports, weigh evidence, produce logical/analytic arguments, interview key informants, demonstrate professional impartiality and diplomacy Persistent at uncovering sensitive information	Skills in more than one of fieldwork, analysis, management or teaching	Copy of previous assessment reports highlighting own contributions
Other	Familiarity with the region or sector associated with the proposal	Has worked in locality	List of projects undertaken

HIA probably requires a combination of specialist and generic skills. A competency mapping process is required, and has been proposed in the UK, in order to identify and separate the two. To date, resources for completing such a competency analysis have not been available. In addition, each setting requires specialist skills. For example, when working in warm climates a good working knowledge of tropical diseases can be crucial. See also Chapter 9.

4.7.7 Terms of reference

The terms of reference (ToR) represent the contractual agreement between client and consultant about the process, resources and content of the HIA. The ToR may be based directly on the bid document and the brief, or may be developed later. For example, the consultant may participate in the scoping process and the ToR may be one of the outputs. It is an important document because the HIA report should be judged, or appraised, according to how well it meets the requirements of the ToR. The HIA report can only be as good as the ToR. Managers should only expect the consultant to do what is in it.

A ToR is likely to include the following items:

- the competence required of the consultant;
- an outline of the scope of the work, i.e. what is to be included and excluded;
- the methods and tools to be used in the assessment;
- the form and content of the outputs and any conditions associated with their confidentiality, production and publication;
- the nature and frequency of meetings with the steering group;
- the budget and source(s) of funding;
- a timeline, including deadlines and milestones;
- the manager or agency to which the consultant is reporting;
- contractual details such as insurance and cancellation.

I would like to provide a template ToR for management to use. However, the task has proved too difficult. Together with the IAIA health section, I drafted a detailed generic ToR using an oil project ESHIA as an example. The document was about 30 pages long and remains unfinished.

4.8 DURING THE ASSESSMENT

The content and method of assessment is explained in detail in Chapters 5–9. Once the consultant is contracted and the assessment is under way, there are still management duties for the commissioner.

One of these is to ensure that the consultant has full access to all relevant documentation, key informants, stakeholders and health statistics. Access to available information is often restricted by political or commercial considerations; it can be facilitated by ensuring that the right people are supporting the assessment. This might include inviting government officers onto the steering group or running a local HIA training course for key stakeholders (Birley, 2007).

4.8.1 No surprises

The commissioner should plan for regular communication with the consultant during the preparation of the draft HIA report. The commissioner should be kept informed as impacts are identified and prioritized, and recommendations are developed. There should be no surprises. The commissioner should not learn about difficult issues for the first time when they read the draft report. The assessor should not learn of changes in the proposal, decisions that have already been taken or external influences after they have delivered the report. Impact assessment should be a consensus-building process and this requires that the commissioner's team is engaged. See Chapter 8 for further discussion.

4.9 INTEGRATION

In some cases there is an intention to integrate the HIA with EIA and SIA, often called ESIA or ESHIA. The management of an integrated assessment provides a number of additional challenges, including report content and budget.

Health impacts are the consequence of interactions among a wide range of health determinants. The health determinants are broadly biological, environmental and social. Many of the outputs of environmental and social assessments can be regarded as statements about changes in the determinants of health and, therefore, as inputs to HIA. For example, the impact on air quality is considered in the EIA. Air quality is a determinant of health of asthmatics. As another example, the impact on income distribution is considered in the SIA. Income distribution is a determinant of health inequalities. It follows that there are both parallel and sequential components to integration of impact assessments.

A well-integrated report would contain cross-references between the EIA, SIA and HIA sections. If the outputs of EIA and SIA are often inputs to HIA, it follows that the number of cross-references is likely to be greater in the HIA. This is illustrated by the following example from the completed ESHIA of Sakhalin Energy (Birley, 2003; Sakhalin Energy Investment Company, 2003). See Table 4.9.

Figure 4.4 illustrates a possible scheme for managing the information flow (Birley, 2003). First, the scope of the assessment must be allocated to the three sub-components. As there is considerable overlap, the allocation is likely to be made on pragmatic grounds and with appropriate budgeting. Second, the three assessments should take place in parallel. The larger box for HIA in the figure does not imply that it is a larger or more important task. It implies, simply, that completion of the HIA may have to await the outputs of the EIA and SIA. Next, a set of integrated recommendations would be generated for safeguarding and enhancing the health of stakeholders, environmental quality and social well-being. The executive summary would provide the final opportunity for integration. A critical appraisal determines whether all the elements identified during scoping have been addressed in one or other of the reports, and provides an opportunity for iteration. Appraisal and budgeting are described in more detail below.

Table 4.9 *Number of cross-references in an ESHIA report*

From	HIA	EIA	SIA
To			
HIA	–	6	3
EIA	37	–	15
SIA	46	15	–

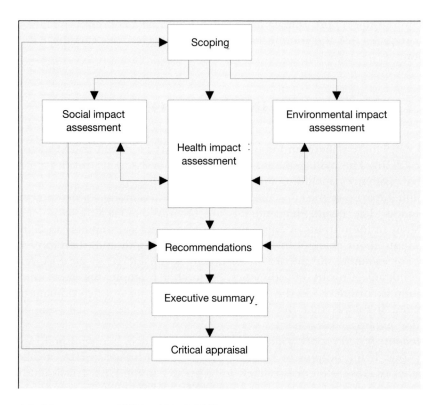

Figure 4.4 *Management of HIA within ESHIA*

As an example, Table 4.10 summarizes how the ESHIA produced for the Sakhalin Energy project approached the issue of water supply from the environmental, social and health perspectives (Sakhalin Energy Investment Company, 2003).

Table 4.10 *Example of drinking water supply in the ESHIA for Sakhalin Energy*

EIA	Projects should be designed to avoid contaminating potential drinking water supplies. Safe drinking water supplies should be provided for staff.
SIA	Drinking water was identified as a major concern of the community. An inventory of existing infrastructure was made.
HIA	An inventory of contaminated supplies was made. The prevalence of medical conditions associated with contaminated water was listed.
Executive summary	The summary acknowledged that many community water supplies would not meet company safety standards without upgrading or repair. It also suggested that some company facilities should use public supplies if there was sufficient surplus.

This report went further in achieving integration than other reports available at the time. However, despite acknowledging the importance of a safe water supply for community and project, the executive summary did not capture the conclusions from the three assessments and the recommendations in the report did not suggest how this joint requirement could be achieved.

4.10 APPRAISAL, REVIEW OR EVALUATION

There are multiple terms in use for this step. I use the term 'appraise', but others prefer review or evaluate. Commissioners of HIA studies have both a right to expect a quality product, and a responsibility to ensure that resources are properly spent. As such it is important to appraise the HIA to ask such questions as whether:

- the evidence presented in the HIA is rigorous, adequate and appropriate;
- appropriate conclusions have been drawn from the evidence;
- the HIA is a quality product;
- the recommendations are practical and appropriate;
- suitable procedures and methods were used.

The appraisal enables the commissioner, or reviewer, to determine the quality of the report and to decide to:

- accept the report as it is;
- require minor amendment;
- return the report for a major rewrite;
- reject the report.

A quality assurance stage is critical. As with all other aspects of management, there is a need to have explicit management systems that ensure quality. The task of managing appraisal is the responsibility of the commissioner. A peer review process should be included whenever possible; this should be budgeted from the outset and should require 1–4 days, depending on the complexity of the proposal.

In many cases, appraisal will be undertaken by clients who have no knowledge of health or HIA. A recent example derived from a motorway project in a poorly developed region with various deficiencies in driver behaviour. It would require relatively large numbers of construction workers, recruited in part from a different country. The feedback from the client, summarized in the box, was anonymous but appeared to be from civil engineers. The report highlighted management of worker–community interactions, driver behaviour, obesity associated with private motorized transport and provision of accident and emergency services. In addition, the client did not believe that the motorway's contribution to climate change was relevant. The feedback illustrates the vast gulf separating the perspectives of the assessor and the client. It was uninformed but indicated where additional explanation was required.

As yet, there are no universally accepted tools for appraisal. However, a tool has recently been proposed for use in UK development planning consent (Fredsgaard et

---

Examples of real feedback received for a scoping report

'Assessor has made a negative statement suggesting project can have an impact on STIs which appears unfounded.'
'I believe there are no problems with workers living in local communities – why assume problems where there is no evidence for this?'
'Link between motorway and nutritional issues (obesity) seems far fetched.'
'The report implies that the motorway will lead to more accidents whereas the motorway will be built to highest safety standards.'

Table 4.11 *Outline of the HIA appraisal tool*

Areas	Categories
1 Context	Site description and policy framework Description of project Public health profile
2 Management	Identification and prediction of potential health effects Governance Engagement
3 Assessment	Description of health effects Risk assessment Analysis of distribution of effects
4 Reporting	Discussion of results Recommendations Communication and layout

al, 2009). The authors suggest that it can be extended to other contexts. This admirably concise tool has been extensively peer reviewed and draws on published literature. It subdivides the appraisal into four areas, 12 categories and 36 subcategories: see Table 4.11. It provides a worksheet for scoring each subcategory. It represents the ideal situation and this may rarely be achieved in practice. It does, however, provide a valuable framework for the appraisal process.

Additional elements to include in an appraisal follow (Birley, 2003; Bos et al, 2003):

- the suitability and adequacy of the terms of reference;
- potential conflicts of interest including financial obligations;
- the potential bias and traps in the HIA and whether these might have affected the findings and recommendations;
- the skills needed to conduct the HIA;
- the validity of the data used as evidence to support the HIA's conclusions;
- the accessibility of the HIA report;
- an evaluation of the feasibility and acceptability of the recommendations;

- consideration of the measures needed to implement and monitor HIA recommendations.

4.10.1 Method quality

In order to appraise the method it is advisable to consider whether an explicit classification process has been used, such as the determinants of health. The literature review should be focused, rigorous, valid and systematically undertaken. There are separate tools for evaluating the literature review and these are described in Chapter 6. Evidence cited should be appropriate and gaps should be clearly identified. The text should support the conclusions reached. For example, assumptions should be made explicit; appropriate conclusions should be drawn from the evidence; and recommendations should be practical, acceptable and achievable.

The objective of HIA

The objective of HIA is not the discovery of absolute scientific truth or publication in peer-reviewed journals. Rather it is to make practical and acceptable recommendations to proponents and stakeholders for safeguarding and promoting human health based on a reasoned judgement about health outcome. The report should be appraised with this objective in mind.

4.10.2 Process quality

A competent assessor may produce a poor report because of management deficiencies. Ideally, the appraiser should be able to investigate this. The following are examples of process issues:

- The suitability and adequacy of the ToR are crucial, as the report should address all the issues identified in it. It should be available to the appraiser.
- Consultation with stakeholders should be included. Was public participation adequate? Were vulnerable groups and differential impacts included?
- There may be evidence of bias associated with conflict of interest and financial obligations. For example, did the assessor report directly to the commissioner or to a steering group?
- The timeliness of the assessment may be inappropriate with regard to the project cycle. For example, an assessment that starts after the concrete has been poured is untimely.
- The competence and experience of the assessment team may be inappropriate. The assessors may be specialists in a small component of the problem and give this excessive attention, at the expense of less familiar and more important components. For example, infantile diarrhoea is far more important than outdoor air pollution in

---

Example of interface

A sewage treatment plant in the Middle East was designed to discharge its product into a channel that was in the jurisdiction of the municipal authority. The plant manager observed illegal abstraction of the product from the channel for irrigation of salad crops, but could not take action because the product had left his jurisdiction.

---

many low-income countries; but outdoor air pollution may receive more attention because it is a standard component of EIA.
- The budget for the HIA may be inadequate and far smaller than the EIA budget (see below).
- The report may be written in a style that is inaccessible to a specific audience.
- The recommendations may be unrealistic, imprecise and unachievable.
- The interface with associated assessments and reports should be clear. These may include, for example, EIA, SIA, health risk assessment, transport assessment, sustainability assessment, inequality assessment, health systems assessment, health needs assessment and social investment plans.
- The jurisdictional boundaries should be clear.

In cases of integrated impact assessment, the appraisal should have the additional elements introduced in Section 4.9. These could include the following:

- The HIA report should have many cross-references to the other impact assessment reports.
- Results of the social and environmental assessments should be referred to in the analysis of health impacts.
- There should be evidence that health, social and environmental assessors worked together.
- Health recommendations should be related to other recommendations.
- Stakeholder groups identified should be consistent with those in the social assessment.
- Geographical and time-phase boundaries should be identified consistently in all the assessments.
- The executive summary should draw together results from all the assessments.

In addition, the appraisal should consider how the budget was distributed between the components. This is discussed below.

4.11 BUDGET

Time and other resources required for an HIA need careful budgeting. A common tendency is using far too much of the budget on baseline studies (health profiles

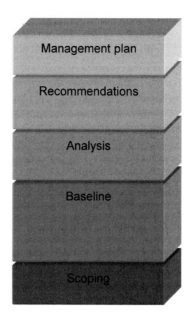

Figure 4.5 *Distribution of the budget between different activities*

and literature review) and leaving far too little for analysis and development of recommendations. As a rule of thumb, it is a good plan to allocate less than 50 per cent of the total HIA budget to baseline studies.

The purpose of the HIA is to make justified recommendations for safeguarding and enhancing health. It follows that analysis, development of recommendations and the management plan are the most important parts of the impact assessment.

Figure 4.5 illustrates how resources could be subdivided between some of the main activities. The diagram is only intended to be indicative. It places baseline studies in the larger context of what is required.

4.11.1 Budgeting for integrated assessment

Integrated assessments (ESHIA) have an additional dimension to the budget distribution. The main budget holder for the ESHIA is usually an environmental consultancy company. The managers in that company usually have a strong EIA bias. They are familiar with delivering comprehensive EIA reports as part of a statutory requirement. Such reports often have ten or more main sub-components including biodiversity, archaeological heritage, pollution modelling, transport modelling and water management. They may regard the HIA as an additional sub-component, requiring only 5–10 per cent of the total ESHIA budget. A health specialist, by contrast, can envisage ten or more sub-studies on health issues alone. As an example, these may include health and social care delivery facilities at primary, secondary and tertiary level; health information systems and epidemiology; communicable disease control and environmental health; inequality; and psychosocial well-being.

Ensure a reasonable budget

The team responsible for the HIA are advised to ensure that a reasonable percentage of the budget is allocated before they agree to be part of the bid.

A more appropriate division of the budget is illustrated in Figure 4.6. The division is indicative and depends on circumstances. If no human community could be affected then the health component would be minimal. The division of the budget assumes that the environmental component, EIA, receives the largest slice of the budget because it is a statutory requirement, not because it is more important. If health is only allocated 5 per cent of the budget, as frequently happens, then a good product is unlikely.

4.12 TIMING

Proposals go through a process of evolution from rough outlines through to detailed specifications. The objectivity and effectiveness of the report could be limited by the stage in this process at which the assessment is conducted. Impact assessments that are started too soon have too little information to draw upon. Impact assessments that are started too late, after the ground has been cleared, cannot make recommendations associated with the siting and design of infrastructure.

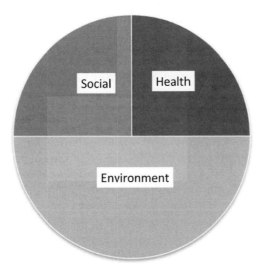

Figure 4.6 *Indicative distribution of the budget between different kinds of assessment*

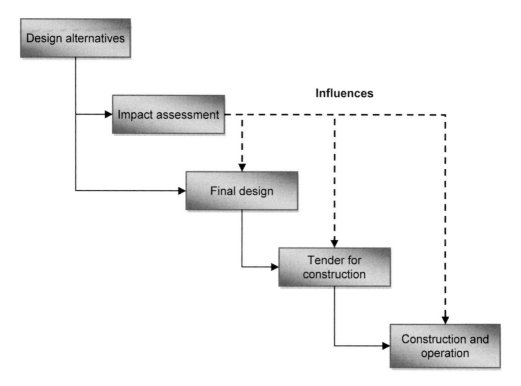

Figure 4.7 *Points of influence*

Project impact assessment is best performed when there are design alternatives to consider and before the construction contract (or equivalent) has been sent out for tender. For example, there may be a need to include recommendations about the clauses in the construction contract. In the case of programmes and policies the timing details are different but similar considerations apply.

Figure 4.7 illustrates the stages that an HIA should seek to influence in the case of a project with a construction component. This ideal is rarely achieved and it is common for an HIA to be undertaken concurrently with construction, when opportunities for influencing design are no longer possible.

The timing of an HIA in relation to other planning and design activities in the project cycle determines its value. Often this is out of the consultant's control. Inappropriate timing may easily render the entire study useless. The critical appraisal of an HIA should determine when the assessment was undertaken and how this timing affected recommendations.

4.12.1 Examples of timing

In our training courses we often provide the following three examples of project timing and ask participants to discuss which example provides best practice. In case you wish to think about it, one possible answer is at the end of the chapter.[1]

4.12.1.1 Example A

The consultant carried out an HIA of a new project based on a two-page outline of a proposal that was being submitted by the investment group. The site was identified but there were no details about potential numbers of users or types of activities.

4.12.1.2 Example B

The consultant carried out an HIA of a new project while it was being constructed. Decisions had already been made about the facilities to be offered. The site had been surveyed and an EIA report was available.

4.12.1.3 Example C

The consultant carried out an HIA of a new project at the time the proposal was being first prepared for consideration by a planning authority. Two alternatives were available and the content of each had been outlined.

4.13 MANAGEMENT PLAN

Other chapters in this book consider the baseline studies, analysis, prioritization and development of recommendations in detail. These are all elements of the HIA report. The report is received by the proponent, appraised using peer and community review, edited, modified and accepted. The final step is to develop, and cost, a management plan for implementing the recommendations and monitoring that implementation. See Chapter 8 for further discussion of recommendations and management plan.

4.14 EVALUATION AND MONITORING

After the HIA has been accepted, management plans have been agreed and implementation has started, there is sometimes an opportunity to evaluate the whole process. This is discussed in more detail in Chapter 12. The results of such evaluations provide an opportunity to improve the effectiveness of HIA. There are a number of different forms of evaluation and these include process, impact and outcome.

Process evaluation is used to answer questions like: was the timing and content of the assessment appropriate? It provides an opportunity to evaluate how successful the process was in practice. It also provides an opportunity to ask the community whether they felt involved – whether participation was achieved (Kemm, 2007).

Impact evaluation monitors the acceptance and implementation of recommendations. Outcome evaluation monitors indicators and health outcomes after the proposal has been implemented (Mindell et al, 2003). Time may reveal the accuracy with which health impacts were identified. However, there is a counterfactual argument: certain outcomes may have been avoided as a result of the assessment and these cannot be observed. Outcome evaluation is complex and there are many methodological difficulties. There may be long latency periods. The composition of the affected population may change. Adjustments may need to be made for confounding factors.

As we cannot know the future, monitoring of health indicators provides an opportunity to adjust the proposal as the future unfolds. Plans and policies often have predetermined review points. Well-implemented projects often have general monitoring stages as part of project management. This is discussed in more detail in Chapter 8.

4.15 CASE STUDY OF HIA PROJECT MANAGEMENT

A few years ago I oversaw an HIA of a gas field development in a remote desert in northern China. The process was challenging and during its course I made a decision to accept a substandard report that I have since regretted. This is the main learning point of the example. The challenges included local inexperience of HIA; language and communication; logistics; entrenched cultural differences; scarcity of data; and limited budget.

The project was under the control of a Chinese company that was part owned by a transnational corporation. The transnational corporation had established a policy and programme for HIA at corporate level and was rolling this out to its associated national companies. The Chinese company had no knowledge or experience of HIA, but understood that it had to comply with corporate requirements. Following discussions with senior management in Beijing, it was clear that these managers could not distinguish between HIA and workforce health and safety management. The company doctor was assigned as the local manager of the HIA and his duties were to request my assistance, control the budget, learn from the experience and to consider my suggestions.

A local Chinese consultancy group was needed to undertake the HIA at the same time as the EIA. A consultancy company responded to the ToR and submitted a competitive bid. The financial side of the bid was managed by the Chinese company and I was not consulted. The local environmental consultancy had considerable experience of EIA and was based at a prestigious state university located several hundred kilometres from the project site. They had no knowledge or experience of HIA and no health professionals on their team. However, there was a health department within the university from which public health professionals could be recruited. For example, one of the professors in the health department was an international expert on Kaschin-Beck, a serious bone disease caused by selenium deficiency that was endemic in the area. One of the health concerns, identified during the scoping stage, was whether the proposal would cause sufficient agricultural disturbance to exacerbate the environmental conditions responsible for this disease. The professor agreed to be part of the team and assigned a junior research assistant to undertake the work. There were substantial language difficulties and I relied on the company doctor to translate my suggestions and requests.

A steering group was established that included representatives from the local health department in the town nearest to the project site. I ran a three-day training course on HIA in the town and this was attended by the consultancy team of environmental experts, the health professor and his research assistant, the public health officials and the company doctor. Prior to the course, the consultancy team and I made a quick inspection of the site and the surrounding area and held meetings with some key informants. My initial impression was that although there were many public health issues in the area, the proposal would only have a minor impact on them.

Following the training course, the consultancy team undertook the HIA. Later it prepared draft HIA reports in Chinese and translated them into English. The quality of the drafts was very poor. For example, the health baseline relied on national rather than local data.

After four iterations, including much rewriting by myself, the report was barely passable but further improvement did not seem practical. A final steering group meeting was scheduled in the local town to consider the report. Two and a half days of air travel were required to reach the locality from Europe. During the meeting, the representatives from the local public health department objected to some of the contents of the report. Their objections seemed rather minor at the time and I proposed that a small amount of editing following acceptance of the report would be sufficient.

In retrospect, this is the decision that I regret: if the local steering group members were not happy with the report then I should not have accepted it. After the meeting, I learned that the consultancy team had made no further visits to the area after the training course because they had no budget to do so. They had no meetings with key informants or with the community and had not made any significant attempt to identify and report local health data. The health professor had not been involved in the assessment and the knowledge and understanding of the junior research assistant was extremely limited. The report had been written by the environmental specialists as a desktop exercise and at minimum cost. As a result of distance, language and cultural differences I was ill informed about the actual situation. I had made far too many allowances for the challenges listed above and had to accept, as an achievable goal, that the HIA would merely provide the first step in a capacity-building process.

4.16 ACCESSIBILITY OF HIA REPORTS

HIA reports should be simple, clear and comprehensive because they may be read by different audiences. Examples of these audiences include local residents' groups, voluntary sector agencies, primary care groups and/or health authorities, local and central government officers and elected members, boards of directors, public inquiries, press and media, and academic researchers. Accessibility is enhanced by language, layout, clarity, presentation, affordability and means of dissemination. The report might need to be adapted for specific audiences. For example:

- A board of directors may just need an executive summary.
- The press and media need a press release.
- Academic researchers need the entire report and any other documentation.
- Local residents need a synopsis of the report in their own language.
- A public inquiry needs all the supporting documentation.
- Publication on relevant websites.

4.17 EXERCISES

See if you can answer the following questions (the answers are at the end of the chapter):

- What is the main purpose of scoping?[2]
- Which stakeholder group is most important in HIA?[3]
- What is the main purpose of engaging with stakeholders?[4]
- What competence is required to lead and undertake an HIA?[5]
- What percentage of the assessment budget should be devoted to HIA for an offshore platform with no additional onshore components?[6]

Take another look at the San Serriffe exercise that was introduced in Chapter 2. Please identify some appropriate geographical boundaries for the assessment and write a sentence justifying each one.

4.18 REFERENCES

Birley, M. (2003) 'Health impact assessment, integration and critical appraisal', *Impact Assessment and Project Appraisal*, vol 21, no 4, pp313–321

Birley, M. (2005) 'Health impact assessment in multinationals: A case study of the Royal Dutch/ Shell Group', *Environmental Impact Assessment Review*, vol 25, no 7–8, pp702–713, www.science direct.com/science/article/b6v9g-4gvgt8v-1/2/01966b5af4f9ae9ecd390e4dd382a5a3

Birley, M. (2007) 'A fault analysis for health impact assessment: Procurement, competence, expectations, and jurisdictions', *Impact Assessment and Project Appraisal*, vol 25, no 4, pp281–289, www.ingentaconnect.com/content/beech/iapa

Bos, R., M. Birley, P. Furu and C. Engel (2003) *Health Opportunities in Development: A Course Manual on Developing Intersectoral Decision-Making Skills in Support of Health Impact Assessment*, World Health Organization, Geneva

EC (European Commission) Health and Consumer Protection Directorate General (2004) *European Policy Health Impact Assessment: A Guide*, http://ec.europa.eu/health/index_en.htm, accessed July 2009

Equator Principles (2006) 'The Equator Principles', www.equator-principles.com, accessed October 2009

Fredsgaard, M. W., B. Cave and A. Bond (2009) *A Review Package for Health Impact Assessment Reports of Development Projects*. Ben Cave Associates, Leeds, www.apho.org.uk/resource/item.aspx?rid=72419

ICMM (2010) *Good Practice Guidance on Health Impact Assessment*, International Council on Mining and Metals, London, www.icmm.com/document/792

IFC (International Finance Corporation) (2006) *Policy and Performance Standards on Social & Environmental Sustainability*, www.ifc.org/ifcext/enviro.nsf/content/envsocstandards, accessed April 2008

IPIECA/OGP (International Petroleum Industry Environmental Conservation Association, International Association of Oil and Gas Producers) (2005) 'HIV/AIDS management in the oil and gas industry', IPIECA, London, www.ipieca.org

Kemm, J. (2007) 'What is HIA and why might it be useful?', in M. Wismar, J. Blau, K. Ernst and J. Figueras (eds) *The Effectiveness of Health Impact Assessment*, European Observatory on Health Systems and Policies, Brussels, www.euro.who.int/en/home/projects/observatory/publications

Mindell, J., E. Ison and M. Joffe (2003) 'A glossary for health impact assessment', *Journal of Epidemiology and Community Health,* vol 57, pp647–651, http://jech.bmjjournals.com/cgi/reprint/57/9/647.pdf

Sakhalin Energy Investment Company (2003) 'Health, social and environmental impact assessments', www.sakhalinenergy.com, accessed March 2003

Scott-Samuel, A., M. Birley and K. Ardern (2001) 'The Merseyside Guidelines for health impact assessment', www.liv.ac.uk/ihia/impact%20reports/2001_merseyside_guidelines_31.pdf, accessed January 2007

Scott-Samuel, A., K. Ardern and M. H. Birley (2006) 'Assessing health impacts on a population', in D. Pencheon, C. Guest, D. Melzer and J. Grey (eds) *Oxford Handbook of Public Health Practice,* Oxford University Press, Oxford

4.19 NOTES

1 Example C is best practice because there is sufficient information available and design options are still available.
2 To produce a ToR that ensures all health concerns are addressed and analysed.
3 The most vulnerable: if they are safe then everyone is safe.
4 To ensure that most health concerns are identified prior to analysis; to empower stakeholders to own the assessment.
5 Attend HIA course, experience of doing HIA, and have health background.
6 Close to zero.

CHAPTER 5

Methods and tools

This chapter discusses:

- A method of HIA in the context of objectives, uncertainties and evidence.
- Various models used in HIA.
- The nature of evidence.
- Sources of information about health concerns.
- Stakeholder engagement.
- Some elements of analysis including epidemiological tools.

5.1 INTRODUCTION

There is no single method of undertaking an HIA and different practitioners may well advocate different methods. The method advocated in this book has the following components:

1　Collect as many health concerns as possible from as many different stakeholders as possible concerning the proposal.
2　Undertake a literature review of the evidence base and use this to identify health concerns
3　Describe the health profile of the proposal-affected communities and use this to identify health concerns.
4　Describe the policy context in which the HIA is being conducted.
5　Arrange the health concerns into categories and subcategories.
6　Summarize the evidence in support of each health concern.
7　Prioritize the health concerns.
8　Develop justifiable recommendations to manage the health concerns.

There are other steps and these can be regarded as part of the process (discussed in Chapter 4) rather than the method. For example:

9　Seek consensus among stakeholders and decision-makers.
10　Enable decision-makers to prepare a practical management plan.

Some of the methods used in HIA are introduced in this chapter. Additional discussion of methods may be found in subsequent chapters, especially Chapters 6 and 7.

5.2 WHAT HIA IS NOT

The following concepts influence the HIA method.

5.2.1 It's not meant to be perfect

The recommendations made in an HIA do not have to be the best possible recommendations that could be made if resources were unlimited. Although optimization is the ideal, the best should not be made the enemy of the good. The proposal simply has to be better for human health than it would otherwise have been. HIA, and other forms of impact assessment, are practical management tools. This means that impact assessment is necessarily imperfect.

5.2.2 Not a search for truth

In this book, HIA is not regarded as a search for scientific or absolute truth. It is not an opportunity to validate a scientific hypothesis. It is an occasion to influence the design and operation of proposals in order to safeguard and enhance the health of human communities.

5.2.3 Not a thesis

An HIA that is conducted on the back of an envelope and that leads to real improvements in human health safeguards is far more valuable than a three volume thesis that sits on a bookshelf. This is a real issue, as many of the decision-makers who commission an impact assessment do not know what to do with the report once they have received it. Many impact assessment reports seem to have no function once a proposal has been authorized by a licensing or lending authority.

5.2.4 Not a prediction

In my view, an impact assessment (of any kind) is not a prediction about the future. This view is contested. 'Prediction', to me, implies a far greater level of rigour and accuracy than is warranted. As stated above, it's not meant to be perfect and it's not a search for the truth. The truth of impact assessments cannot be tested. This is the counterfactual argument and it runs as follows. Any statement about the future made in the course of an impact assessment will influence those who are responsible for planning and managing that future. Consequently, they will change their plans and management ideas to a greater or lesser extent. This, in turn will change the outcome. The future

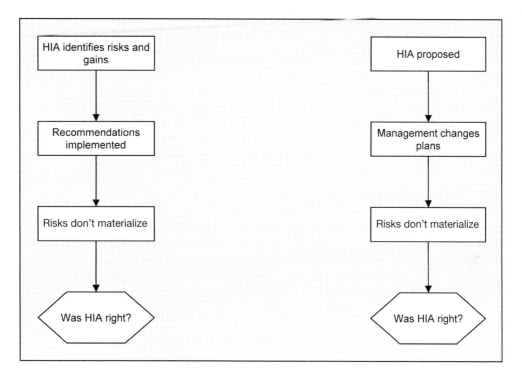

Figure 5.1 *Two views of the counterfactual argument*

that unfolds will be different to the future that was 'predicted'. So the accuracy of HIAs cannot be tested in the traditional scientific sense. This is an expression of the 'uncertainty principle': the observer changes that which they observe. No manager wishes to be accused of damaging people's health. Any kind of inquiry into whether a proposal can do so is likely to change some of the decisions that the manager will make. Figure 5.1 makes these points graphically.

5.2.5 Not a before-and-after tool

HIA is not about making before-and-after comparisons. Some managers commission HIAs because they want to guard against being sued by people who may claim that the proposal has damaged health. The commissioners would like to be in a position where they can demonstrate, in a court, that the prevalence of all diseases is no worse before and after the proposal has been implemented. While commissioners of impact assessments may legitimately wish to conduct before-and-after comparisons, it is not the role of HIA to do so. Section 5.15.7 discusses the resources and approaches required for conducting comparisons. See also Chapter 6. The resources required are usually much larger than those required for an HIA.

5.2.6 Not a positivist approach

HIA is based on an interpretive rather than positivist approach to science (Veenstra, 1999). The positivist approach assumes that there are natural laws governing the behaviour of large human groups and that these can be discovered through science. The interpretative approach supposes that the observer imposes meaning and values and that social reality varies between groups. Sets of practices or techniques used for the purpose of revealing intelligible meaning are referred to in philosophy as hermeneutics. Work on the underlying hermeneutics of HIA is probably needed.

5.2.7 Not for stopping proposals

HIA is not about stopping proposals, however inappropriate the proposal may seem. It is simply about pointing out to the owner of the proposal what the likely consequences are for human health. It is the job of the decision-maker, not the HIA specialist, to weigh the consequences of their actions. If decision-makers thought that HIA was about stopping their favourite proposal they would simply avoid commissioning HIAs.

5.3 THE NATURE OF EVIDENCE

Many different kinds of evidence must be used in HIA. These include, but are not limited to, scientific evidence. Other evidence may be derived from informants and unpublished reports. The evidence used may be incomplete, inconclusive, imprecise, not completely credible, biased or uncertain. The evidence should be gathered and used in an ethical and unbiased manner to guide, support and justify the conclusions and recommendations of the HIA (Scott-Samuel et al, 2001; Birley, 2002; Mindell et al, 2006). See also Chapter 12.

Table 5.1 suggests a hierarchy for the strength of different sources of evidence used in HIA (Pennington et al, 2010). The strength is relative, and weaker levels of evidence may still be valid and reliable. A literature review for an HIA will normally draw on relevant evidence from all levels in the hierarchy and the strength of each piece of evidence can be stated.

For example, we reviewed the association between housing interventions and health in order to support an HIA of housing regeneration.[1] It included information from a systematic review (Thomson et al, 2001), a single HIA (Gilbertson et al, 2006) and a single study of a housing regeneration project (Critchley et al, 2004). There was also information from officers and from housing residents' personal experience of regeneration.

The HIA method should establish the credibility of the evidence available. Credible evidence can have the following attributes (Audi, 1998; Schum, 1998):

- tangible – open to direct inspection;
- authoritative – accepted as originating from an unbiased source such as a reference book;
- testimonial – asserted by an informant.

Table 5.1 *A hierarchy of evidence*

Rank	Strength	Type	Source
6	Strongest	From academic literature	Reviews of reviews or meta analyses
5			Systematic reviews or reviews of HIAs
4			Reviews of single HIAs
3			Single studies
2		From people	Evidence from experts (key informants)
1	Weakest		Evidence from stakeholders

In HIA, we often rely on the testimony of informants. Evidence acquired through testimony has two components: the testimony itself and the credentials and credibility of the informant. The credibility of informants is based on their competence and sincerity. This, in turn, depends on their objectivity, veracity and observational sensitivity (ability to make careful observations). It also depends on the nature of the association between the informant and the topic.

In addition to credibility, evidence has properties of inferential force (can be used to make a strong argument) and relevance. The inferential force of evidence is dependent on individual judgement. Evidence is relevant if it makes the existence of a material fact more or less probable. Relevant evidence may be:

- direct – leads straight to a conclusion;
- circumstantial – provides a step towards a conclusion; or
- ancillary – supports or refutes the strength of other evidence.

In the complex systems encountered during HIA, there are many separate pieces of evidence and they may reinforce or contradict each other. The evidence is used to choose between alternative conclusions. To use a simplified example[2], a particular proposal may have:

- little or no impact on health: no action is required;
- some negative impacts on health: action is required;
- or only positive impacts on health: action may be required.

The conclusion is reached by establishing the truth of proximate causes, such as the three principal categories of health determinants, expressed as questions. Is the proposal likely to change:

1 Individual/family health determinants in such a way as to increase or decrease the vulnerability of a community?
2 Physical, social, or economic health determinants in such as way as to change the exposure of the vulnerable community?
3 Institutional health determinants in such as way as to increase or decrease the ability of health-protection services to safeguard the vulnerable community?

The answer to these questions, in turn, depends on examination of the subsidiary health determinants. In this way, a chain of reasoning is constructed: from changes associated with the proposal, to changes in health determinants, to likely changes in health outcomes.

The evidence used in HIA, as in much other public health, is likely to be incomplete. Action must often be based on the precautionary principle (WHO, 1999). The objective is to make rational decisions under conditions of uncertainty by using available evidence in a constructive manner. As the assessments are prospective, the evidence is not derived from the outcome and prospective assessments are not verifiable (see the counterfactual argument above) (Mcintyre and Petticrew, 1999). All the same, the best assessment of the future impacts of a new proposal is often based on knowledge of what happened when a similar proposal was implemented in a similar region. The same mistakes are likely to be repeated. The evidence base for such impacts is small, but growing.

I believe that the objective of HIA is to make justifiable recommendations for safeguarding and enhancing the health of communities affected by a specific proposal. The recommendations should be able to withstand challenge either by decision-makers or in a court of law. Such recommendations are dependent on a balanced and reasonable interpretation of available evidence.

5.4 AVAILABLE EVIDENCE

General scientific evidence is published in peer-reviewed journals, reports and books. It should be gathered and reviewed in a systematic, ethical and unbiased manner. Pragmatism is required: time and resources are limited and the objective is to inform decision-makers, even if evidence is sparse. Sometimes there is scientific controversy, e.g. about the health impacts of high-tension power lines. In many cases, there is no controversy, e.g. the species and behaviour of local malaria mosquitoes may be well established. Additional published evidence is often found in internal reports, referred to as the grey literature. There may be reviews about the health impacts of proposals of a similar type. For example, the systematic review about the effect of housing improvement on health mentioned above (Thomson et al, 2001).

Objective, rigorous, quantitative scientific evidence may, or may not, be available. For example, if the toxicology of a pollutant is well understood then the number of new cases of particular diseases arising from a specific rate of emission can be modelled. Much of the evidence will not be quantitative, but this does reduce its value.

There is a guide available for reviewing published evidence for use in HIA (Mindell et al, 2004, 2006). This guidance is based on the ready availability of information in high-income economies and may not apply so well in low- and middle-income economies, where information is often sparse. In summary, the guidance suggests that a literature review should include the following features:

- The aim is to summarize the evidence, and support the analysis of potential health impacts and the recommendations that are made.
- Reviews should be reported in such a way that those responsible for their critical appraisal can understand them clearly.

- The questions that the review seeks to answer should be clear, focused and relevant. Set and state the inclusion and exclusion criteria, which may include countries and languages.
- The population groups, or stakeholders, should be identified.
- Reviews of the effectiveness of mitigation measures should consider reversibility and inequality. Reversibility means that if exposure to A causes B, then less exposure to A will result in a reduction in B, but this should not be assumed.
- Check for existing reviews, especially systematic ones, and report how you checked.
- State the search terms used. Report the dates on which the literature search began and ended and the range of years included. Report the total number of articles identified by the search and the number included and excluded.
- List key informants contacted.
- Identify information gaps.
- Rank the quality of the evidence reported. Identify conflicting results.
- Make sure that citations and references reported are complete.
- State your conclusions and justify them from the evidence presented.
- Provide a 'lay summary'.
- Ideally, ask someone who was not involved to comment on the draft review.

The guide identifies nine steps required in a quality literature review and these are summarized in Table 5.2. For each of the steps, the guide includes explanations and advice.

In some cases the evidence will be contradictory and a process of triangulation should be used. This simply means comparing independent sources of data. If two sources disagree, then a third is found. Independence can be hard to establish, as many sources refer to each other. Table 5.3 provides an illustration; two of the sources may have obtained their data from a common, but unknown, source. The evidence may be open to different interpretations and a consensus-building process may be required (NICE, 2008).

5.5 HEALTH CONCERNS

The starting point for the IIIA method is the collection of health concerns. Health concerns, in this context, are unstructured, non-prioritized, broad, vague or specific, and realistic or unrealistic. They may be positive – how the proposal could increase health gains – or negative. The intention is to obtain a comprehensive list of concerns that covers every eventuality. The analysis proceeds by examining, classifying, rejecting and prioritizing these concerns. Figure 5.2 indicates some of the sources that may be used.

Concerns may be elicited from the community or key informants using techniques like mind-mapping ('brainstorming'). The mind-mapping process is used to generate health concerns from a group of participants. During the process all concerns are regarded as equally valid and all participants are considered equally competent. The concerns are sorted and categorized at the next stage. Health profiles identify health needs associated with the existing community and its physical and social environment

Table 5.2 *Summary of guidance on the review of published literature*

Step	Additional comments
Framing the questions that the review seeks to answer	The questions should be clear, focused and relevant. The population groups and demographic factors should be stipulated, including vulnerability. Evidence should be sought for effectiveness of mitigation strategies; reversibility; effect on inequalities; economic appraisal.
Determining whether a literature review is required, and its scope	Check for existing reviews and appraise their quality.
Purpose, organization and structure	State instigator and purpose. The structure should include the review question; details of the literature search conducted; findings; references for all articles and reports included; conclusions; date of completion of literature review. Obtain independent comments on the draft.
Setting inclusion and exclusion criteria	State the types of studies sought to answer the review questions. More than one study design may be included. State inclusion/exclusion criteria such as countries/languages, designs, interventions and populations.
Literature search	State the search strategies used, e.g. range of years, databases searched, search terms used, languages included, experts contacted, whether grey literature was included, and how it was identified, number of articles or reports identified. Comment on any constraints, e.g. time, access to databases, or inability to obtain copies of papers.
Critical appraisal	Note weaknesses in a study that may affect confidence in its conclusion, e.g. lack of impartiality of sources; suitability and rigour of research methods; how far conclusions are supported by results. List articles or reports excluded on the grounds of quality.
Interpretation	Detail any process or methods used to combine and synthesize the findings. Identify gaps. Identify factors affecting the quality, e.g. bias and confounding. Identify economic appraisals. Summarize all studies and provide full references. Discuss the comparability of the studies reviewed with the specific context of the HIA. Consider whether studies address cause and effect. Report exposure-effect/dose-response relationships. State principles used where there is conflict, e.g. weighting.
Conclusions	Provide clear conclusions based on and justified by the evidence. Summarize the gaps, biases, conflicts, quality issues and other limitations. State the relevance to the specific HIA.
Reporting	Include a 'lay summary' that is easy to read and points to main report.

Table 5.3 *Triangulation of international data regarding HIV/AIDS prevalence rates in Nigeria in 2005*

Source	Website	Adult HIV/AIDS prevalence rate (aged 15–49) (%)
Global health facts	www.globalhealthfacts.org	3.1
CIA World Fact Book	www.cia.gov	5.4
UNAIDS	www.unaids.org	3.1

(see Chapter 6). Experience of similar proposals in similar regions identifies health concerns that were relevant elsewhere and may be relevant to the new proposal. The experience may be anecdotal; only rarely will there be scientific studies of before-and-after effects. There should be many general scientific reports that are of partial relevance.

The impact assessment process should be iterative and open ended, so there will always be opportunities to add new health concerns to the list. Showing the list to key informants and stakeholders may prompt them to think of additional items to include.

The categories of health outcomes and health determinants described in Chapter 2 provide a prompt. The health concerns identified should cover a broad range of categories. For example, if the only concerns identified are non-communicable diseases associated with pollution then it is probably necessary to reconsider. Could any aspect of the proposal affect the nutritional status of the community? Could barriers to physical activity be created or removed? What about anxiety?

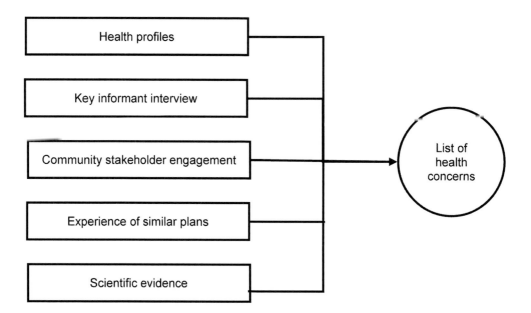

Figure 5.2 *Sources of information used to identify health concerns*

5.6 EXAMINING THE HEALTH CONCERNS

During the analysis, each health concern is examined and classified for each proposal stage and each proposal alternative: see Figure 5.3. The analysis distinguishes the consequence (degree of benefit or severity) of the impact on different community groups (see prioritization in Chapter 7). In each case, a conclusion must be reached and a recommendation made and justified. The justification will depend, in part, on the evidence reviewed. Consideration should be given to the extent to which mitigation will reduce the consequence; there may be residual or cumulative effects (see Chapters 7 and 12).

The classification scheme should separate:

- the proposal alternatives, if there are any, such as location A versus location B;
- the proposal stage, such as construction, operation and decommissioning;
- the community groups affected, such as the most vulnerable;
- the nature of the health concern, such as health outcome or health determinant;
- direct and cumulative effects.

For each health concern, the analysis should consider:

- objectivity, whether there is a plausible association with the proposal;
- likelihood, consequence and priority (see Chapter 7);
- strength of the evidence;
- recommendations required to safeguard or enhance human health;
- effect of recommendations (the residual impact);
- community perception of risk;
- certainty of evidence.

5.7 ENGAGEMENT WITH INFORMANTS

Health concerns are gathered by engaging with as many different informants, or stakeholders, as possible. Stakeholder engagement meets a number of objectives. In countries that respect democracy it provides an opportunity for the community to take control of the factors that affect their own lives. It is of value in managing the perception of risk. Stakeholders can be a useful source of baseline data. For example, a professional stakeholder may remember an unpublished report that is sitting on a bookshelf.

Perceptions of risks arise in part from a feeling of lack of control. Local stakeholders are the involuntary recipients of a plan that may not bring them much personal benefit. They must inevitably regard the proposal with some suspicion. They may fear that they and their children will be poisoned or lose their homes. These fears can persist regardless of the attention to detail that is inherent in a modern proposal. They may not believe that emissions are at a safe level or that their personal interests are respected. Management of risk perception includes enabling the community to take control over their own lives. For example, it may be advisable to provide the community with an

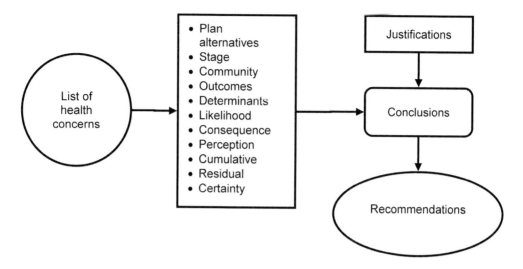

Figure 5.3 *Classifying health concerns*

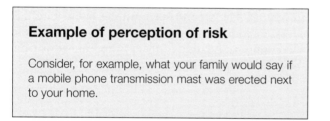

Example of perception of risk

Consider, for example, what your family would say if a mobile phone transmission mast was erected next to your home.

independent means of monitoring the levels of air pollution in their neighbourhood. See also Chapter 7.

Engagement is complex and there are many pitfalls. For example, one must avoid making promises or raising expectations. One-to-one interviews can be arranged with some of the key informants. These are individuals with special knowledge as a result of their profession or standing in the community. There can be a number of key informants who claim to represent their communities but do not; for example, traditional leaders are not always recognized by the community as representatives. In order to reach a larger number of people and hear their views, community workshops are often required. In some contexts, people are reluctant to attend such meetings and surveys will have to be arranged. These may involve telephone interviews or knocking on doors. In many cases a local consultant, translator, specialist facilitator or other respected person should undertake the work. Care should be taken to ensure that all groups are well represented in order to avoid gender, ethnic, age or other biases. Separate workshops may be needed for this purpose. Try and include representatives of the medical community in the discussions, interviews and workshops. A recent report reviewed the challenges of effective community engagement in extractive and infrastructure projects (Herbertson et al, 2009). The following principles were proposed:

- prepare communities before engaging;
- determine what level of engagement is needed;
- integrate community engagement into each phase of the cycle;
- include traditionally excluded stakeholders;
- gain free, prior and informed consent;
- resolve community grievances through dialogue;
- promote participatory monitoring by local communities.

Effective engagement means involving the local community in the actual decision-making process. Proponents are naturally hesitant to do this and may seek to reduce the level of engagement to the minimum. For example, engagement may become a one-way communication of information. This is short-sighted and can engender community distrust. Engagement should be built on the principle of free, prior and informed consent. Many plans have failed because of community mistrust.

5.8 KEY INFORMANTS

Stakeholder engagement will normally take place in two stages. The first stage will be through discussions with key informants. The second stage will be by inviting the key informants, and other stakeholders, to community workshops, or by visiting their homes. Managers often appoint a community relations officer who can advise on identifying key informants and the best way to approach them.

Typical key informants include the following:

- project staff;
- local doctors and nurses in public and private clinics and hospitals;
- managers of local public utilities and services;
- community representatives;
- academics;
- department of health officers at national and district level;
- environmental health officers;
- public health officers;
- health statistics departments;
- NGOs;
- police.

Each stakeholder will have legitimate, specialist concerns. The concerns will cover many different themes; some will be subjective or intangible. For example, some stakeholders may express a general fear that the proposal will contribute to cancer or crime. Other stakeholders may identify specific issues based on their expert knowledge of the region, the community, or the proposal. For example, a regional health specialist may point out that there is a reservoir of bubonic plague or a high incidence of traffic injury in the locality. A local health-clinic manager in a rural area may be concerned that an influx of population would overwhelm services. Each of these concerns must be recorded and acknowledged.

Engagement provides an opportunity to secure the support of powerful and influential people for the impact assessment. It is an opportunity to explain to them why the impact assessment is taking place and how it is likely to proceed. In some cases, an HIA training course may be scheduled in order to increase the competence of local consultants. Government officers can be invited to attend such a training course and they may subsequently feel enthusiastic about supporting the impact assessment and participating in the engagement process. See Chapter 4.

One of the duties of the community relations officer is to map the importance of stakeholders according to their power and influence. The officer may be primarily interested in the powerful and influential. By contrast, HIA is primarily concerned with the most marginalized and vulnerable. If that community is safe, then everybody is likely to be safe. Community relations officers may be reluctant to provide introductions to marginalized people.

Key informant interviews should ideally be conducted using a semi-structured process. The objective of the interview should be determined in advance together with the kind of information that is being sought. Several key informants may be asked the same questions and a structured approach will assist with the subsequent analysis.

5.9 COMMUNITY WORKSHOPS

Community workshops are often part of the stakeholder engagement process. Instead of meeting stakeholders on a one-to-one basis, they are met as a group, providing an opportunity for focus group discussions and more detailed elucidation of health concerns.

The primary purpose of the community workshop, from an HIA perspective, is to capture and record the health concerns of the different community groups that are represented. The secondary purpose has to do with management of the perception of risk (see Chapter 7). Table 5.4 illustrates the range of participants and some possible concerns for a proposal in a rural low-income economy context. For example, community leaders may be concerned that the proposal will reduce their legitimate role (see involuntary resettlement in Chapter 9).

5.9.1 How to do a workshop in a low income country

Communities in low-income countries do not always benefit from written information or sophisticated PowerPoint presentations. They may have difficulty understanding the size, scale and time requirements of the proposal. They do appreciate clear graphical materials. Principles to consider include the following:

- Communicate in a manner that is familiar to them.
- Go to them, rather than asking them to come to you.
- Accept their hierarchy with regard to the seating arrangements and do not impose your hierarchy on them.
- Accept their meeting protocols, which may include a time of prayer.
- Gather, record and report all their concerns.

Table 5.4 *Typical workshop participants and their concerns, in a low-income country context*

Participant	Concern
Community representatives	Difficult to speak out in the presence of powerful people. Range of health concerns, often associated with poisoning and with job opportunities.
Community leaders	Loss of legitimacy. Interested in obtaining medical services and other utilities for their communities.
Government officials	Suspicious of the proponents. Unwilling to share information. Unclear about the aims and objectives of the impact assessment. Interested in the supply of medical services.
Representatives of the medical community	Ensuring improved medical resources. Concern about increasing workload.
Proponents	Wish to maintain community support; likely to make unwise promises. Cannot see the association between their plan and community health.
Translators	Ensuring future contracts.
Professional facilitators	To be seen to do a good job. Ensuring that different community groups all have an opportunity to share their concerns.
Local consultants	Gathering and recording health concerns. Ensuring future contracts.
Community relations officer	Management of community expectations and management of powerful people.

- Hold separate workshops for powerful/influential people and local communities.
- Identify and follow up new sources of baseline information.
- Plan for a duration of one day per workshop.
- Repay participants for their lost time.
- Provide feedback and demonstrate that you have heard.

In some contexts, participatory or rapid rural appraisal techniques may be appropriate (Chambers, 1983). This may include tools like map-making using twigs and stones.

Figure 5.4 illustrates the wrong way to hold a community workshop. In this example:

- the community were transported to a large hotel far from the project site;
- the proponents sat on a raised dais;
- the village community were marginalized at the back of the room;
- the rich and powerful sat near the front of the room;
- a formal presentation was made using PowerPoint with a diagram of the project that had no size reference on it.

Figure 5.4 *Wrong way to hold a workshop*

Copyright Tropix.co.uk

By contrast, Figure 5.5 illustrates a community workshop that was held sitting on mats under a woven grass shelter near the project site. There were many women and children present and the women spoke assertively through an interpreter.

Many of the stakeholders who will participate in HIA will also participate in environmental or social assessment and these may also have community engagement processes. There is an opportunity for integration. It is unfair to the local community to survey them twice. You risk 'survey fatigue' There will also be overlap with professional stakeholders. For example, the environmental assessment may wish to consult the local water supply and waste management utilities. The social assessment may wish to consult with the local public transport utility. In all these cases, discussion and integration can be planned in advance. See also Chapter 4.

5.10 MORE ON 'MIND-MAPPING'

There are several ways of gathering health concerns from the community. One way is called mind-mapping, using a focus group, introduced above. Mind-mapping simply refers to an uncritical approach to gathering ideas. Participants should be told that all ideas are equal at this stage and that sorting them out will take place later. They

Figure 5.5 *Better way to hold a workshop*

Copyright Tropix.co.uk

may be encouraged to write ideas on a piece of paper and post them on the board. Alternatively, they may be encouraged to speak so that their ideas can be written down by a facilitator. They are asked the question: 'How do you think that this proposal could affect the health of people you know?'

Some of the stakeholders who are asked this question may not be familiar with the concept of health. They need help to understand that there is no limit to the kind of concerns that they can raise. I have found that the question can usefully be rephrased as: 'What would your mother say if you asked her how this project could affect her health?'

The mind-mapping session generates themes for the HIA. After the mind-mapping is finished, the health concerns can be sorted into categories. For example, people may refer to sexually transmitted infections, pollution, traffic noise, loss of livelihood, strangers coming into their community, loss of tranquillity, disturbance of sacred land, fear of crime, fear of getting sick, fear of being poisoned and fear of having one's brain damaged. These could be categorized as in Table 5.5.

These concerns can then be compared with the existing conditions experienced by each of the stakeholder communities. For example, there may already be high levels of traffic noise. The proponent is strictly only responsible for changes to the existing

Table 5.5 *Categories of health concerns from a mind-mapping meeting*

Category	Example
Health outcomes	Sexually transmitted infections
Health determinants	Pollution, traffic noise, loss of livelihood, strangers coming into the community, loss of tranquillity, disturbance of sacred land
Health risk perceptions	Fear of crime, getting sick, being poisoned and having one's brain damaged

conditions, such as additional traffic noise, that are attributable to the proposal. An acceptable total noise level may be exceeded. The health concerns will need prioritizing and then recommendations will be developed. Some of the concerns may have no factual basis other than as a perception by the community. Such community concerns are important because they cause stress, anxiety and opposition, and they should be mitigated.

Additional survey work in the community may be required to ensure that the focus group was representative. For example, there may be groups that are seldom heard because they are difficult to reach or because they are silenced by social controls. In some cultures, speaking up at public meetings is limited to the rich and the powerful and these are usually male.

5.11 TELEPHONE INTERVIEWS – A CASE STUDY

In some countries it is appropriate to conduct stakeholder interviews by telephone. There are specialist survey companies available to do this. They select a stratified random sample of households using, for example, postcodes. Interviews are conducted in the evening as well as during the day in order to ensure that the working population is accessed.

The following example concerned a waste-to-energy incineration proposal on the island of Jersey, in the UK (Birley et al, 2008). The proposed new incinerator would replace an older one that had pollution and traffic issues. A literature survey confirmed a consensus view among scientists and many environmental groups that modern 'clean burn' incinerators are much safer than earlier models, although caution is still required. The new incinerator would be located on an industrial peninsular on the far side of a bay from the main town. A waste management hierarchy of reduce, reuse, recycle was inherent in the proposal. Waste calculations demonstrated that even with maximum practical recycling, there would be a growing amount of waste that would need to be buried or incinerated. Incineration would be used to produce electricity and hot water.

A telephone interview was conducted with a stratified random sample of the island population. A team of trained interviewers provided a brief description of the proposal before asking eight main questions: see Table 5.6. Some of the questions had subsidiary components and this is illustrated by question 3j. The demographics of the interviewees were also recorded and they were asked if they would be interested in attending a

Table 5.6 *Example of telephone interview for an incineration project*

	Question	Instructions for interviewer
1	Considering the information you've just heard, what do you think the top three potential positive and top three potential negative impacts of the incinerator will be on you, your friends and family?	Multiple choice – code maximum of three positive and three negative – unprompted
2	How confident are you that the introduction of a new incinerator will be managed and run by the local government to ensure the waste facility and associated traffic doesn't affect the health or safety or well-being of you and your family?	Single choice – prompted
3	How important do you think each of the following are in ensuring the project is acceptable to you, your family and friends?	Grid – prompted
3j	Are there any other things which would help to ensure the incinerator is acceptable to the community?	Open – unprompted
4	Which groups of people in the community, if any, do you think need most protection from any possible negative impacts of the incinerator?	Multi-choice – unprompted
5	Do you have any concerns about how the construction of the new incinerator could affect you, your family, or your friends?	Multi-choice – unprompted
6	What do you think the potential effects of the incinerator are likely to be on your family's health and well-being?	Single choice – unprompted
7	Who do you think will benefit most from the incinerator?	Multi-choice – unprompted
8	Is there anything else to do with waste management and health that you'd like to tell me about?	Open – unprompted

follow-up meeting. For many of the questions the survey tried to anticipate the range of answers and these were included in the survey form to assist the interviewer with rapid coding. There was always an opportunity to record responses that had not been anticipated. Most of the questions were unprompted and asked in an open-ended manner. In some cases, the question was presented as a range of choices with a rank of importance for each. The interview form was agreed with the proponent in advance. Follow-up interviews and workshops also took place.

There were 456 respondents and as multiple responses were requested, there were over 1000 replies to question 1: see Tables 5.7 and 5.8. The respondents were 264 women and 192 men.

Table 5.7 *Negative responses to question 1*

Coded answers	Count	%
The view from the waterfront – get worse	171	38
Traffic congestion – get worse	157	34
Air quality – get worse	101	22
Odour/smells – get worse	85	19
Noise pollution – get worse	60	13
Standard of living – get worse	49	11
No impact/Don't live near enough for it to impact on me	48	11
Other (negative)	48	11
Risk of traffic accidents – increase	47	10
No positive impacts – do not build it	44	10
Don't know/not sure/no opinion	36	8
Risk of explosions – increase	32	7
Negative impact – cost	28	6
The amount of rubbish around – get worse	23	5
New location – poor	22	5
Value of our property – decrease	21	5
Ease of walking or cycling – get worse	21	5
Job opportunities – decrease	9	2

Table 5.8 *Positive responses to question 1*

Coded answers	Count	%
Air quality – get better	218	38
No negative impacts – it was time it was built	213	37
Odour/smells – get better	117	20
The amount of rubbish around – get better	105	18
Standard of living – get better	73	13
Other (positive)	70	12
Noise pollution – get better	53	9
No impact/Don't live near enough for it to impact on me	48	8
Job opportunities – increase	48	8
Don't know/not sure/no opinion	36	6
Traffic congestion – get better	26	5
New location – good	19	3
Risk of explosions – decrease	17	3
Positive impact – electricity generation	17	3
Ease of walking or cycling – get better	12	2
Risk of traffic accidents – decrease	11	2
The view from the waterfront – get better	10	2
Positive impact – encourage recycling	9	2
Value of our property – increase	1	0

Table 5.9 *Community opinion about the health impact of an incinerator*

Coded answer	Count	%
No potential effects/will not affect me	375	53
Don't know/not sure	91	13
Change in well-being or quality of life	68	10
Respiratory disease, e.g. asthma	68	10
Improvement/positive effect	66	9
Other	52	7
Stress/depression or mental illness	47	7
Increase in cancer rates	30	4
Accident rates	21	3
Circulatory disorders, e.g. heart attacks, strokes	17	2

Some of the responses were 'other'. These typically referred to the need for recycling and composting of waste, the visual appearance of the incinerator or that it should be sited elsewhere. The most important issues for the community were: opportunities to recycle/compost, traffic and the need for more information. There was little concern about construction as the site was not in a residential area. Many people thought that it would not affect their health although there were some concerns.

Table 5.9 indicates the responses to question 6. While the majority did not seem particularly concerned, there were clearly a number of specific issues that needed to be addressed.

Less than 50 per cent of the population were 'very' or 'reasonably' confident in the Jersey government. However, opportunities to be involved in air pollution monitoring or to be on a management committee were not strongly supported. The most important messages that came through were the need for more regular communication about the proposal, including public exhibitions and opportunities to visit the working facility. All of this would help to manage the perception of risk issues and increase confidence.

5.12 ANALYSIS

The analysis stage is where disparate information and knowledge is brought together and used to formulate statements about future changes that are attributable to the proposal.

The analysis proceeds from a broad list of health concerns, through a process of classification, to a narrower list of health issues that require additional attention and prioritization. Decisions have to be made and justified. Health priorities should be an output of the assessment and not prejudged because of administrative or political convenience – this is discussed further in Chapters 4 and 7.

As discussed in Section 5.2, the analysis does not have to be perfect; it simply has to be plausible and comprehensive. The best should not be an enemy of the good. All the important issues must be included and there should be an emphasis on health safeguards and health gains. Avoid focusing on popular issues regardless of their

Table 5.10 *Three kinds of gap*

Type of gap	Examples
Health profile	Health status of stakeholder communities
	Determinants of health in stakeholder communities
Knowledge base	Association between types of proposals and health outcomes
Management system	Poor communication interfaces between the management systems involved

relative importance. For example, in an infrastructure development proposal in a poor country there may be a source of industrial emissions. The risks associated with these emissions may be of order $1{:}10^6$ while the local risk of childhood illness from diarrhoea may be of order $1{:}10$. The analysis should consider whether the proposal could affect the rates of childhood diarrhoea. Prioritization is discussed in Chapter 7.

5.13 GAP ANALYSIS

Gap analysis is an important component of HIAs. The available information is almost always incomplete and the time allocated to acquiring new information is too brief. The objective of a gap analysis is to determine whether enough data have been collected and analysed to meet the objectives of the HIA.

There are three main kinds of gaps and these are illustrated in Table 5.10 and described below: health profile, knowledge and management.

Gaps in the baseline data are discussed in Chapter 6. In many settings, there will simply be insufficient information to characterize accurately the existing health conditions and needs of the community. Examples of baseline data gaps include:

- percentage of homes without safe water supply;
- percentage of homes with a chronically disabled tenant;
- insecticide resistance rate in local vector population.

There may be gaps in knowledge about how different types of proposals affect health. Examples of knowledge-base gaps could include:

- the relationship between green space and health;
- the relationship between irrigation systems and malaria;
- the causal pathways linking the proposal and health outcomes.

There are an increasing number of reviews of such relationships and these are referenced throughout this book.

Third, there may be gaps in the proposal management system. These are often at the interface between different management systems or administrations. Examples of management system gaps include:

- the interface between the proponent and government services;
- the interface between client and contractors;
- occupational health services and community health services;
- intersectoral collaboration between ministries, for example agriculture and health;
- contractual requirements between the proponent and the construction company;
- the interface between HIA and EIA.

Identification of some of these gaps may lead to specific recommendations. Gaps usually remain and then assumptions must be made. These should be stated so that the reader is free to agree or disagree.

Example of a management system gap

A social landlord maintained a detailed database about the location and condition of homes. However, there was no information about the occupants. There was a plan to upgrade the conditions of the homes, but no system to identify vulnerable residents.

5.14 POLICY ANALYSIS

HIA is usually conducted within a specific policy context. At an international level, this may be the Equator Principles or the Treaty of Amsterdam, discussed in Chapter 1. There are also standards for air quality and much else. There may be specific policy statements at national or local level. Private companies also have relevant policies. Policy statements provide a framework to which the proposal must conform and can provide justification for the recommendations made in the HIA report.

In an industrialized economy, such as the UK, there are likely to be many policy documents to analyse. Policy changes fast and the number and diversity of documents are large. Spatial planning provides a good example. At national level there are policy planning statements. At local level there are local development frameworks and core strategies. See Chapter 11 for more information.

5.15 CAUSAL MODELS

Causal models provide a tool for analysing the chain of reasoning between proposal and outcome. A causal model of health impacts was introduced in Chapter 2. In summary, a proposal changes the determinants of health and this, in turn, changes health outcomes. The model acknowledged that other changes in the environment and society change health determinants, independently of the proposal.

Table 5.11 *Components of the DPSEEA model*

Hierarchy of interactions	Explanation and examples
Driving forces	Economic development, population growth and technology, poverty
Pressures	On production, consumption and waste
States	States of the environment that include pollution levels, natural hazards and resource availability
Exposures	Exposure of communities to the risk states
Effects	Changes in well-being, morbidity and mortality

Some additional models are briefly reviewed below.

5.15.1 DPSEEA model

The driving forces, pressures, states, exposures, effects, actions (DPSEEA) model provides an alternative approach to organizing the association between proposals and health outcomes (WHO, 2005). It is based on a hierarchy of interactions and these are summarized in Table 5.11. External driving forces in society change the pressures on production and this leads to changes in the state of the environment. Changes in exposure to hazards lead to positive or negative changes in health effects. The model highlights the different kinds of remedial action that are possible at each level. It seems to be used most frequently in discussions of biophysical determinants of health, such as pollution. For example, it has been used in a comparative risk assessment of transport policies (Kjellstrom et al, 2003).

The model has also been used in an analysis of the health impacts of European employment policy (EC Health and Consumer Protection Directorate General, 2004). The driving force was the policy. The pressure was work flexibility. The states were physical work environment, such as hazards in the workplace; psychosocial work environment, such as job control; and work–life balance, such as caring for others. The exposures included accidents, bullying and substance abuse. The effects were physical, psychological and social well being, including psychosomatic diseases. The authors reported that it was not possible to quantify the entire model. A modified form of the model is being developed in Scotland to guide various policy initiatives (Margaret Douglas, pers. comm.).

5.15.2 Equity models

In Australia and New Zealand, a method of addressing equity in each step of HIA has been established. This has been referred to as equity-focused HIA (CHETRE, undated). The early examples were all within the health sector itself, where health equity was a specific objective of the proposal. There is increasing interest in improving the methods available for ensuring that HIA is equity focused, see Chapter 1 for more details.

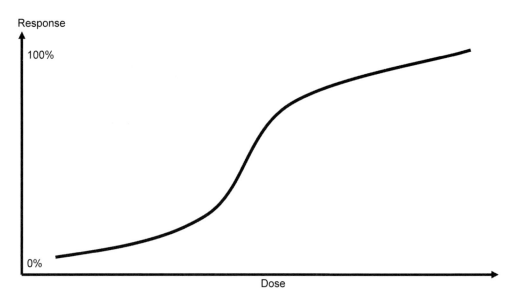

Figure 5.6 *Example of a dose-response function*

5.15.3 Quantification

Quantification based on predictive modelling is not discussed in detail in this book. It is a specialist subject and there is a growing literature (for example Veerman et al, 2005; Bhatia and Seto, 2011). There are three main approaches to quantitative HIA:

1 statistical regression methods;
2 quantitative risk assessment;
3 dynamic population health models.

5.15.3.1 Regression

Regression methods are based on establishing a statistical dose-response using a linear or non-linear function. The S-shaped curve in Figure 5.6 is one common example of a mathematical relationship between the dose of a toxic substance and the number of people whose health is affected. Data are usually available from animal experiments and other sources to determine the general shape of the curve. A mathematical formula is then derived and used to extrapolate the dose required to produce a negligible response. In many cases, there is no minimum dose at which the response is zero. The shape of this curve follows from the fact that most people respond to an average dose of the toxin. There are always some people who are much more sensitive than the average and some who are much less sensitive. HIA should take account of the distribution of effects.

Similar mathematical functions are currently unavailable for many of the determinants of health. Regression methods can demonstrate association but not causality.

Additional tools are used in epidemiology to establish causality. Typical examples of well-established, causal associations include:

- associations between airborne particulate concentration and incidence rate of cardiovascular diseases;
- associations between national alcohol sales and fatal accident rates.

5.15.3.2 Quantitative risk assessment

Quantitative risk assessment (QRA) combines the dose-response relationship with an exposure distribution in the population to estimate the disease burden attributable to the risk factor. The four main steps in QRA are:

1 hazard identification;
2 dose-response relationship;
3 exposure assessment;
4 risk characterization.

An additional step of economic valuation is often added.

A recent review examined the strengths and weaknesses of using QRA in HIA (O'Connell and Hurley, 2009). The review suggested that, where possible, HIA should use a range of quantitative and qualitative evidence. At the policy level, quantitative evidence may be given more consideration. QRA methods are used to provide point and range estimates for the health risks associated with a variety of hazards, such as setting standards for air quality. The methods depend on estimating how a proposed policy will affect population exposures. QRA estimates are often derived from dose-response, exposure-response or concentration-response functions. When a large range of health outcomes may be affected, a summary measure of health may be used, such as DALYs discussed in Chapter 2. These can be combined with economic costs to produce cost effectiveness or cost-benefit analysis: see below. There are often many uncertainties, and these should be made explicit. These methods are most useful when the epidemiological evidence base is strong. Examples include outdoor pollution and all-cause mortality; elevated blood lead levels and neuro-developmental impairment; and the risk associated with various other heavy metals. However, it can be difficult to obtain good-enough data, even on rates of hospital admissions and other indicators of morbidity. A plenary address at the International HIA conference on the current state of quantitative HIA reiterated that there is a shortage of reliable data even for key health issues in the EU (Mackenbach and Llachimi, 2009).

There have been a number of large-scale European studies that used QRA. For example, the Apheis programme calculated the health impact of long-term exposure to particulates in terms of attributable number of deaths and potential life expectancy across 23 European cities (Medina et al, 2004). The CAFE programme (Clean Air For Europe) determined the economic impact of air pollution reduction across the European Union (CAFE, 2008). The benefit–cost ratio was in the range 4–14. This estimate, together with other factors, informed the European Commission's work in proposing a new directive for controlling very fine particulate matter ($PM_{2.5}$).

There are concerns that quantitative methods could allow policy-makers to ignore the underlying complexities and many other qualitative judgements used to derive the quantitative measures and functions. For example, monetary value has to be assigned to premature mortality. One of the most significant limitations of the QRA approach is the small number of risk factors with well-defined dose-response relationships. Most policy issues do not have these characteristics and a mixture of both qualitative and quantitative methods is likely to remain necessary.

5.15.3.3 Dynamic modelling for health impact assessment

There are a number of predictive quantitative models used in HIA. Four of them were described by McCarthy and Utley (2004) and more have been developed since then. One recent example is the DYNAMO-HIA programme (DYNAMO-HIA, 2007). The aim of this programme is to develop a web-based tool to assess the health impact of European Union policies through their influence on health determinants. According to the website, the tool will be used to estimate the impact of smoking, obesity and alcohol consumption policies. The focus is thus on health policy, rather than the impact of non-health policy on health.

A recent paper illustrates the use of quantitative modelling for two projects in the UK – an airport and an incinerator (Phillips et al, 2010). The paper draws on the normal outputs of the EIA that estimate the concentration of an environmental variable in the locality of a human community. The method was used to assess the health impacts of air pollution, noise and road traffic accidents associated with project operation.

The incinerator was located in an urban area and some 5.5 million people were exposed to the airborne pollution and additional traffic. The airport was located in a rural area and some 55,000 people were exposed to additional air pollution, road traffic and noise. The air pollutants considered were PM_{10}, $PM_{2.5}$, NO_2 and airborne carcinogens. For several health measures, the percentage change in risk associated with a unit change in exposure in populations had been established in earlier studies. The baseline rate of the health effects was known for the average population. These included chronic bronchitis, cardiovascular hospital admissions, respiratory GP consultations, lower respiratory symptoms, mortality, restricted activity days, years of life lost, non-traumatic deaths, annoyance, awakenings and reading ability delay. The new exposures to air pollutants and noise for the population were drawn from the environmental statement. The change in health effect in the exposed population could then be estimated. New road accidents were calculated directly from the extra number of vehicle trips.

The modelling outcome indicated that the health effect associated with air pollution from the incinerator was very small and would be undetectable in normal circumstances. The paper contains a number of caveats and these are also important in understanding the current value and limits of such methods for use in project level HIA.

Table 5.12 *Forms of capital asset*

Form of capital asset	Explanation	Examples from a low-income country perspective
Natural capital	The natural resource stocks from which resource flows useful for livelihoods are derived	Environmental health determinants such as natural vector breeding sites, animal herds, drinking water sources and wastewater sinks, food supplies, distance of travel for wild food and fuel collection
Financial capital	The financial resources which are available to people and which provide them with different livelihood options	Treatment-seeking behaviour, medicine purchases, food security, purchasing barriers to infection, insurance, reserves to counter lost production associated with illness, remittances from outside, banks and savings
Physical capital	The basic infrastructure and the production equipment and means which enable people to pursue their livelihoods	Drinking-water delivery, communication routes, health centres, machinery, boats, diversionary structures, irrigation systems, housing quality
Human capital	The skills, knowledge, ability to labour and good health necessary to pursue different livelihood strategies	Good health; freedom from fear, pain and suffering; well-being; educational achievement; empowerment of women and minorities; capacity and capability of personnel in institutions responsible for protecting health including health centres; health-promoting knowledge, beliefs, attitudes and behaviours; seasonal work migration
Social capital	The social resources (networks, social claims, social relations, affiliations, associations) upon which people draw when pursuing different livelihood strategies requiring coordinated actions	Conflicts over traditional water, wild foods and land rights leading to traumatic injury, malnutrition and uncertainty; distributional mechanisms; neighbourhoods, community support

5.15.4 Economic models

There are many economic models that may be appropriate for use in HIA. This is a large topic and it will not be explored in detail in this book. See Chapter 7 for a discussion of cost benefit, cost effectiveness and willingness to pay.

The livelihood model may provide a useful approach for conceptualizing the health impact of proposals. In this model, there are five forms of capital asset (Carney, 1998; Birley, 2002): see Table 5.12. Proposals have the function of transferring these capital assets from one form into another. For example, men, money and machinery are used to transform natural assets, such as minerals, into financial assets. Proposals can change livelihoods and increase assets. This model provides an alternative way of classifying

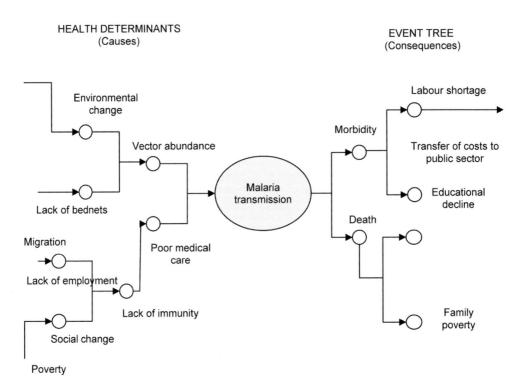

HEALTH DETERMINANTS
(Causes)

EVENT TREE
(Consequences)

Figure 5.7 *Malaria example as cause and consequence tree*

health determinants that may be valuable in economic models. It helps to ensure that economic analysis takes account of externalities such as environmental damage and the disruption of social support structures.

5.15.5 Causal trees, networks and flow diagrams

Causal models trace the influence of a chain of health determinants. It is perfectly plausible that a determinant changes another determinant: and these are sometimes referred to as determinants of determinants or 'wider determinants'. They can be captured by the use of flow charts and conceptual or causal trees. These help with the formulation of hypotheses.

For example, the Foresight project has mapped the complex web of health determinants that are the cause of obesity (Butland et al, 2007; Vandenbroeck et al, 2007). The map was developed in order to help understand the complex systemic structure of obesity and to help policy-makers to generate, define and test policy options.

Figure 5.7 sketches a cause and consequence tree associated with an increase in malaria transmission using a 'bow-tie model', introduced in Chapter 2. There are two wings representing cause and consequence, and a central point referred to as the top, main, or primary event. The primary event is something to be prevented – such as malaria transmission. It is the result of a web of secondary causes and many of these have to happen at the same time in order for the primary event to happen. If the primary event happens, there is a complex web of consequences. The model is used to make recommendations to prevent all the secondary causes happening at the same time, and to make recommendations about the emergency action that would be required if prevention fails. An example like this might be built to justify a conclusion that there was an increased malaria risk and unacceptable consequences. The proposal changes some of the determinants of health and these are indicated on the left hand side. The combined effect of those changes is a change in malaria transmission. If malaria transmission increases there are a number of consequences for the community and these include morbidity, mortality, educational impairment, loss of productive work and costs to government medical services. There are also consequences for the proposal including reputational damage, treatment costs and labour shortages.

Figure 5.8 illustrates some of the links between domestic solid waste management and health outcomes in a poor, peri-urban, tropical environment (Birley and Lock, 1999). It was built to analyse the health impacts of solid waste management. Filariasis and dengue are mosquito-borne communicable diseases found in warm climates. The risk of poisoning and explosions is present in poorly managed landfills. Waste often attracts an itinerant population of waste pickers who depend on waste recycling for their livelihoods but are exposed to various health risks.

The HIA Gateway website (APHO, undated) lists causal diagrams for a diverse range of links between major determinants and health. At the time of writing, these include noise, climate change, runway extensions, traffic intensity, a rapid transport scheme, employment, built environment, cycling, peri-urban natural resource development, travel infrastructure, transport policy, new supermarkets, the living wage ordinance, hospital construction and alcohol policy. There is also an example available for the mining sector (ICMM, 2010).

5.15.6 Epidemiology

Epidemiological tools provide comparative data about the past, present and future health status of human populations. This is a large subject and there are many specialists. HIA practitioners do not have to be experts in epidemiology, but should grasp the basic principles.

The science of epidemiology has descriptive, analytic and predictive components. An example of descriptive epidemiology is the life expectancy of a population. This is a simple statistic that requires information about the proportion of the population that die in each age group from 0 to the oldest possible age. It depends on accurate recording of age of death by the national government. There are many sources of inaccuracy. For example, rural communities may not record infant deaths.

Analytic epidemiology uses statistical tools to compare the health status of different communities, or the same community at different times. It answers question like: 'is there any significant evidence that people exposed to emissions from this factory are sicker than those who are not exposed?' In order to obtain scientifically robust results, many sources of potential bias have to be identified and controlled and relatively large sample sizes have to be used. It is expensive, specialist and non-routine. Baseline data collected for comparative purposes would have to meet such standards.

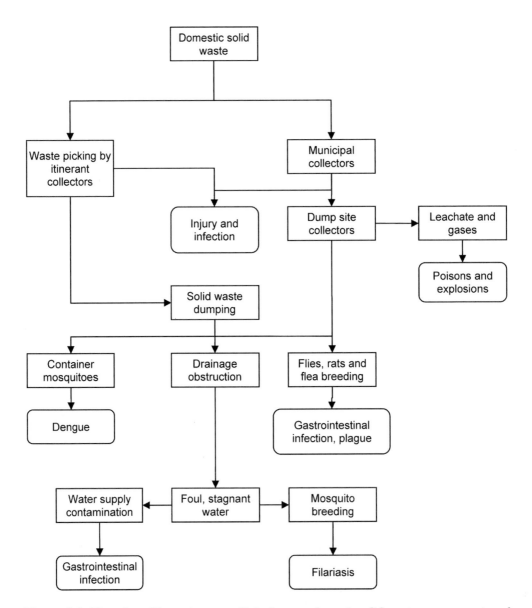

Figure 5.8 *Flow chart illustrating some links between domestic solid waste management and health outcomes in a tropical environment*

Table 5.13 illustrates a common statistical tool that could be used to make a before-and-after comparison of the consequences of a proposal. The four cells contain the numbers of people from two sample communities. One community has been exposed to a proposal and the other has not. In each community, the number of people with and without a certain health outcome has been counted. Do the data indicate that the proposal has affected community health? The null hypothesis would be that all four surveys had the same frequency of the test variable. In other words, we start by assuming

Table 5.13 *Example of 2 × 2 table of association*

	Proposal	
Health outcome	With proposal	Without proposal
People with outcome	12,549	4312
People without outcome	891	32,765

that neither the location nor the proposal is associated with differences in disease rates. A statistical test, such as odds-ratio or chi-square, is used to check the significance of the association between the proposal and the outcome. The significance is described by the probability of observing the same event by chance. If the probability is less than 5 per cent then the association is usually regarded as significant. Association does not prove causality and epidemiologists use a different set of tools to decide whether there is a causal association. There could be other explanations. For example, the people affected by the proposal may be from a different ethnic group or social class.

Predictive epidemiology uses models to forecast the future prevalence rate of a disease in the population. The forecast is based on statistical analysis of previous information plus a mathematical relationship linking cause and effect. The mathematical relationship may be empirical, such as a linear regression, or deductive, such as a set of coupled differential equations. Alternatively, a computer simulation model may be used. In some cases, models have a strong predictive value and have proved useful in policy-making, when large populations are involved.

There are many standard texts on epidemiology including a guide to the evaluation and use of epidemiological evidence for environmental health risk assessment (WHO, 2000).

5.15.7 Designs for before-and-after comparisons

HIA is not intended for before-and-after comparisons, as discussed in Section 5.2.5. Nevertheless, commissioners sometimes ask for it and may need an overview of statistical design.

5.15.7.1 Principles of statistical design

Sample surveys have to be designed to meet their objectives. Design has to address sample size, controls and confounding variables. Appropriate designs that are able to detect significant changes in disease rates over the short term or in small populations are usually complex and expensive. A good design should also address questions such as how the sample was selected and possible selection bias. In addition, comparability of the groups in age and gender should be considered.

The design should be able to determine whether the observed change is a consequence of the proposal, or the consequence of other, parallel, changes in society or environment.

Table 5.14 *Required sample size (significance level = 1%, power = 0.95)*

Rate in first population (%)	Sample size if rate in second population is 5%
10	1064
20	195
25	129
30	94
40	39
50	40
60	29
70	22

5.15.7.2 Sample size

Populations are sometimes sampled for the explicit purpose of deciding whether a proposal has changed health risks. Assuming that the samples are obtained randomly, the following principle applies: to detect a small change you need a large sample size and vice versa.

Statisticians produce formulae to determine the sample size required under various conditions. The formulae take account of the importance of two kinds of errors: finding a change when there isn't one and not finding a change when there is one. The first kind of error would benefit a campaigning organization seeking to hold a company responsible for changes in disease incidence. The second kind of error would benefit a company seeking to deny responsibility for changes in disease incidence.

Table 5.14 indicates the sample sizes required under some simple assumptions (the significance level, the design effect and the power of the test) (Fleis, 1973; Antcliffe, 1999). If there was a base disease rate of 5 per cent and there was a concern that the proposal might increase it to 10 per cent, then a sample size of 1064 would be required in each survey in order to detect that change. For effects of order 1:1,000,000 the sample size is correspondingly larger. In cases where the sample size is relatively large compared with the size of the community, it is sometimes possible to compensate by aggregating the data (Hansell and Aylin, 2003). An expert epidemiologist should be consulted.

A good survey report will specify in advance the size of change that is to be detected, the assumptions being made and the sample size required. Such statements are frequently missing from HIA reports, indicating that insufficient attention has been given to survey design.

5.15.7.3 Alternative designs

Case-control studies provide an alternative method of detecting differences associated with near and far from the plan site. For each case of disease detected in a proposal-affected population, a matched control set (between 1 to 3 controls) is selected from a non proposal-affected population. These studies also require the assistance of an expert epidemiologist.

5.15.7.4 Significance levels

The examples above demonstrate that making a robust scientific assessment of the health impact of a proposal can be complex, time consuming and expensive. Scientific studies often use a significance level of 5 per cent or 1 per cent to ensure that observed changes in disease rates are not simply the result of chance events.

By contrast, HIA is not intended to provide a similar level of scientific robustness, or to provide before-and-after comparisons. It is a pragmatic process that is intended to make justifiable recommendations for safeguarding and enhancing human health. Therefore, there are circumstances where a survey carried out as part of an HIA should consider using a significance level of 10 per cent. In this case the sample size is smaller and the survey is less time-consuming.

5.15.7.5 Stratified sampling

People who are included in statistical samples are chosen at random in order to avoid bias. However, random sampling from the whole population can mean that important groups are under-sampled. To overcome this bias, the population is first divided into groups and then each group is sampled randomly. Table 5.15 provides examples of the groups that may be used, together with an example of the context in which this may be relevant.

Table 5.15 *Examples of stratified sampling and relevance*

Stratification	Example
Age	Child survival
Sex	Anaemia in women
Location	Urban versus rural water supply, administrative district
Occupation	Cancer rates by industry
Ethnicity	Body mass index between ethnic groups

5.15.8 Incorrect analysis

There are many ways to draw incorrect conclusions from epidemiological data. For example, the following data are from a real health profile report.

The authors wanted to present the data in Table 5.16 as rates in order to make comparisons. Table 5.17 indicates how the data were analysed.

Table 5.16 *Raw data on malaria and other diseases by age group*

Age group	0–5	6–15	>15	Total
Malaria	30	109	169	308
Other diseases	53	82	175	310
Total	83	191	344	618

Table 5.17 *Poor presentation of the data as rates*

Age group	0–5	6–15	>15	Total
Malaria	10	35	55	100
Other diseases	17	26	56	100

Table 5.18 *Better presentation of the data as rates*

Age group	0–5	6–15	>15
Malaria	36	57	49
Other	64	43	51
Total	100	100	100

This analysis suggests that malaria and other diseases are less common in children than adults. In fact, it simply shows that fewer children were sampled than adults. In poor countries, about half the population are under 15 years old so it is unlikely that there were fewer children than adults in the sample population.

Table 5.18 is a better representation of the data. It indicates that malaria is less commonly diagnosed in 0–5 year olds compared to other diagnoses. The rate may be lower in children or there may be poor diagnosis. Malaria is often diagnosed incorrectly on clinical signs alone. In the end, the data have not contributed much understanding.

5.16 CHANGES ASSOCIATED WITH THE PROPOSAL

There are likely to be many changes in specific health determinants attributable to the proposal. The changes may be positive or negative and large or small. There are also differences in the level of certainty that can be attributed to those changes. Most frequently, the changes will be identified in qualitative terms and based on uncertain evidence. Table 5.19 provides a convenient way to summarize the argument. An example is included for an infrastructure proposal in a tropical location, where malaria has been identified as a concern. During the construction stage, the physical environment will be changed in some way that promotes malaria mosquito breeding sites. In Africa, for example, one species of malaria mosquito breeds in many kinds of rain pools, including tyre tracks and excavations. The community living adjacent to the construction site may be affected – the mosquitoes have a flight range of several kilometres. The certainty of the conclusion depends on whether construction takes place during the rainy season.

This summary should be supported by a detailed explanation. The explanation should follow the summary, so that the users do not have to read it unless they choose to do so. The explanation has to consider the combined influence of the changes in the determinants of health on the health outcomes. For example, the risk of malaria may increase if the plan changes the environment and creates more vector mosquito breeding sites, if there is a susceptible community and if the health agencies cannot provide adequate protection.

About mosquitoes

Infrastructure development in warm climates can change the abundance of malaria vectors. There are about 50 important species of malaria mosquito in the world and about two species in each malarious country. Each species has different habits. The habits include: fresh versus saline water, shaded versus sunlit water, still versus moving water, outdoor versus indoor biting, early evening versus late evening biting, outdoor versus indoor resting. So a clearing in the middle of a forest could stimulate local mosquito breeding, or not. Expert knowledge is needed here.

Table 5.19 *Summary table for changes in health determinants, with example*

Category of health determinant	Specific health determinants	Proposal stage	Stakeholder community affected	Direction of change and measurability (calculable, estimable, qualitative)	Certainty (definite, probable, speculative)
Individual/ family					
Physical environment	Malaria mosquito breeding sites	Construction	Peripheral communities	Increases in vector abundance expected and measurable	Probable
Institutional					

In this example there might be:

- sensitive receptors – susceptible people in peripheral villages, often children;
- limited resources – a health agency that cannot protect the community and a community that is too poor to protect itself;
- an overall context – a country unable to afford malaria control, with a vector mosquito and increasing drug resistance;
- magnitude of change – increased prevalence rates of malaria, of order 1–10 per cent;
- absence of mitigation – no prior design consideration concerning community malaria;
- uncertainty – about whether the mosquito abundance will increase.

We make the assessment based on the assumption that no special mitigation measures have been included in the proposal design or operation to take account of the health concern. We then make a judgement as to whether the normal measures are sufficient

Table 5.20 *Summary of changes associated with the proposal for completion as an exercise*

Category of health determinant	Specific health determinants	Project activity	Stakeholder community affected	Direction of change and measurability (calculable, estimable, qualitative)	Degree of certainty (definite, probable, speculative)
Individual/family					
Environmental					
Institutional					

to prevent the concern from escalating. If they are insufficient, the predicted conclusion will be an adverse change in health. If we evaluate that change to be significant then we have to make recommendations for special mitigation measures. Recommendations are discussed in a separate chapter.

The analysis will identify three main kinds of association between the proposal, the health determinants and the health outcomes:

1   No significant change is anticipated.
2   Associations that are so complex that it is not feasible to relate changes in health determinants with changes in health outcomes. In these cases, the changes in health determinants themselves are all that can be used.
3   Associations for which there is a clear link between changes in health determinants and changes in health outcomes.

5.17 SAN SERRIFFE EXERCISE CONTINUED

In the exercises at the end of Chapters 2 and 4 you were introduced to an infrastructure development project in the imaginary country of San Serriffe.

Use the information available to you about this project, including the lists of health determinants that you constructed in an earlier exercise, to fill in part of Table 5.20. For example, what health determinants may change in individual families during the resettlement of the fishing village? Assume that the project will supply the resettlement community with good quality housing, latrines and running water in accord with the World Bank standard for involuntary resettlement (World Bank, 2001).

Advanced exercise: reach a conclusion about one positive, one no-change and one negative change in health associated with the San Serriffe project.

5.18 NOTES

1 Thanks to Andy Pennington of IMPACT for reminding me of this.
2 This example is simplified because in reality there may be a mixture of positive and negative health impacts and trade-offs may be needed.

5.19 REFERENCES

Antcliffe, B. L. (1999) 'Environmental impact assessment and monitoring: The role of statistical power analysis', *Impact Assessment and Project Appraisal,* vol 17, no 1, pp33–43

APHO (Association of Public Health Observatories) (undated) The HIA Gateway, www.apho.org.uk/default.aspx?qn=p_hia, accessed July 2009

Audi, R. (1998) *Epistemology,* Routledge, London

Bhatia, R. and E. Seto (2011) 'Quantitative estimation in health impact assessment: Opportunities and challenges', *Environmental Impact Assessment Review,* vol 31, no 3, pp301–309

Birley, M. H. (2002) 'A review of trends in health impact assessment and the nature of the evidence used', *Journal of Environmental Management and Health,* vol 13, no 1, pp21–39

Birley, M. H. and K. Lock (1999) *The Health Impacts of Peri-urban Natural Resource Development,* Liverpool School of Tropical Medicine, Liverpool, www.birleyhia.co.uk/Publications/periurbanhia.pdf

Birley, M., D. Abrahams, A. Pennington, F. Haigh and H. Dreaves (2008) 'A prospective rapid health impact assessment of the energy from waste facility in the States of Jersey stage 2', University of Liverpool, Liverpool, www.liv.ac.uk/ihia/impact%20reports/energy_from_waste_stage_2_-_final.pdf

Butland, B., S. Jebb, P. Kopelman, K. McPherson, S. Thomas, J. Mardell and V. Parry (2007) *Tackling Obesities: Future Choices,* www.foresight.gov.uk, accessed September 2009

CAFE (2008) 'Clean air for Europe, cost benefit analysis', www.cafe-cba.org, accessed June 2010

Carney, D. (ed.) (1998) 'Sustainable rural livelihoods, what contributions can we make?'. Department for International Development, London

Chambers, R. (1983) *Rural Development: Putting the Last First,* Institute of Development Studies, University of Sussex, Brighton

CHETRE (Centre for Health Equity Training, Research and Evaluation) (undated) 'HIA Connect, building capacity to undertake health impact assessment', www.hiaconnect.edu.au/index.htm, accessed September 2009

Critchley, R., J. Gilbertson, G. Green and M. Grimsley (2004) *Housing Investment and Health in Liverpool,* Centre for Regional Economic and Social Research, Sheffield Hallam University, Liverpool

DYNAMO-HIA (2007) 'Dynamic Modelling for Health Impact Assessment', www.dynamo-hia.eu/root/o14.html, accessed December 2009

EC (European Commission) Health and Consumer Protection Directorate General (2004) *European Policy Health Impact Assessment: A Guide,* http://ec.europa.eu/health/ph_projects/2001/monitoring/fp_monitoring_2001_a6_frep_11_en.pdf, accessed July 2009

Fleis, J. L. (1973) *Statistical Methods for Rates and Proportions,* Wiley, New York

Gilbertson, J., G. Green and D. Ormandy (2006) 'Sheffield decent homes health impact assessment', Sheffield Hallam University, Sheffield, www2.warwick.ac.uk/fac/soc/law/research/centres/shhru/sdh_hia_report.pdf

Hansell, A. and P. Aylin (2003) 'Use of health data in health impact assessment', *Impact Assessment and Project Appraisal,* vol 21, no 1, pp57–64

Herbertson, K., A. Ballesteros, R. Goodland and I. Munilla (2009) *Breaking Ground: Engaging Communities in Extractive and Infrastructure Projects,* World Resources Institute, Washington, DC, http://pdf.wri.org/breaking_ground_engaging_communities.pdf

ICMM (2010) *Good Practice Guidance on Health Impact Assessment,* International Council on Mining and Metals, London, www.icmm.com/document/792

Kjellstrom, T., L. van Kerkhoff, G. Bammer and T. McMichael (2003) 'Comparative assessment of transport risks: How it can contribute to health impact assessment of transport policies', *Bulletin of the World Health Organization,* vol 81, pp451–457, www.scielosp.org/scielo.php?script=sci_arttext&pid=S0042-96862003000600016&nrm=iso

Mackenbach, J. and S. Llachimi (2009) 'Quantitive health impact assessment: Where do we go from here?', Presented at HIA Conference 2009, Rotterdam.

McCarthy, M. and M. Utley (2004) 'Quantitative approaches to HIA', in J. Kemm, J. Parry and S. Palmer (eds) *Health Impact Assessment: Concepts, Theory, Techniques and Applications,* Oxford University Press, Oxford

Mcintyre, L. and M. Petticrew (1999) 'Methods of health impact assessment: A literature review'. University of Glasgow, Glasgow

Medina, S., A. Plasencia, F. Ballester, H. Mucke and J. Schwartz (2004) 'Apheis: Public health impact of PM10 in 19 European cities', *Journal of Epidemiology and Community Health,* vol 58, no 10, pp831–836

Mindell, J., A. Boaz, M. Joffe, S. Curtis and M. Birley (2004) 'Enhancing the evidence base for health impact assessment', *Journal of Epidemiology and Community Health,* vol 58, no 7, pp546–551, http://jech.bmjjournals.com/cgi/reprint/58/7/546.pdf

Mindell, J., J. P. Biddulph, A. Boaz, A. Boltong, S. Curtis, M. Joffe, K. Lock and L. Taylor (2006) *A Guide to Reviewing Evidence for Use in Health Impact Assessment,* www.lho.org.uk/viewResource.aspx?id=10846, accessed February 2011

NICE (National Institute for Clinical Excellence) (2008) *Social Value Judgements: Principles for the Development of NICE Guidance,* second edition, NICE, London, www.nice.org.uk/aboutnice/howwework/socialvaluejudgements/socialvaluejudgements.jsp

O'Connell, E. and F. Hurley (2009) 'A review of the strengths and weaknesses of quantitative methods used in health impact assessment', *Public Health,* vol 123, no 4, pp306–310, www.sciencedirect.com/science/article/b73h6-4vxjw0j-3/2/230e9cb507bb95c878fb4cfa05b1b991

Pennington, A., H. Dreaves and F. Haigh (2010) 'A comprehensive health impact assessment of the City West housing and neighbourhood improvement programme', IMPACT, University of Liverpool, Liverpool

Phillips, C., M. McCarthy and R. Barrowcliffe (2010) 'Methods for quantitative health impact assessment of an airport and waste incinerator: Two case studies', *Impact Assessment and Project Appraisal,* vol 28, no 1, pp69–75, doi:10.3152/146155110X488808

Schum, D. A. (1998) 'Legal evidence and inference', in *Routledge Encyclopedia of Philosophy, Version 1.0.* Routledge, London and New York

Scott-Samuel, A., M. Birley and K. Ardern (2001) 'The Merseyside Guidelines for health impact assessment', www.liv.ac.uk/ihia/IMPACT%20Reports/2001_merseyside_guidelines_31.pdf

Thomson, H., M. Petticrew and D. Morrison (2001) 'Health effects of housing improvement: Systematic review of intervention studies', *British Medical Journal,* vol 323, no 7306, pp187–190, www.bmj.com/cgi/content/abstract/323/7306/187

Vandenbroeck, P., J. Goossens and M. Clemens (2007) *Foresight Tackling Obesities: Future Choices – Building the Obesity System Map,* Foresight, London, www.foresight.gov.uk

Veenstra, G. (1999) 'Different wor(l)ds: Three approaches to health research', *Canadian Journal of Public Health*, vol 90, no S18–S21, ppS27–S30

Veerman, J. L., J. J. Barendregt and J. P. Mackenbach (2005) 'Quantitative health impact assessment: Current practice and future directions', *Journal of Epidemiology and Community Health*, vol 59, no 5, pp361–370, http://jech.bmj.com/content/59/5/361.abstract

WHO (World Health Organization) (1999) 'Environmental health indicators: Framework and methodologies' WHO, Geneva, http://whqlibdoc.who.int/hq/1999/who_sde_oeh_99.10. pdf

WHO (2000) 'Evaluation and use of epidemiological evidence for environmental health risk assessment, guideline document', WHO Regional Office for Europe, European Centre for Environment and Health, Bilthoven Division and International Programme on Chemical Safety, Copenhagen

WHO (2005) 'The DPSEEA model of health-environment interlinks', www.euro.who.int/ ehindicators/indicators/20030527_2, accessed November 2009

World Bank (2001) 'Safeguard Policies, Operational Policy 4.12: Involuntary resettlement', http://web.worldbank.org/wbsite/external/projects/extpolicies/extsafepol/0,,menupk:584 441~pagepk:64168427~pipk:64168435~thesitepk:584435,00.html, accessed October 2006

CHAPTER 6

Baseline report

- The aims and contents of baseline reports are described.
- A distinction is made between international and national proposals, HIAs and health need assessments.
- Sources of information are illustrated.
- Methods for managing the gaps in available information are explained.
- Challenges of primary data collection are discussed.

6.1 INTRODUCTION

Different kinds of evidence are used in HIA in order to understand the future consequences of a proposal. These were introduced in Chapter 5. One kind of evidence is a baseline report of existing health, social and environmental conditions in the proposal area and of what is known about the impacts of similar proposals. The primary objective of the baseline report is to contribute to the analysis of health impacts and the formulation of health mitigation and enhancement measures, as described in the previous chapter.

The baseline report is also valuable for understanding the existing health needs of the community, or health needs assessment (HNA), independently of any development proposal (Wright et al, 1998; NICE, 2005). An HNA may be used by a proponent to formulate social investment programmes. These are programmes for improving community health that are independent of the impacts of the proposal. For example, a proposal to build a road may have no impact on the existing domestic water supplies in nearby communities. But the water supplies may be inadequate and a social investment programme might be proposed for improving them. Proponents sometimes include social investment programmes in their plans as part of their business case (discussed in Chapter 1).

Proponents often envisage a third objective for baseline reports. They wish to make a before-and-after comparison of the health of stakeholder communities in order to demonstrate that the proposal has had a positive impact on stakeholder health, or not had a detrimental impact. Unfortunately, the level of detail that would be required to achieve this third objective in a scientific, or legally robust manner falls outside the scope of HIA. It could require an epidemiological study, as discussed in Chapter 5, which is expensive, of long duration and complex. Low-cost, robust, before-and-after

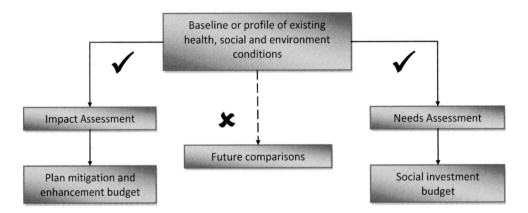

Figure 6.1 *Illustration of confusion of objectives*

comparisons are only practical in countries where routinely collected health data are already of very high quality. HIA consultants, themselves, would like to make before-and-after comparisons in order to understand the effectiveness of their own work. However, after the HIA is complete, the consultant usually has no further involvement in the proposal and is therefore unable to return at a later date and make comparisons.

This confusion of objectives is important, so Figure 6.1 is included to emphasize the point. The baseline report can be used by the impact assessment to plan mitigation and health enhancement measures; it can be used by the needs assessment to plan social investments, but it is unlikely to be useful for making future comparison.

6.2 COMPONENTS OF BASELINE ANALYSIS

The main components of baseline reports are:

- literature reviews;
- health profiles;
- gap analyses;
- surveys;
- studies;
- published and unpublished reports;
- opinions of key informants;
- opinions of the affected community and other stakeholders.

The baseline report combines international, national, regional and local reports of health status and health determinants in order to build a profile of the existing community and its health needs.

The evidence may range in degree of certainty from definite to speculative. It may range in degree of measurability from qualitative to calculable. There will often be gaps in the evidence and assumptions will have to be made. This is acceptable if the assumptions are made explicit. The reader is then able to decide whether to agree or disagree. This, in turn, assists with the analysis of impacts and needs and the subsequent development of justifiable recommendations.

The process varies according to context. To illustrate this, a local proposal in the UK will be contrasted with an international proposal in a low-income country.

6.2.1 Baseline data in the UK

In the UK, an HIA can be commissioned from local HIA specialists who are already familiar with much of the high-quality, routinely collected health data available. Much of the data are available on publicly accessible websites maintained by government bodies and are disaggregated to ward or super output area level (a kind of neighbourhood) (DCLG, 2007a, 2007b, 2007c). Health and planning data often share the same geographic boundaries, creating a common data set. The Public Health Observatories provide detailed health profiles (for example, North West Public Health Observatory, undated). Comparisons can be made over time and between community groups for a range of health indicators and outcomes. See Section 6.3.1 for an example.

Despite the wide range of sources and the large body of scientific endeavour, the data are still incomplete and a number of relevant gaps will usually be identified.

6.2.2 Baseline data in low-income countries

In this book, an international proposal is conceived and managed outside the recipient country, often a low-income country. Public health data for low-income countries is different in quantity and quality from data in high-income countries, such as the UK. There are usually many more gaps. The baseline report may be constructed in three main stages, illustrated in Table 6.1. The first stage is likely to be conducted at an international level. Much of the data available at this level is aggregated to large populations but there are also academic publications about special surveys in specific localities. Similarly, a national proposal refers to one that is conceived and managed within one country.

Table 6.1 *Sources of baseline information for an international proposal*

Level	Examples of sources of information
International	Internet: health and development data; international academic publications.
National	Unpublished reports in ministries, annual statistical summaries and maps in hospitals and clinics, university research projects, NGO reports, national and regional websites; national academic publications.
Local	Public health specialists, medical practitioners, nurses, pharmacies and healers, local government departments including utilities and police.

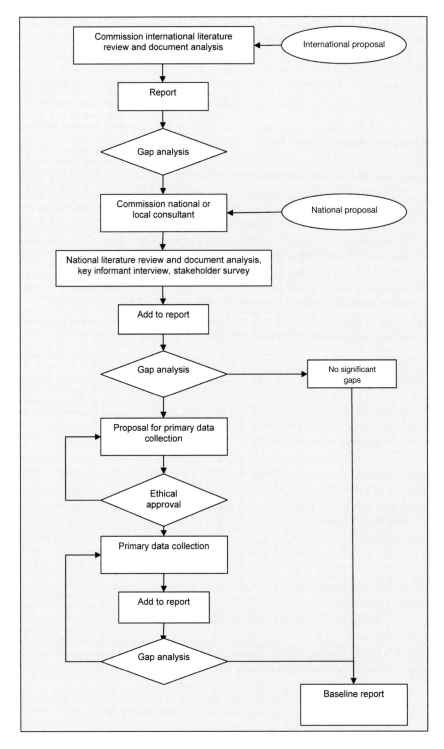

Figure 6.2 *Iterations required to compile a baseline report*

Table 6.2 *Summary of a gap analysis*

Gaps in the literature review	Relevance
Food security and food safety	Determinants of malnutrition and anaemia
Diarrhoea and other gastrointestinal infections	Determinants of childhood morbidity, malnutrition and anaemia
Leishmaniasis vectors and their distribution	To determine if project is in a leishmaniasis transmission zone
Road traffic accidents and other injury	Project generates road traffic
Gender and other inequality	Basis for malnutrition and anaemia
Role of traditional healers	Understanding local health care needs
Mental illness	Understand effect of rapid change on community well-being
Special reports on health of similar communities in the region	Understand special health needs of typical desert communities
A ranked list of top 10 causes of morbidity and mortality at national, regional, district and local levels, divided by sex and adult/children	Understand local priorities
A discussion of causes of inaccuracy or bias in national medical data	Understand reliability of data
Distribution of indicators by age, sex, ecosystem, social class	Population health indicators disguise important differences and inequalities between community groups

Preparation of a baseline report for an international proposal may require a minimum of three corresponding iterations (see Figure 6.2). Preparation of the baseline report for a national proposal is similar, except that the first iteration is not required. The steps in Figure 6.2 are described below.

6.2.3 Literature reviews and gap analysis

A common tool for all baseline reports is a review of the published and unpublished literature. This is also required for other components of the analysis and was discussed in Chapter 5. There will usually be gaps in the available information. Gap analysis, introduced in Chapter 5, is required in order to decide if the gaps are significant and need corrective action. Gaps occur where reports have not been obtained or no data have been collected. The gap analysis lists these gaps and recommends the subsequent action required. The action might be a further literature review, surveys or field studies.

6.2.4 Example gap analysis

Table 6.2 is a summary gap analysis that I produced when working as an international consultant. I helped commission a baseline literature review by a national consultant for

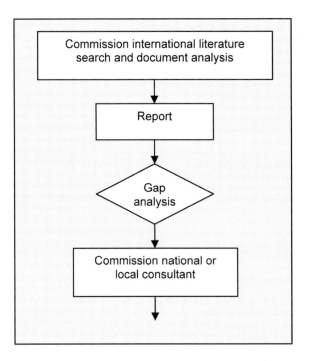

Figure 6.3 *International information review*

a project located in Pakistan. Judging by the number of gaps, it may be concluded that the scope of the literature review was inadequate, perhaps because the commissioning process was inadequate. This has to be anticipated and managed.

One of the gaps was associated with a vector-borne disease called leishmaniasis. The literature review carried out by the national consultant made no mention of leishmaniasis. An Internet search of 'leishmaniasis Pakistan' produced 103,000 hits. Leishmaniasis was a health concern because disturbance of the habitat by the project might affect prevalence rates.

6.2.5 International information

The first step in the international baseline report is a desk study using the Internet, summarized in Figure 6.3.

There are many Internet-based sources of national data and many different indices that can be used to compare and contrast health and its determinants between nations. For example, WHO maintains comparative statistics as part of the global burden of disease estimates (WHO, 2002). This page links to spreadsheets that aggregate national-level data collected by ministries of health. These data indicate the disease priorities in each region according to measured incidence and known severity. A common indicator of population health status is life expectancy at birth. Note that WHO data are a compilation of official reports by governments. These, in turn, are a compilation

Life expectancy

A typical value of the life expectancy statistic in low-income countries is 48 years. This does not mean that people tend to die when they are 48. It means that the death rate in children and young people is relatively high.

Table 6.3 *Typical data compiled by WHO – male mortality in the African region*

Age	All causes	Infectious and parasitic diseases
0–4	2,308,049	1,395,082
5–14	273,064	114,398
15–29	490,406	257,639
30–44	791,860	548,194
45–59	604,795	290,412
60–69	403,532	101,869
70–79	368,274	46,959
80+	176,959	16,847

of data collected from public medical centres. Like all data sources, there are inherent inaccuracies, but they have comparative value.

Table 6.3 illustrates some of the African regional data. The data are not expressed as rates because population size is unclear.

6.2.6 Inherent inaccuracies

There are many reasons why government health data are inaccurate. The data are only captured from patient attendance at public clinics. In some low-income countries, the national health budget is so small that the public has little confidence in the service and use alternative practitioners in the private sector. These include formal doctors as well as traditional healers. The private sector usually does not provide any statistical information. Data may also be deliberately under-reported in the case of politically

Example of data inaccuracy from the US

In the US, the AIDS Healthcare Foundation stated in 2008:

'... the numbers they've been reporting for years have been inaccurate, and have incorrectly portrayed the US epidemic as static ...'

Reuters, 27 March 2008

sensitive diseases such as HIV/AIDS and cholera. Because of irregular drug supplies, people may only use public clinics when they know that a drug delivery has been made – biasing statistics.

There are often no reliable demographic census data. Without this, the number of cases reported has little meaning because the data cannot be expressed as a rate. There may have been 10,000 deaths recorded from malaria, but was the population size 1,000,000 or 2,000,000? Not all deaths are reported. For example, newborn babies may be buried unnamed and in unmarked graves.

Government clinics are often poor record-keepers because they do not have time or staff to fill in forms accurately or promptly. The forms that are filled in may be collected and analysed too slowly. Countries do reform their health systems and then health data reporting improves.

6.2.7 International indicators

Common health indicators are reported by ministries of health and include infant mortality rates and health determinants such as access to water supply or sanitation, and educational achievement. Less well-known determinants of health include human rights, corruption, economic factors and inequality. There are many possible indicators. Table 6.4 contains examples. In most cases, the indicator is compared with the best and worst value.

The health indicators for Nigeria are generally poor, although not the worst. Some of the indictors that I have chosen may require additional explanation.

6.2.8 Human rights

Human rights abuses affect the HIA process because they interfere with the freedom of stakeholders to express an opinion, as well as affecting mental and physical well-being. This reduces the likelihood that public concerns can be identified, an unbiased HIA can be produced, or that well-being can be safeguarded and enhanced.

Human rights indicators do not have reliable statistical measures, making comparisons between countries complex. They are usually expressed in descriptive terms. The following example is a summary of what Amnesty International said about Nigeria (Amnesty International, 2008):

- Elections were marred by widespread violence and widely criticized by observers.
- Security forces continued to commit human rights violations in the oil-rich Niger Delta with impunity and few among the local population benefited from the region's oil wealth.
- Police and security forces executed hundreds of people extra-judicially.
- Religious and ethnic tensions persisted.

Table 6.4 *Examples of international indicators of health: Nigeria*

	Nigeria	Best country	Worst country	Relevance/explanation
Infant mortality rate	10%	<0.2%	18%	Capacity of country to care for infants
Under-five mortality rate	19%	<0.3%	28%	Risk of child death
Expectation of life at birth (male/female)	47/48	80/86	37/37	Risk of early death
Adult HIV prevalence rate	3%	<0.1%	26%	Epidemic risk
Households without water	52%	0	78%	Risk of communicable disease
Households with sanitation	44%	100%	9%	Risk of communicable disease
GDP per capita	US$1500	US$71,400	US$600	Relative poverty of nation
Health expenditure per capita	US$51	US$5711	US$14	Available medical services
Corruption index	2.2	9.4	1.4	Affects HIA process
Adult literacy rate male	78%	>98%	31%	Empowerment
Adult female literacy rate	60%	>98%	15%	Empowerment
Adult literacy ratio of female to male	77%	99%	31%	Gender inequality
Human rights	See Section 6.2.8			Affects HIA process
Inequality in income or expenditure (GINI)	43	24	74	0 = absolute equality 100 = absolute inequality
Human Development Index rank	158	1	177	Composite index indicating relative development of nation
Political rights and civil liberties	4	1	7	Surveying public opinion
Press freedom ranking	110	1	195	Availability of quality assured public information

Sources: Health data: www.globalhealthfacts.org; corruption: www.transparency.org; human development reports, many indicators including health data: http://hdr.undp.org/en/; economic data: http://worldbank.org/; political and civil liberties, freedom of the press: www.freedomhouse.org

6.2.9 Corruption index

High levels of corruption can undermine the HIA process. Corruption can affect the availability of reliable health data, the provision of services and the ability to implement health management measures that require partnerships with national institutions. For example, government officers may require payment (effectively a bribe) to supply public data. The budget for implementing recommendations may get diverted. It may affect the provision of many public services and consequently the health, safety and well-being of the community. For example, the former Azerbaijani health minister was jailed for corruption in 2007 (RFE/RL, 2007) and accessing data from the ministry was problematic during his period in office.

Table 6.5 *Country rankings in a corruption index produced by Transparency International*

Score range	Countries achieving
9–10	Finland, New Zealand, Denmark, Iceland, Singapore, Sweden, Switzerland
8–9	Norway, Australia, Netherlands, UK, Canada, Austria, Luxembourg, Germany, Hong Kong
7–8	Belgium, Ireland, US, Chile, Barbados, France, Spain
6–7	Japan, Malta, Israel, Portugal, Uruguay, Oman, United Arab Emirates, Botswana, Estonia, Slovenia
5–6	Bahrain, Taiwan, Cyprus, Jordan, Qatar, Malaysia, Tunisia
4–5	Costa Rica, Hungary, Italy, Kuwait, Lithuania, South Africa, South Korea, Seychelles, Greece, Suriname, Czech Republic, El Salvador, Trinidad and Tobago, Bulgaria, Mauritius, Namibia, Latvia, Slovakia
3–4	Brazil, Belize, Colombia, Cuba, Panama, Ghana, Mexico, Thailand, Croatia, Peru, Poland, Sri Lanka, China, Saudi Arabia, Syria, Belarus, Gabon, Jamaica, Benin, Egypt, Mali, Morocco, Turkey, Armenia, Bosnia and Herzegovina, Madagascar, Mongolia, Senegal
2–3	Dominican Republic, Iran, Romania, Gambia, India, Malawi, Mozambique, Nepal, Russia, Tanzania, Algeria, Lebanon, Macedonia, Nicaragua, Serbia and Montenegro, Eritrea, Papua New Guinea, Philippines, Uganda, Vietnam, Zambia, Albania, Argentina, Libya, Palestinian Authority, Ecuador, Yemen, Republic of Congo, Ethiopia, Honduras, Moldova, Sierra Leone, Uzbekistan, Venezuela, Zimbabwe, Bolivia, Guatemala, Kazakhstan, Kyrgyzstan, Niger, Sudan, Ukraine, Cameroon, Iraq, Kenya, Pakistan, Angola, Congo Democratic Republic, Cote d'Ivoire, Georgia, Indonesia, Tajikistan, Turkmenistan
1–2	Azerbaijan, Paraguay, Chad, Myanmar, Nigeria, Bangladesh, Haiti

The NGO Transparency International provides an annual indicator of the level of perceived corruption in each country of the world (TI Secretariat, undated). The magnitude of the indicators in a specific country varies slightly from year to year and Table 6.5 is not necessarily up to date. Nigeria was in the lowest rank when I compiled it.

Unfortunately, countries with large supplies of natural resources frequently rank poorly on corruption and human rights indices. Exploitation of natural resources may increase corruption and human rights abuses – see Chapter 12.

6.2.10 Inequality

Data aggregated at national level hide major variations between communities. There are now data available on the Internet that disaggregate some of the indicators by socio-economic quintile, age and sex, and by geographic region. In all countries, there are major differences in health indicators between groups. Examples were described in Chapter 2 and many more have been published (CSDH, 2008). Inclusion of a description of such differences is important for understanding the impact of projects on different groups and for allocating mitigation measures in the most equitable manner.

The GINI coefficient, listed in Table 6.4 is a comparator of economic inequality. It measures the distribution of income between economic groups using a complex formula (for a quick overview, see Wikipedia, 2010).

6.2.11 Trends

The United Nations Development Programme publishes a Human Development Index that can be compared between countries and regions over time (UNDP, 2008). This index enables comparisons to be made on the basis of more realistic variables than just income. For example, the Human Development Index for the Democratic Republic of the Congo has declined compared to other countries in Africa, indicating that this country is not keeping up with its peers.

6.2.12 Reporting adverse data

Most of us do not like others to criticize our own countries, although we may do so ourselves. Most countries contain unfavourable elements. Much of the data discussed above, although factual, would be regarded as critical. Such data need to be handled with diplomacy. They may be too sensitive to include in a public report, even though it is all in the public domain. They may also be too sensitive to include in a private report that would be seen by government officials or national partners. On the other hand, it is information that affects the planning and design of an HIA. In some cases, it may form part of an annex to an HIA report that has restricted circulation.

The ethics of undertaking HIA in countries with poor human rights and corruption records is discussed elsewhere in this book.

6.2.13 Standard components of a health baseline

For international proposals, the scope of work provided by the client can include a list of specific items that should be reviewed in the baseline report. The following set of subheadings is an example from a transnational oil company:

- health legislative and institutional framework;
- national health accounts;
- health policy and implementation programmes;
- community health risk factors or health determinants;
- morbidity and mortality;
- health care delivery services;
- private/traditional health workers;
- other health organizations.

There is an unfulfilled opportunity to develop standard, fit-for-purpose, guidance on what to include in baseline studies in a range of settings.

6.3 NATIONAL INFORMATION

There is an urgent need to build local capacity for HIA in many countries and a local consultant should be provided with opportunity and training whenever possible. Some

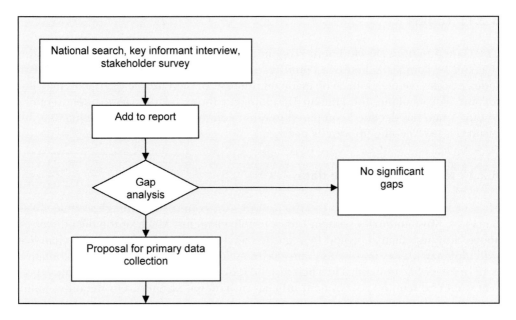

Figure 6.4 *National and local information review*

information cannot be obtained from the Internet. It may only be available in the local language, only in hard copy, internal to local health agencies, regarded as confidential or the property of private medical clinics. It can be hard to find. Government officers may be reluctant to supply standard government health data. They may expect a payment, suspect the inquirer is a spy, dislike external consultants or simply be unused to the free exchange of information. A local consultant who speaks the language and is already known to them is likely to achieve better results than an international 'laptop consultant' on a 'data safari'.

For national proposals that are conceived and managed within one country, the literature review starts at this point and includes both national and international sources. For international proposals, described above, this is the next step. See Figure 6.4.

The national information review should follow a scoping exercise that identifies the health concerns associated with the proposal. This helps to focus the review. Without this focus, the review may contain too much information of too little relevance. The review should address all of the issues identified during scoping as well as those identified during the international desk-study (if there was one). Section 6.2.4 included an example of a gap analysis where the local consultant had failed to investigate the available literature, or had failed to make connections. I have described additional pitfalls elsewhere (Birley, 2007).

In addition to collecting printed reports, key informant interviews will be required. See Chapter 5 for more details. Whenever possible, triangulate the information. For example, during the HIA for a project in China, we wanted to know the local road traffic accident rates. We asked the local police, who said there were no road traffic accidents. We then visited the accident and emergency unit in the local hospital, who

Table 6.6 *Example of health data available at ward, regional and national level*

	Towcester Mill	Towcester Brook	South Northants	East Midlands	England
Total people	4298	4632	79,293	4,172,174	49,138,831
	%	%	%	%	%
General health: Not good	6	5	6	9	9
General health: Fairly good	21	17	20	23	22
General health: Good	73	78	75	68	69
People of working age with a limiting long-term illness	9	8.5	9	14	13
People with a limiting long-term illness	13	11	13	18	18
All people who provide unpaid care: 1–19 hours a week	75	74	78	69	69

said they frequently treated traffic injuries. There were no written reports available to confirm these opinions.

6.3.1 Example of health profiles in the UK

We compiled baseline data for a housing project near the small town of Towcester, in England. One focus of the HIA was active travel. The town was divided between two electoral wards, Mill and Brook, with about 4500 people living in each. The Office of National Statistics provides comparative data on their self-reported health and this is listed in Table 6.6 (ONS, undated). The data suggest that the health of these two communities is higher than for the region. More detailed data were provided by the local health authority. We knew, for example that two doctors' surgeries served the same community and that one had a higher percentage of low-income households than the other. We knew what services each surgery provided and the travel time to the nearest hospitals. We also had data on access to transport including bike use (1 per cent of all journeys). The accidental death rate for the subregion was relatively high as a result of road traffic accidents. This appeared to be associated with the large number of rural roads.

Other examples of UK health profile data are scattered through this book.

6.3.2 Examples of national data

There are often research reports published in the academic literature, or by specialist agencies, that are generally or specifically relevant. In the UK, for example, there are reviews of the association between health and housing, transport, green space and much else and these are accessible via the HIA Gateway (APHO, undated). The following two examples are from Nigeria.

6.3.3 Malaria example

Malaria transmission depends on the presence of certain mosquitoes and this is a specialist topic. For example, a literature review was undertaken during an HIA of an infrastructure project in a coastal location of Nigeria. The literature review identified a published study of potential malaria vectors from a nearby coastal location (Awolola et al, 2002). See Table 6.7. The design and location of the infrastructure could influence the distribution and abundance of the mosquitoes.

The data demonstrated that:

- There were four potential malaria mosquitoes in the coastal zone.
- *Anopheles arabiensis* was never found to be carrying the parasite and only fed on animals. It could be excluded.
- The most important malaria mosquito, *Anopheles gambiae ss*, was only present during the wet season.
- The other malaria mosquitoes were present all year round.
- Some 2–4 per cent of the malaria mosquitoes were infective and able to transmit the malaria parasite.
- The breeding site requirements of the four species were different.

6.3.4 Sexually transmitted infections example

Data on STI prevalence rates are often needed. The data can be used to formulate management plans for reducing transmission between the workforce and the local community.

Table 6.7 *Malaria vectors in coastal Nigeria*

Mosquito species in the genus *Anopheles*	When abundant	Feed on	% annual sample	% infective	Breeding sites
gambiae ss	wet season	human	45	4	Puddles
melas	all year	human	29	2	Saline ponds
moucheti	all year	human	21	2	Riverine forest
arabiensis	dry season	animal	5	0	Ponds

Reproduced by kind permission of Maney Publishing

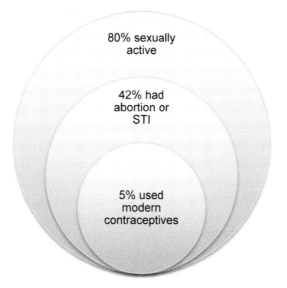

Figure 6.5 *Summary of data regarding sexual behaviour among 17–19-year-old rural women in the Niger Delta*

During an HIA of an infrastructure project in Nigeria, we identified a special study commissioned by an international donor. The report was readily available but it was not formally published in the academic press (Brabin et al, 1995). It was a survey of the sexual behaviour and sexual infections of rural adolescent women in the Niger Delta. Some of the results are summarized in Figure 6.5. It indicated that young women in that community were extremely vulnerable to reproductive ill health. This vulnerability was a function of their poverty and disempowerment. The project would create ideal conditions for an HIV epidemic. Careful management would be required of the construction workforce including the provision of condoms and continuous peer-to-peer health promotion. In addition, the project should seek to reduce gender inequalities by providing training, credit and employment opportunities for women.

6.3.5 Existing health services

Descriptions of existing health services are often required in HIA. This is a specialist topic best undertaken by those engaged in provision of the services or those engaged in health sector reform and health systems research. Reports are usually available in ministries of health at national and local level. These have usually been prepared for the purpose of planning additional services or obtaining additional funding.

In the case of rural locations in low-income countries, a great deal can be learned by visiting a primary care centre. In the worst cases, one may observe that drug supplies and staff salary payments are poor and erratic, records are not kept properly, water supplies and sanitation facilities are limited, equipment is broken or poorly maintained and users travel large distances to wait in long queues for little benefit. Such facilities are not able to cope with the influx associated with a new proposal.

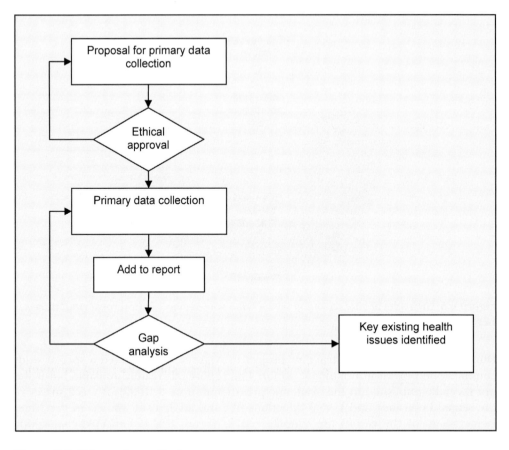

Figure 6.6 *Primary data collection*

6.4 PRIMARY DATA COLLECTION

Primary data can be defined as data collected solely for the purposes of the HIA itself (Harris et al, 2007). By contrast, secondary data can be defined as data collected for another purpose that is used in the HIA. When gaps are identified in the availability of essential data, there may be proposals put forward for primary data collection. Health surveys can be conditional on ethical approval procedures, depending on the kind of data to be collected (see Figure 6.6).

Table 6.8 illustrates some of the different kinds of primary data that may be collected.

Examples of social surveys include concerns about the hazardous nature of emissions from a proposed plant and opinions about the quality of existing services. Experiences may include earlier phases of the proposal, similar projects in the vicinity or current service delivery. Behaviours could include medical care-seeking behaviour, substance abuse and sexual behaviour. Social surveys can provide information about

Table 6.8 *Types of primary data*

Type of primary data	Examples
Social	Concerns, perceptions, opinions, experiences and behaviours
Biomedical	Blood samples, height and weight measurements, respiratory function
Environmental	Mosquito densities, noise levels
Direct observation	Service provision, behaviour, current land uses

socio-economic status, sources of employment, difficulties associated with water supplies and much similar information. These should be commissioned from specialists.

Biomedical data can include rates of morbidity and mortality for defined medical conditions. It also includes any information about the medical conditions of individuals, biomedical samples taken from those individuals and behaviours that may be considered private – such as sexual behaviour. The collection of biomedical data is complex and sensitive. It requires the approval of a national ethics committee and a wide range of conditions must be fulfilled. See Section 6.7 Ethical issues.

By contrast, the collection of health determinant data may be relatively straightforward. Valuable information can often be obtained by an expert from a simple walk through the proposal-affected area. A trained observer may note:

- service provision – such as the location of clinics, pharmacies and community centres, the contents of local stores and food outlets and the quality of public transport;
- behaviour – such as the prevalence of graffiti and vandalism and waste management practices;
- current land use – such as location of footpaths, green space, existing projects and waste management.

A wide range of indicators of health determinants can be devised. For example, proposals in low-income countries may affect the price of basic foods in the local market, leading to changes in the rate of malnutrition. It is relatively simple to establish a method of monitoring and comparing these prices. This, in turn, can be used to determine how much food can be sourced locally.

Table 6.9 summarizes some of the differences between measuring health outcomes and health determinants.

Primary data are likely to be collected using statistical techniques and will require careful analysis. See Chapter 5 for examples. The results are added to the report and a final gap analysis will then be required to determine whether sufficient information has been collected to meet all the health concerns. If not, a further round of data collection may be necessary.

There may be an overlap between the environmental, social and health surveys. For example, the environmental survey in a rural developing country setting may identify and count a range of different animal and plant species. If there are vector-borne diseases in the locality, it may be necessary to survey the vector species. Who should do the mosquito survey: the health group or the environmental group? The

Table 6.9 *Comparison of measuring health outcomes and health determinants*

	Health outcome	Health determinant
Challenges	Hard to measure	Easy to measure
	Well defined	Less well defined
	Requires ethical approval	Fits in with the environmental and social
	Sensitive	surveys
	Expensive	Relatively cheap
		Suitable for monitoring and comparison
		Often not sensitive
Examples of what to measure	Morbidity and mortality rates, disaggregated by age, sex, location and socio-economic group	Water supplies
		Market prices
		Condition of medical services
		Care-seeking behaviour
		Wealth distribution
		Risk perceptions
		Crime rates
		Traffic density
		Vector densities

social survey will pick up information about the perceptions of the community and their health concerns. Who should do the social survey: the health group or the social group? The division of labour between the different impact assessments should be made pragmatically, according to local circumstances.

As an example of the complexities of field surveys, obtaining meaningful information about mosquitoes can require the following:

● A medical entomologist should be consulted.
● The survey must take account of seasonal variation.
● Population density should be determined from light traps, human landing catches, house resting catches and larval dipping.
● Breeding sites should be mapped using GPS.
● A sophisticated laboratory is needed for analysing:
  – species and species complexes;
  – parasite infection rates;
  – feeding preferences;
  – insecticide resistance.

6.5 EXAMPLES OF GOOD AND BAD SURVEYS IN DEVELOPING COUNTRIES

6.5.1 Bad example

The community were subsistence farmers and fishing folk. The primary causes of morbidity were communicable diseases including malaria, diarrhoea and respiratory infection. No accurate prevalence rates were available. There was no information about mosquito vectors and their breeding sites in the area. The local consultant proposed to do surveys of hypertension, diabetes and stroke – non-communicable diseases that were becoming increasingly prevalent among wealthy urban dwellers. There was no literature review or gap analysis to support their proposal. No ethical clearance was sought. The project manager did not wish to obtain medical ethical clearance (see Section 6.7 Ethical issues).

6.5.2 Good example

The community were semi-nomadic pastoralists. The study sought to know about the effect of the project on local malnutrition rates. A randomized survey of adult female anaemia rates was undertaken and socio-economic data were collected. Anaemia rates were disaggregated by socio-economic quintiles and distance from the project site. A 10 per cent confidence interval was used to determine sample size. A GPS unit was used to fix the position of survey households. The study design was reviewed by a national medical ethics committee and free prior informed consent was obtained from the participants. See also Section 6.8.

6.6 PLANNING A FIELD SURVEY OF MEDICAL CONDITIONS

A field study may be commissioned in order to determine the prevalence rate of priority medical conditions. Such a study has the following components:

- a clearly defined objective;
- ethical requirements;
- choice of sample size;
- randomized sampling process;
- questionnaire;
- clinical examination protocol;
- data management protocol;
- data analysis protocol;
- report writing.

When a decision has been made to collect this kind of primary data, it should be regarded as a last resort. It will only tell you what is, not what will become as a result of the proposal. There may be seasonal variations, as with environmental data, and the ideal survey should include at least one full year with work during each season. Additional variation will be associated with socio-economic conditions and geographical location. The services of a professional epidemiologist will be required to advise on sample size, randomization, confidence levels and control groups. Costs will be relatively high. The data will be meaningless unless disaggregated by age, sex, socio-economics and geographical location.

Example of intrusive medical survey

Imagine that a stranger knocks on your door and asks to take samples of blood from your children. How would you respond? We should not subject other people to procedures that we are not prepared to accept for ourselves.

6.7 ETHICAL ISSUES

All medical data are sensitive and collecting them from individuals requires a sound ethical procedure. The proposal should be submitted to a medical ethics committee for approval. This will generally not be necessary when health determinant data are being collected, although there are cases when it will be. For example, it is unnecessary to obtain ethical approval to survey the breeding sites of vector mosquitoes. On the other hand, it is necessary to obtain ethical approval to survey the use of condoms in the community.

An ethical health survey has the following features, among others:

- Free prior informed consent – people must not feel obliged to take part in order to gain some personal advantage.
- Confidentiality protocols – sensitive information is held in a safe, confidential format.
- Statistical robustness – you know in advance of the survey exactly what statistical tests will be used in order to make specific comparisons.
- Medical treatment – what are you going to do if your survey detects a particular illness in a particular individual? Will you pay for their treatment or will you refer them to a clinic for further examination?
- Clear goals and objectives.
- Appropriate sample sizes and stratification.
- Unbiased sampling.

A considerable period of time may be required to obtain the approval of the national medical ethical committee.

Whatever primary data collection technique is used, it should have the following properties:

- concise: well focused, time and size limited;
- effective: providing precisely the information that is needed;
- fit for purpose: of the minimum size necessary to obtain that information;
- ethical: respecting the right of privacy as well as all other human rights.

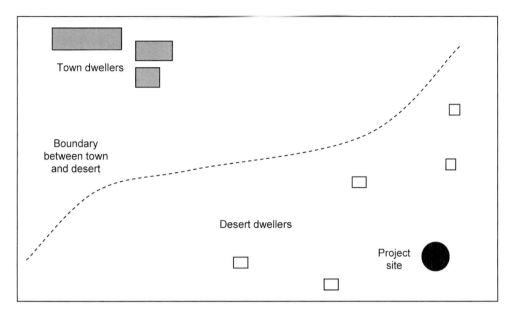

Figure 6.7 *Map indicating project site in relation to town and desert communities*

6.8 EXAMPLE OF BASELINE FIELD DATA FOR A DESERT PROJECT IN A DEVELOPING COUNTRY

Figure 6.7 illustrates a large infrastructure project associated with the energy sector located in a desert (based on Viliani and Birley, 2010). The data in the following graphs are illustrative; they are based on the real case but simplified for use as an example. There were two main communities: the primary community of desert dwellers near the project site; and the secondary community of town dwellers who were further from the project site. The community near the project site were substantially poorer than in the nearby town. They were desert dwellers and lived in scattered settlements, grazing their animals and moving when necessary. They belonged to different ethnic groups and clans. They lacked water, food, grazing, veterinary services, medical services, education, employment opportunities, markets and roads. The town was up to two days' walk away. A field study was undertaken to understand the baseline wealth and health status of the two communities. The study design was subject to ethical committee approval, as discussed above. Figure 6.8 illustrates the distribution of wealth, ranked into quintiles. In the town, many of the population were in the upper quintiles of relative wealth. In the desert, many of the population were in the lower quintiles.

The health of the two communities was measured using adult female anaemia rates as an indicator. Anaemia was measured using field kits that analysed finger-prick blood samples. See Figure 6.9. Anaemia rates were very high in all the wealth quintiles indicating that the community as a whole had significant health needs. The anaemia rates of women living in the desert were strongly associated with their wealth quintile.

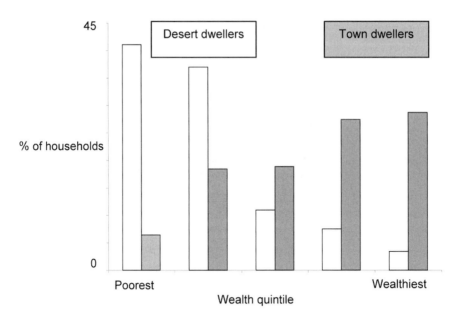

Figure 6.8 *Illustration of the distribution of wealth between desert and town communities by wealth quintile*

One partial explanation is that women in the poorer desert community were only allowed to eat meat about once per month while men ate meat more often. Women subsisted on milk and grains. Family planning was very limited.

The assessors had to decide if the project could affect these anaemia rates. For example, job opportunities created by the project would lead to an input of wealth into the household and this could contribute to an improvement in nutritional standards for women. However, the wealth would belong to the men and they might not share it with the women. The skills needed for working on the project might be more common in the town than the desert. The community might benefit from the project through local sourcing of food or supply of trucking services. It was concluded that the main opportunity to affect the women's anaemia rates would be through social investment projects, which the client was already supporting. Micro-finance initiatives that targeted women had been established using the services of a specialist local NGO.

6.9 SCOPING THE BASELINE: A CASE STUDY IN AN INDUSTRIALIZED ECONOMY

A large infrastructure project was located inside the border of European country 'A'. Neighbouring country 'B' was accessible via a number of country roads and there were no border traffic restrictions. The construction was expected to be undertaken by

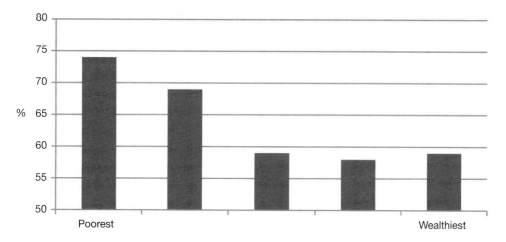

Figure 6.9 *Illustration of female anaemia rates by wealth quintile*

temporary migrant workers from Eastern Europe: the classic 'mobile men with money' who are expected to be vulnerable to, and to create epidemics of, STIs. Country A had an enlightened regime for regulating, protecting and taxing commercial sex workers (CSWs) and promoting safe sex practices. Country B lacked a similar regime. Anecdotal information suggested that some CSWs preferred the untaxed regime and were located in country B, close to the project site.

A health profile was prepared, with the consent of the client, of the affected communities in country A and this included STI rates and management procedures. The client did not authorize resources to prepare a similar profile for country B, or to investigate cross-border public health procedures. The client refused to acknowledge that part of country B was within the affected area, even though the border was quite open. The scope of the baseline study was inappropriate.

6.10 EXERCISE

- Conduct an Internet search of the relationship between health and transport in Eastern Europe.
- Explore the WHO and UNDP websites seeking health and health indicator data for your own country.
- Search for a public health annual report for the administrative area in which you live and find the current priorities for action.

6.11 REFERENCES

Amnesty International (2008) *Annual Report*, http://archive.amnesty.org/air2008/eng/home page.html, accessed February 2011

APHO (Association of Public Health Observatories) (undated) The HIA Gateway, www.apho.org. uk/default.aspx?qn=p_hia, accessed July 2009

Awolola, T., R. Hunt, A. Ogunrinade and M. Coetzee (2002) 'Dynamics of the malaria-vector populations in coastal Lagos, south-western Nigeria', *Annals of Tropical Medicine and Parasitology*, vol 96, no 1, pp75–78, www.maney.co.uk/journals/atm and www.ingentaconnect. com/content/maney/atmp

Birley, M. (2007) 'A fault analysis for health impact assessment: Procurement, competence, expectations, and jurisdictions', *Impact Assessment and Project Appraisal*, vol 25, no 4, pp281–289, www.ingentaconnect.com/content/beech/iapa

Brabin, L., J. Kemp, O. K. Obunge, J. Ikimalo, N. Dollimore, N. N. Odu, C. A. Hart and N. D. Briggs (1995) 'Reproductive tract infections and abortion among adolescent girls in rural Nigeria', *The Lancet*, vol 345, no 8945, pp300–304, www.ncbi.nlm.nih.gov/entrez/query.fcgi? cmd=retrieve&db=pubmed&dopt=citation&list_uids=7837866

CSDH (Commission on Social Determinants of Health) (2008) 'Closing the gap in a generation: Health equity through action on the social determinants of health. Final Report of the Commission on Social Determinants of Health', www.who.int/social_determinants/the commission/finalreport/en/index.html, accessed July 2009

DCLG (Department for Communities and Local Government) (2007a) *The English Indices of Deprivation 2007 – Full Report*, www.communities.gov.uk/documents/communities/pdf/ 733520.pdf, accessed February 2009

DCLG (2007b) *The English Indices of Deprivation 2007 Summary*, www.communities.gov.uk/ documents/communities/pdf/576659.pdf, accessed February 2009

DCLG (2007c) 'Thematic mapping of index of multiple deprivation', www.imd.communities.gov. uk, accessed February 2009

Harris, P., B. Harris-Roxas, E. Harris and L. Kemp (2007) *Health Impact Assessment: A Practical Guide*, Centre for Health Equity Training, Research and Evaluation (CHETRE), University of New South Wales, Sydney, www.health.nsw.gov.au

NICE (National Institute for Health and Clinical Excellence) (2005) *Health Needs Assessment: A Practical Guide*, www.publichealth.nice.org.uk/page.aspx?o=513203, accessed November 2006

North West Public Health Observatory (undated) 'Health profiler North West of England', www. nwph.net/healthprofiler/#, accessed November 2009

ONS (Office for National Statistics) (undated) Various UK statistics, www.statistics.gov.uk, accessed April 2010

RFE/RL (2007) 'Azerbaijan: Former Azerbaijani health minister jailed on corruption charges', Caucaz, Europenews, www.caucaz.com/home_eng/depeches.php?idp=1625&PHPSESSID= df4e939cd1a29c

TI Secretariat (undated) Transparency International, the global coalition against corruption, www.transparency.org, accessed February 2010

UNDP (2008) Human development reports, http://hdr.undp.org/en/statistics/data, accessed 2008

Viliani, F. and M. Birley (2010) 'Desert gas project', in *HIA Conference: Urban Development and Extractive Industries: What Can HIA Offer? 7 April, Geneva*. World Health Organization, www. who.int/hia/conference/posters/en/index.html

WHO (World Health Organization) (2002) *Revised Global Burden of Disease (GBD) 2002 Estimates*, www.who.int/healthinfo/global_burden_disease/estimates_regional_2002_revised/en, accessed November 2009

Wikipedia (2010) GINI coefficient, http://en.wikipedia.org/, accessed 2010

Wright, J., R. Williams and J. R. Wilkinson (1998) 'Health needs assessment: Development and importance of health needs assessment', *British Medical Journal*, vol 316, pp1310–1313, www.bmj.com/cgi/content/full/316/7140/1310

CHAPTER 7

Prioritization

- Four approaches to prioritization are explained: risk assessment matrix, economic analysis, win-win, and values and standards.
- Perception of risk is discussed.
- Residual risk is explained.

7.1 INTRODUCTION

During the HIA, a set of positive and negative health concerns is identified and recommendations are made. These require action. Any action that follows for either safeguarding or enhancing health will have resource implications. As the budget is limited, priorities will have to be agreed. This chapter explores some of the different approaches to setting priorities. It focuses more on the reduction of negatives rather than enhancement of positives because my experience lies there.

Prioritization is an aspect of risk management. One general approach to risk management distinguishes four responses to an identified risk: take, terminate, transfer or treat. A proponent may decide to take a risk and not implement a proposed recommendation. In doing so, they impose an involuntary risk on the local community who may not be so willing to accept it. The decision to terminate a risk is equivalent to deciding not to go ahead with the proposal, or to make significant changes to the location or technology used. The decision to transfer a risk does not transfer responsibility. It might include actions such as insurance, sharing the risk with other government agencies or contracting out. The decision to treat a risk involves action such as implementing the recommendations made in the HIA.

There are a several approaches to risk prioritization. These include:

- risk assessment matrix;
- economic methods;
- win-win;
- values and standards.

7.2 RISK ASSESSMENT MATRIX

One common tool for risk management is called the risk assessment matrix (RAM), risk analysis matrix or risk prioritization matrix. See Figure 7.1. It is used for analysing all kinds of risks and it is based on probability and consequence. This method assumes that health impacts can be categorized in these two dimensions. Probability is also referred to as frequency or likelihood of occurrence. Consequence is the magnitude of the effect.

The probability ranges from 0, meaning the event will not occur, to 1, meaning it will definitely occur. Most probabilities will lie between these extremes. The consequence varies from very beneficial through neutral to very harmful. In the case of harmful consequences, terms such as severity are appropriate. For example, the severity of air pollution may range from an annoying smell to acute poisoning. A person drowning during water sports is a low probability event that is very harmful to the person involved.

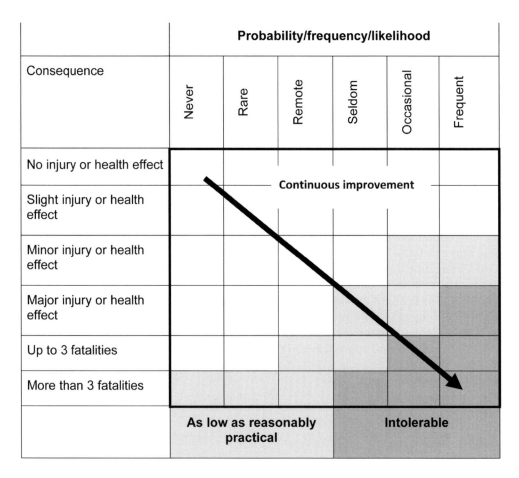

Figure 7.1 *A risk assessment matrix (the arrow indicates increasing significance)*

Table 7.1 *A matrix with beneficial as well as detrimental consequences*

Consequence	Probability/frequency/likelihood				
	Very low	Low	Medium	High	Very high
Very Beneficial					
			7		
No effect					
Very harmful	1				

The following are examples of health impacts with different degrees of likelihood and consequence. The examples are based on an industrial plant to be constructed in a low-income country:

1 Explosive release of poisonous gas that engulfs a settlement and kills many people.
2 Increase in HIV infections during construction.
3 Local merchants benefit from selling goods and services.
4 People are killed, or seriously injured, by transport moving along the access roads to a project site.
5 Some people have to move home and subsequently suffer psychosocial or mental health problems.
6 Increase in malnutrition from loss of wild foods.
7 Quality of life is improved as a result of gaining meaningful employment in the new plant.
8 Increase in access to medical facilities.
9 Increase in people who can transport their own produce to market and sell it as a result of the upgraded transport infrastructure.

These impacts can be plotted in a matrix that has likelihood and consequence as its axes. Each axis is best represented by a small number of ranks. See Table 7.1. There are usually insufficient data to plot the events objectively. The accuracy of the plot may not be high and a number of assumptions may be required. Plotting provides an opportunity for discussion and the development of consensus. This method of analysis may be sufficient to prioritize different kinds of risks and to trigger appropriate action. As an exercise, the nine impacts listed above could be plotted. The approximate position of events 1 and 7 are indicated, as an example.

In order to use the matrix for prioritization, the significance of the risk must be determined and this is a function of consequence and likelihood:

Significance = Consequence + Likelihood

The most common use of this method is the prioritization of negative impacts, or risks. The likelihood and consequence classifications vary with context and different descriptors are used. In this example, significance is divided into three bands labelled continuous improvement, as low as reasonably practical (ALARP), and intolerable. These are defined in Table 7.2. For example, a risk that is common and could cause a fatality is considered intolerable by most rational people. This implies that immediate action is required to reduce it to a lower band. A risk that is ALARP may be acceptable to the proponent. Systems are in place to reduce the risk as much as possible, but some risk remains. Emergency plans are required for treatment and additional risk reduction measures should be developed in the medium term. Risks that are rare or minor are accepted in the short and medium term, but further risk management plans should be developed whenever the opportunity arises. Again, the classification of significance and the action implied varies with context.

The descriptors of likelihood used in the RAM are based on the assumption that sufficient data are available from previous proposals of a similar type to calculate the frequency of occurrence of a particular outcome. This is rarely the case in project-level HIA at present, but may apply to occupational health and safety in large industries with good global record-keeping. It may also apply for the HIA of national policies where good public health data have been collected and analysed for many years. For example, much is known about the effect of airborne particulates on urban populations.

Table 7.3 and Table 7.4 are examples of classification of consequence using different descriptors, with an explanation of their meaning. The first example uses

Table 7.2 *Example of ranking significance*

Significance	Explanation
Intolerable	Short-term, interim risk reduction required. Long-term risk reduction plan must be developed and implemented.
As low as reasonably practical	Additional long-term risk reduction required. If no further action can reasonably be taken, management approval must be sought to continue the activity.
Continuous improvement	Risk is tolerable if reasonable safeguards/management systems are confirmed to be in place. Risk reduction at management discretion.

Table 7.3 *Example of classification of consequences of occupational risks*

Class of likelihood	Consequences for occupational health and safety	Examples
No injury or health effect	None	
Slight injury or health effect	Not affecting work performance and not affecting daily life activities	First aid cases and medical treatment cases. Exposure to health hazards that give rise to noticeable discomfort, minor irritation or transient effects reversible after exposure stops
Minor injury or health effect	Affecting work performance, such as restriction to activities or need to take up to five days to fully recover, or affecting daily life activities for up to five days, or reversible health effects	Restricted work-day cases or lost work-day cases resulting in up to five calendar days away from work. Illnesses such as skin irritation or food poisoning
Major injury or health effect	Affecting work performance in the longer term, such as absence from work for more than five days, or affecting daily life activities for more than five days, or irreversible damage to health	Long-term disabilities Illnesses such as sensitization, noise-induced hearing loss, chronic back injury, repetitive strain injury or stress.
Permanent total disability or up to three fatalities	Resulting from injury or occupational illness	Illnesses such as corrosive burns, asbestosis, silicosis, cancer and serious work-related depression. Car accident resulting in one to three fatalities
More than three fatalities	Resulting from injury or occupational illness	Multiple asbestosis cases traced to a single exposure situation. Cancers in a large exposed population. Major fire or explosion resulting in more than three fatalities

the context of occupational health and safety, where the system is well developed. The second example uses different descriptors and relates these to clinical illness.

More robust approaches to using the RAM in HIA are under development. For example, improvements have been proposed in the context of mining projects in developing economies (Winkler et al, 2010). Consequence was classified by extent, intensity, duration and health effect. Each of these was given a score of 0–3. The scores were summed to define an impact severity (or total consequence). See Table 7.5. This example is described further in Chapter 10.

Table 7.4 *Example of classification of medical consequences*

Class	Consequence for public health
Incidental	No impact
Minor	Illness or adverse effect with limited or no impacts on ability to function and medical treatment is limited or not necessary
Moderate	Illness or adverse effects with mild to moderate functional impairment requiring medical treatment
Major	Serious illness or severe adverse health effect requiring a high level of medical treatment or management
Severe	Serious illness or chronic exposure resulting in fatality (<10) or significant life-shortening effects
Catastrophic	Serious illness or chronic exposure resulting in fatality (>10) or significant life-shortening effects

Table 7.5 *Four-step procedure for determining impact severity*

Impact level and score	Consequence components			
	Extent	Intensity	Duration	Health effect
Low (0)	Rare	Minor	<1 month	Not perceptible
Medium (1)	Local, small and limited. A small number of households is affected	Easy adaptation to the health impact. Maintenance of pre-impact level of health	1–12 months. Low frequency	Annoyance, minor injuries or illness that does not require hospitalization
High (2)	Project area, medium but localized. Village level	Adaptation with some difficulty. Maintenance of pre-impact level of health with support	1–6 years. Medium or intermittent frequency	Moderate injury or illness that may require hospitalization
Very high (3)	Extends beyond the project area. Regional level	Unable to adapt to the health impact or to maintain pre-impact level of health	>6 years. Long-term/ irreversible. Constant frequency	Loss of life, severe injuries or chronic illness that may require hospitalization

Reproduced by permission of the publisher

Risks that have a high probability and are very harmful are clearly unacceptable. Risks that are low frequency and minor are probably acceptable. The challenge lies in classifying risks that lie between these extremes.

Consequences will be classified differently by different stakeholders; proponents may take a different view from local communities. For example, if your child developed a minor skin rash after a processing plant was located upwind of your home – would

Table 7.6 *Alternative prioritization matrix*

	High importance	Low importance
High modifiability	✓✓	✓
Low modifiability	✓	✗

you consider this a minor consequence? People have different risk perceptions. For example, which of the following risks is more important?

● Outdoor pollution associated with industrial and vehicle emissions OR indoor pollution associated with tobacco smoke?
● Exposure to industrial contaminants OR exposure to contagious diseases?

These examples suggest that it is difficult to rank consequence. It is determined by the number of people affected, the amount of pain and suffering and whether they live or die. Some diseases cause acute effects – within a few days, weeks or months – while others may not occur for many years. It is difficult to value a life. Some partial solutions are suggested by the use of economic methods, described below, and DALYs, described in Chapter 2.

One example of an alternative matrix approach to prioritization is illustrated in Table 7.6 (Harris et al, 2007). This approach gives high priority to impacts that are modifiable. The impacts that have high importance and that are also highly modifiable have priority. Impacts with low importance and low modifiability have least importance. Impacts that are in-between are hardest to resolve, as is the case with other prioritization systems. The matrix is used to focus a discussion and the original text has suggestions for managing disagreements.

Another example of an alternative approach is to combine consequence and severity into a single set of significance ranking: major, moderate, minor and neutral (ICMM, 2010). This may be useful when defining consequence and severity is too complex or gives an unwarranted impression of scientific accuracy.

7.3 PERCEPTION OF RISK

In order to obtain consensus for prioritizing, stakeholders need to share a perception of a risk. However, the risks that proposal managers perceive are usually different from the risks perceived by the community. The managers are voluntary risk takers while the community are involuntary risk receivers (see below).

7.3.1 An example of risk perception issues

Figure 7.2 is a photograph of the babies' cemetery in Sumgayit, Azerbaijan. Sumgayit has been described as one of the most polluted places on the planet (Islamzade, 1994; Wikipedia, 2009). Many chemical industries were concentrated there. Most of the industry was created during the Soviet era and now lies abandoned (see Figure 7.3). Unknown chemicals have contaminated, and continue to contaminate, the air, water and soil. Asbestos sheets lie crumbling on the ground. It is widely believed that children born in Sumgayit had, and continue to have, a high risk of birth defects. Cancer rates are elevated (UNECE, 2003). The babies' cemetery is a vast field of little headstones. Some of the headstones carry photographs of the dead children. This tragedy is known to many Azeris. Sumgayit lies on the north side of a peninsula. Not far away lies the capital city of Baku and to the south lies the modern oil and gas processing plant at Sangachal (see Figure 7.4). The modern plant has a gas flaring chimney. It receives raw oil and gas from a pipe that stretches out under the Caspian Sea. It cleans the oil and gas and pumps them into pipes that take the product to the Mediterranean. The plant is situated several kilometres from the nearest homes, in a desert habitat, and emissions are monitored and controlled to international standards.

Some of the local inhabitants at Sangachal expressed concern about the emissions from the new plant. They feared for the health of their children and themselves. The fear persisted despite reassurances from the staff of the new plant. Many of those staff were unaware of the Sumgayit story and asked what relevance it had to them. The risk perceived by the local inhabitants was at variance with the statistical risk perceived by the engineers and plant operators. Both perceptions of risk were valid.

7.3.2 Classifying risk perceptions

There is an important difference between the risk perceptions of the informed specialist and the risk perceptions of the informed layperson (BMA, 1990; Funtowicz and Ravetz, 1992; Silbergeld, 1993). The expert, the engineer and mathematician use a single dimension to measure risk: statistical probability; a number between 0 and 1. The layperson, by contrast, seems to perceive and value risks in many dimensions such as those in Table 7.7. As individuals, we are all prepared to take quite high voluntary risks. For example, we may engage in dangerous sports, work in hazardous environments or have unhealthy lifestyles. At the same time, we are not prepared to accept involuntary risks. For example, we may object to a low concentration of chemical pollutant but happily smoke cigarettes. We also seek to protect our children from risks that we would accept for ourselves. For example, people consider the death of a child in a plane crash to be less acceptable than the death of an adult. Both may have the same frequency per distance travelled. Mass death is less acceptable than one-by-one death. We are more concerned with unfamiliar risks than with familiar risks. We regard some manners of dying with more dread than other manners of dying, though this varies between individuals. We are more concerned about risks that may be contagious – such as a plague.

These dimensions of risk influence the manner in which the community perceive a risk. So the perception of risk is also real. Perceived risks give rise to stress, anxiety, anger and concern. Perceived risks need to be managed just as effectively as any other kind of risk. Management strategies include empowerment, communication and acknowledgement.

Table 7.8 is an example that compares the risk ranking of an expert panel with public opinion in the US (Silbergeld, 1993). Both groups were asked to rank the four

Figure 7.2 *Sumgayit baby cemetery, Azerbaijan*

Copyright Tropix.co.uk

Table 7.7 *Some dimensions of risk and acceptability*

Dimension of risk	Examples of perceived relative importance
Voluntariness	Rock climbing versus pipeline explosion
Dread, aversion, horror	Cancer versus drowning
Protecting the innocent	Children versus adults
Many versus few	Plane crash versus car crash
Familiarity	Car crash versus chemical leak
Contagion	Plague versus 'flu

issues that are indicated. The results illustrate some of the differences in the perception of risks between different communities. This implies that priorities for mitigating health impacts could be different.

The literature on the perception of health risk and environmental problems has recently been updated by a study from northwest England (Luria et al, 2009). The study examines perceptions of risk in the context of waste management, odour and air contamination, land contamination, non-ionizing radiations, chemical and hazardous substances, flooding and cancer. The list of reported symptoms is impressive.

Figure 7.3 *Abandoned Soviet chemical industry at Sumgayit*

Copyright Tropix.co.uk

For example, some residents in the vicinity of mobile phone masts reported some of the following: insomnia, fatigue and poor concentration, skin conditions, dizziness, migraines, general musculoskeletal and mental suffering, aggravation of heart and neural diseases, tachycardia, noise annoyance and ear problems, nose bleeds, general concerns about cumulative effects, general fear of utilizing open spaces in proximity of a mast, and loss in property values. The prevalence rate of 'non-specific health symptoms' associated with proximity to radio transmitters may be as high as 1–2 per cent of the total population, although there is still no evidence of a causal pathway associated with this kind of radiation.

7.3.3 Voluntary risk takers versus involuntary risk receivers

Table 7.8 *Perceptions of risk in an example from the US*

Issue	Expert panel	Public opinion
Hazardous waste sites	Low–medium	High
Pesticide residues on food	High	Medium
Indoor air pollutants	High	Low
Consumer exposure to chemicals	High	Low

Figure 7.4 *Modern processing plant at Sangachal*

Copyright Tropix.co.uk

Proponents are voluntary risk takers and beneficiaries, as are the associated financial institutions. The community affected by the proposal are involuntary risk receivers (World Commission on Dams, 2000). They are often poor, vulnerable and marginalized. For members of the community, the proposal may be an unnecessary hazard better placed elsewhere, unless they expect to receive direct benefit from it, such as employment. Figure 7.5 and Figure 7.6 make this point graphically.

Figure 7.7 sketches the differences between expert and public estimates of risk (based on BMA, 1990; DOH, 1997). The straight line represents the set of points where both communities agree. The shaded ellipse represents points that are off the line, where the two groups disagree. For example, the public may regard the risk of disease associated with smoking as being relatively low compared to the statistical risk estimated by experts. In this case the public will be complacent. On the other hand, the public may regard the risk of a brain tumour from living near to a mobile phone mast to be relatively high compared to the statistical risk estimated by experts, if there is such a risk. In this case, the public may express outrage. The solution is to work towards a shared understanding of risk.

Figure 7.5 *Voluntary risk takers – stock market traders in Sri Lanka*

Copyright Tropix.co.uk

7.4 THE NIMBY SYNDROME

One consequence of the perception of risk issues is the NIMBY (not in my backyard) syndrome. The term is usually applied in a pejorative way to opponents of a development, implying that they have narrow, selfish or myopic views. Local communities, faced with development changes of many kinds, band together to reject the change. They argue that it should be placed elsewhere. This is a widespread social phenomenon and there is a growing analytic literature that is outside the scope of this book (Anon, 2006; Luria et al, 2009; NE, undated). One of the causes of NIMBYism may be the manner in which new development proposals are managed. Perhaps there is insufficient community engagement. Perhaps fears are not solicited, heard, acknowledged and addressed in a timely manner.

Opposition to change is often argued in terms of health impacts: fear of crime, pollution, strangers, traffic and generalized risks. Many of the risks and fears are perceptions, not well supported by evidence. The community tends to suggest that the proposal should be located in a different neighbourhood: one that is less articulate, less powerful, poorer and less able to resist. On the other hand, the community is demonstrating cohesion in the face of a perceived outside threat in order to maintain social and environmental values.

Figure 7.6 *Involuntary risk receivers – a rural family at home in Tanzania*

Copyright Tropix.co.uk

Developments that are rejected in this way are often of benefit to society as a whole, but have perceived local disbenefits. Examples include mobile phone masts, drug rehabilitation centres, incinerators, highways and convalescent homes for wounded military personnel.

Changes in land use are a key issue in the UK and US and there are specialist companies that assist developers to counter opposition. The reasons for opposition and their trends are published annually, though the evidence base for these statistics is not indicated (The Saint Consulting Group, 2009). In 2009 in the UK, the main reasons listed were traffic (25 per cent), protection of green space/environment (23 per cent), protecting community character (21 per cent), proximity to home (17 per cent) and protecting property values (5 per cent). The website also lists the popularity of different kinds of development with schools and private housing at the popular end and waste facilities, casinos, power stations and quarries at the unpopular end. Wind farms were relatively popular.

There are a number of categories of NIMBY response. One of these is to oppose the proposal on behalf of someone else. For example, the local schoolchildren may be cited as a vulnerable group regardless of whether the local schools, or the parents, have been consulted. A related concept is inverse NIMBYism, in which those who are further from a proposed location of some infrastructure have more frequent negative concerns than those who are close to an existing location. This has been observed

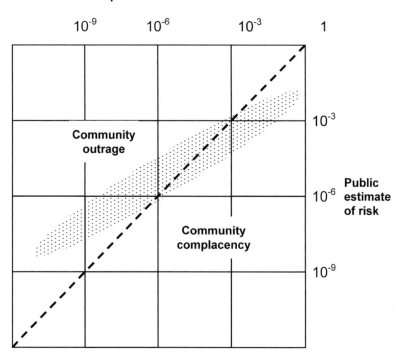

Figure 7.7 *Illustration of differences in expert and public estimate of risk and public response*

in debates about new wind farms in Scotland and Ireland (Warren et al, 2005). For example, a stratified telephone survey of households was conducted by parish for a proposed waste incineration plant as part of an HIA. The new facility would replace an old facility that was about 3 kilometres from it (see also Chapter 5). The results were aggregated by distance from the proposed site. Table 7.9 illustrates some of the data. The sample cannot be tested statistically because the association with distance was not part of the study design, but it may illustrate the inverse NIMBY effect. Some 42 per cent of respondents who lived far from the site (about 10 kilometres), and who would not be able to see it, were concerned that it would affect the view. By contrast, only 34 per cent of those who lived near the site had a similar concern.

More generally, people may react to proposals with either unrealistic expectations or unrealistic fears. Unrealistic expectations are particularly common in poor communities confronted with large oil or mining proposals. People may expect that the proposal will make them rich and provide plenty of employment and infrastructure. They will initially welcome it. But they may become very disappointed and angry when it doesn't. The unrealistic fears, on the other hand, cause people to oppose the proposal. They should find that their fears are unjustified, if the proposal is implemented with proper safeguards. Both sets of expectations need sensitive management and both may be found in the same community. Those conducting HIAs should be careful to avoid creating expectations and to leave management of expectations to the proponent.

Table 7.9 *Belief that the project would have negative impacts according to distance from site*

Factor impacted	Far	Near
Traffic congestion	77 (41%)	80 (30%)
Air quality	48 (26%)	53 (20%)
View	78 (42%)	93 (34%)
Total respondents	186	270

7.5 ECONOMIC ANALYSIS

Any action to safeguard, mitigate or enhance health in association with a proposal is likely to have costs and benefits. As resources are always limited, a decision-maker is likely to request either a cost-benefit analysis or some other form of economic analysis.

Cost-benefit analysis converts both the costs and the benefits to monetary units so that they can be compared. For example, the financial cost of installing air pollution filters is compared with the financial value of the human lives saved. One example of the application of cost-benefit analysis was a programme to estimate the health benefits of cycling (Cavill et al, 2008). The evidence reviewed suggested that the benefit–cost ratio was generally about 5:1. A tool was developed for estimating the ratio in specific settings.

Some of the difficulties with this method are as follows:

- The costs of pain, grief and suffering, diminished quality of life, reduced educational achievement and lost production are hard to quantify.
- There is often a need to value a life lost. Estimates are available, but can be controversial.
- It is often hard to estimate a quantitative measure of the number of people who will be affected at a particular level of severity and probability because of a particular proposal-related health issue. It is probably easiest in the case of poisoning, where a dose-response model can be used.
- Economic analysis measures the effect on a whole society. Health impacts are often experienced inequitably with some communities gaining and some losing. Taking account of such differential impacts adds an extra layer of complexity to the economic analysis.
- How do you compare the economic benefit of reducing long-term disability through the provision of jobs, to the health gain from reducing pollution? What unit should be used?

Cost-effectiveness methods overcome some of the difficulties of cost-benefit analysis by using non-monetary units for health and disease. They enable alternatives to be compared. For example, the financial cost of installing air pollution filters and the number of human lives saved is compared with the alternative of not installing them. Disability adjusted life years (DALYs), discussed in Chapter 2, and quality adjusted life years (QALYs) provide a common metric for quantifying health impact. A number of papers discuss the possible use of these methods in HIA (DOH, 1996/2004; Birley et al, 1998). Examples of application are harder to find.

Willing-to-pay methods provide an alternative approach to measuring the cost of ill health, disability and death. These methods determine the value of anything by observing or asking people what they are prepared to pay. For example, the price that people are prepared to pay for life insurance indicates what value they place on their own life. The difference in the price of property in pleasant and unpleasant environments indicates the value that people place on a pleasant environment. These methods have their own challenges and a detailed discussion lies outside the scope of this book. Examples include the valuation of the health benefits of reduction in air pollution and the valuation of non-fatal road injuries (DOH, 1996/2004).

7.6 WIN-WIN

An ideal outcome from an HIA would not only meet health objectives but would also improve the effectiveness of a proposal. This would be a win-win solution and may be obtained by negotiating solutions for mutual gain. We can imagine a scenario in which the proponent and a government health protagonist are negotiating the actions to take to safeguard or enhance health. See Table 7.10. If negative health impacts have been identified, the health protagonist could seek to block the proposal. Then health would gain but the proposal would lose. Alternatively, the proponent might ignore health concerns and proceed aggressively to complete the proposal. Then the proposal would gain and health would lose.

Priorities for action could be based on mutual gains: the health protagonist promotes opportunities for health enhancement that contribute to the success of the proposal.

Examples of win-win:

- The design of a residential area or construction camp might seek to deter mosquitoes and be easier to maintain.
- An industrial plant might substitute a hazardous process by a less hazardous one and also increase productivity.
- A construction project might improve the skill pool in the local community and so contribute to long-term employment, as well as completing the construction.

7.7 VALUES, STANDARDS AND THRESHOLDS

Another approach to establishing priorities is to refer to a system of values, standards or thresholds that have already been agreed. These may be qualitative or quantitative. Each health impact can then be tested by asking whether it conforms to the system. Examples include:

- equity values;
- cultural values;

Table 7.10 *Explanation of win-win solutions*

		Proponent	
		Gains	Loses
Health protagonist	Gains	Win-win	Diminish proposal
	Loses	Ignore health impacts	Stalemate (stand-off, impasse, deadlock)

- an organization's business principles and standards;
- environmental health standards;
- statutory requirements;
- toxicity thresholds.

An equity value approach might consider an impact unacceptable if it tends to increase inequality. For example a negative impact may affect poor communities more than wealthy communities, women more than men, or the marginalized and powerless more than the influential and powerful. Cultural values may lead to significantly different priorities. For example, in some pastoral communities a cow may be valued more highly than a child, because the short-term survival of the community depends on the health of cows.

As an example of a business principles approach to values, Shell committed to the following (Shell International, 1997):

- pursue the goal of no harm to people;
- protect the environment;
- manage health, safety and environment issues as any other critical business activity.

Some transnational corporations have a principle of avoiding double standards. This implies that a practice cannot be tolerated in a low-income country if the same practice would not be tolerated in a high-income country.

There are many environmental health standards. These include quantitative toxicity thresholds that have been established for various chemicals and are used in EIA to establish emission thresholds for pollutants in air, water and soil. Hygiene standards are used for food preparation, domestic and industrial water quality, and housing design. There are standards for urban street design. Government creates regulations for many aspects of life including driving speeds, noise levels and alcohol consumption.

Standards-based approaches are relatively simple and have the added force of being justified by reference to statutory duties or other obligations.

7.7.1 Case study: Motorway in Europe

Table 7.11 summarizes the policies, guidance and standards that were identified for a motorway project in Europe. The project planned to use loans from the World Bank and US sources, so that IFC and OPIC standards were applicable. Many of the standards included a list of specific recommendations for safeguarding health.

7.8 RESIDUAL IMPACTS

Health impacts can be further classified as pre-mitigation, residual and cumulative.

The pre-mitigation impact is identified by the HIA. It is the impact that is expected if the recommendations made in the HIA are not implemented. The residual impact refers to the unwanted impact that remains after the recommendations are implemented. If the residual impact is still significant, then the proposed mitigation measures are inadequate. The cumulative impacts are concerned with the contribution that the proposal makes to other changes that are taking place in the environment or society. Cumulative impacts are discussed in Chapter 12.

In some cases, HIA reports confuse pre-mitigation and residual impacts. In other words, they report the priority of an impact as though it had already been mitigated. This is bad practice. The initial assessment should be made on the assumption that no special mitigation measures have been implemented. For example, an HIA concludes that there is no significant impact from air pollution, although the proposal design indicates that there are no measures included for removing pollutants. Careful reading indicates that a recommendation is made in the HIA report that emissions should be filtered. The conclusion is based on the assumption that the recommendation has been implemented.

7.8.1 Example of residual impact analysis

Table 7.12 is an example of residual impact analysis. A communicable disease, here called X, that could produce severe morbidity and even mortality, is common in the local community. The assessment team were confident that the project would facilitate further transmission. They proposed mitigations measures including the provision of appropriate medical care. The residual analysis indicates that they were of medium confidence that implementing the recommendations ('with mitigation') would lead to some reduction in transmission and the severity of cases. However, they felt that social investment would be required, in addition to project-related mitigation, in order to reduce the significance to a low level.

The real case, which inspired this example, is described in more detail in Chapter 10 (Winkler et al, 2010).

7.9 EXERCISE

Take another look at the San Serriffe exercise, last seen in Chapter 5. Choose a health impact and assess its significance before and after mitigation.

Table 7.11 *Applicable policies, guidance and standards*

Policy/guidance/standard	Summary
Human rights	Progressive improvement in human health, no backward steps
Treaty of Amsterdam	High level of health protection shall be ensured. Principle of health in all policies
EU charter on transport and the environment (WHO, 1999)	Reduce need for motor vehicles; shift to health-promoting modes; involve environmental and health authorities; identify vulnerable groups; meet challenges of economies in transition; shift freight away from roads
UNECE integrated transport, health and environment pan-European programme (WHO and UNECE, 2009)	Job creation through investment in sustainable transport; reduce greenhouse gas emissions and other transport pollutants; promote healthy transport policies
World Bank/IFC/Equator Principles guidance and standards	Prevent sexually transmitted infections; ensure emergency preparedness; best transport safety practices; manage interaction between workforce and community; improve driving skills; promote access to services; ensure pedestrian and cyclist safety; set speed limits; require health to be a consideration
OPIC standards (OPIC, 2009)	Support projects that reduce greenhouse gas emissions by 30%; public disclosure of projects that emit >100 kilotonnes CO_2 a year
National transport strategy	Improve and develop transport infrastructure and services to support economic development, provide freedom of movement to all communities, provide access for safety and security systems, and ensure that public transport services were available for all citizens, addressing the particular needs of women and minorities.

Table 7.12 *Residual impact for disease X*

	Consequence	Likelihood	Significance	Confidence
Pre-mitigation	High	Probable	High	High
With mitigation	Medium	Possible	Medium	Medium

7.10 REFERENCES

Anon (2006) 'The NIMBY syndrome and the health of communities', http://goliath.ecnext.com/coms2/gi_0199-6992607/the-nimby-syndrome-and-the.html, accessed January 2010

Birley, M. H., A. Boland, L. Davies, R. T. Edwards, H. Glanville, E. Ison, E. Millstone, D. Osborn, A. Scott-Samuel and J. Treweek (1998) *Health and Environmental Impact Assessment: An Integrated Approach*, Earthscan/British Medical Association, London

BMA (British Medical Association) (1990) *Living with Risk*, Penguin, London

Cavill, N., S. Kahlmeier, H. Rutter, F. Racioppi and P. Oja (2008) 'Methodological guidance on the economic appraisal of health effects related to walking and cycling'. World Health Organization Regional Office for Europe, Copenhagen, www.apho.org.uk/resource/item.aspx?rid=78771, accessed 2010

DOH (Department of Health) (1996/2004) *Policy Appraisal and Health*, Department of Health, London, www.apho.org.uk/resource/view.aspx?rid=78635, accessed 2010

DOH (1997) *Communicating About Risks to Public Health: Pointers to Good Practice*, Department of Health, London

Funtowicz, S. O. and J. R. Ravetz (1992) 'Three types of risk assessment and the emergence of post-normal science', in S. Krimsky and D. Golding (eds) *Social Theories of Risk*, Praeger, Westport/London

Harris, P., B. Harris-Roxas, E. Harris and L. Kemp (2007) *Health Impact Assessment: A Practical Guide*, Centre for Health Equity Training, Research and Evaluation (CHETRE), University of New South Wales, Sydney, www.health.nsw.gov.au

ICMM (2010) *Good Practice Guidance on Health Impact Assessment*, International Council on Mining and Metals, London, www.icmm.com/document/792

Islamzade, A. (1994) 'Sumgayit: Soviet's pride, Azerbaijan's hell', www.azer.com/aiweb/categories/magazine/23_folder/23_articles/23_sumgayit.html, accessed July 2009

Luria, P., C. Perkins and M. Lyons (2009) 'Health risk perception and environmental problems: Findings from 10 case studies in the North-West of England'. John Moores University, Liverpool, www.cph.org.uk

NE (NIMBY Experts) (undated), www.nimbyexperts.com/index.html, accessed January 2010

OPIC (Overseas Private Investment Corporation) (2009) 'Greenhouse gas/clean energy initiative', www.opic.gov/sites/default/files/docs/ghg_fact-sheet_070109.pdf, accessed 2010

The Saint Consulting Group (2009) '2009 UK Saint Index – Headline results NIMBYism', http://tscg.co.uk, accessed January 2010

Shell International (1997) 'Statement of general business principles', www.shell.com/home/content/aboutshell/who_we_are/our_values/sgbp, accessed February 2011

Silbergeld, E. K. (1993) 'Revising the risk assessment paradigm', in C. R. Cothern (ed.) *Comparative Environmental Risk Assessment*. Lewis Publishers, Florida

UNECE (2003) 'Environmental performance review Azerbaijan', www.unece.org/publications/environment/epr/welcome.htm, accessed 2011 February

Warren, C. R., C. Lumsden, S. O'Dowd and R. V. Birnie (2005) '"Green on green": Public perceptions of wind power in Scotland and Ireland', *Journal of Environmental Planning and Management*, vol 48, no 6, pp853–875

Wikipedia (2009) 'Sumgayit', http://en.wikipedia.org/wiki/Sumqayit#cite_note-2, accessed July 2009

Winkler, M. S., M. J. Divall, G. R. Krieger, M. Z. Balge, B. H. Singer and J. Utzinger (2010) 'Assessing health impacts in complex eco-epidemiological settings in the humid tropics: Advancing tools and methods', *Environmental Impact Assessment Review*, vol 30, no 1, pp52–61, www.sciencedirect.com/science/article/b6v9g-4wfppc7-1/2/a9621176a138c680fd2e77e594b806d1

WHO (World Health Organization) (1999) *Charter on Transport, Environment and Health*, World Health Organization, Copenhagen, www.euro.who.int/en/who-we-are/policy-documents/charter-on-transport,-environment-and-health

WHO and UNECE (World Health Organization and United Nations Economic Commission for Europe) (2009) 'Amsterdam Declaration, making the link: Transport choices for our health, environment and prosperity', www.unece.org/thepep/en/hlm/documents/2009/amsterdam_declaration_eng.pdf, accessed May 2010

World Commission on Dams (2000) *Dams and Development: A New Framework for Decision-making*, Earthscan, London

CHAPTER 8

Recommendations and management plan

- Developing justifiable recommendations is the principal aim of HIA.
- Different methods of appraising recommendations are described.
- A hierarchy of recommendations is explained.
- Some challenges are identified.
- Examples are provided of good and bad recommendations.
- Recommendations accepted by management should be incorporated in a management plan.

8.1 INTRODUCTION

One of the final steps in an HIA is to recommend management measures that address the health concerns that have been identified. The intention of the recommendations is to safeguard health, maximize opportunities for health enhancement and mitigate (alleviate) deteriorations in health. This is arguably the principal aim of HIA and the most important component. It requires the following sub-tasks:

- prioritizing impacts (see Chapter 7);
- making recommendations for action;
- justifying each recommendation;
- indicating the residual impact (see Chapter 7) and the cumulative impact (see Chapter 12);
- drafting the management plan.

An HIA report will need to make recommendations that reflect the aims and objectives of the HIA, as listed in the terms of reference. Recommendations need to be focused on the health impacts of the proposal, and where health gains can be maximized and health loss minimized. Strictly, the proponent is only responsible for the health changes attributable to the proposal. It is not responsible for the existing health needs of the community and should not be seeking to replace the role of national or local government (or other sections of these). However, there are many cases where the proponents will also choose to include social investments that will bring community benefit independently of their proposal (see Chapter 1). In some cases, there will be other agencies that have an indirect interest in the success of the proposal. These will often include health improvement programmes.

For example, in the UK there is a concept of planning gain. A private developer is required to provide amenities for the benefit of the community that are not part of the original development objectives. In the case of a residential development, these may include parks, playgrounds and clinics. In a low-income country, a corporation may provide schools and primary health care facilities, micro-finance initiatives or public transport. Such facilities improve the developer's reputation, ensure its licence to operate, add value to the local workforce and contribute to local development. There is a risk that such development can be unsustainable, or create new costs for local public services, and this risk should be evaluated.

The HIA recommendations should be distributed between the construction, operational and decommissioning stages of the proposal. For example, during the construction stage of an infrastructure proposal there are many more temporary community groups in the locality than during the operational phase. The health issues are likely to be different and responsibilities are likely to be devolved from the proponent to construction companies. The construction companies must ensure that their workforces do not abuse the local community, or introduce new communicable diseases. During the operational stage, there is likely to be a small but stable workforce that is directly under the control of the proponent, plus a range of small but stable service providers in the local community. These may include food growers, retailers, haulage companies, cleaning and gardening services, security and so forth.

Recommendations must be assessed for their feasibility, the resources required to implement them and how they might fit in with or influence existing programmes, e.g. health improvement, health promotion or social investment programmes. They should align with and respect existing government health policies and programmes. It may be appropriate to form partnerships with local government and NGOs in order to deliver a service more effectively. These are often referred to as public–private partnerships (World Bank Institute, 2007). For example, in a poor rural community, there may be concern that prices of basic foods will increase during the construction phase as a result of the procurement requirements of the catering subcontractors. There may be local agricultural NGOs that are working to improve food production, storage and marketing facilities. The proponent may be able to support the NGO with equipment or materials as well as providing a guaranteed market for produce at a fair-trade price. Such partnerships can improve the sustainability of the recommendation.

Recommendations will need to be specific and practical, and examples of good and bad practice are provided below. The recommendation should be presented so that the reader of the report can determine which health impacts it addresses, why the impacts are expected and how the priority has been determined. There should also be discussion of alternative options.

The recommendations are presented to the proponent, who will then decide which to accept and which to reject. The proponent is responsible for cost control and will need to justify any expenditure required. Ideally, the cost of implementation should be estimated for each recommendation proposed. In practice, the data required to do so may not be available at this stage. In any case, the cost may need to be set against the cost of ill health, pain, suffering and death averted, which is discussed further in Chapter 7.

Table 8.1 *Example mitigation hierarchy*

Hierarchy	Explanation
Avoid, reduce at source, minimize	Design the proposal so that a feature causing an impact is removed or altered (e.g. relocate).
Abate on site	Add something to the design. An example is pollution control. In the case of emissions this is often called 'end-of-pipe'.
Abate at receptor	If an impact cannot be abated on site then measures can be implemented off site. For example, install double-glazed windows to minimize noise impact at a nearby residence.
Repair or remedy	Examples include medical care, different water supplies and road realignment.
Compensate in kind or through other means	Where other mitigation approaches are not possible or fully effective, then compensation for loss, damage and general intrusion might be appropriate. This could include financial compensation, or providing community facilities for loss of recreation and amenity space.

Once the recommendations have been accepted, a management plan will be required for implementation. An outline of that plan can be provided in the HIA report. It should indicate who must do what and when. It should include milestones and monitoring.

8.2 MITIGATION HIERARCHY

There is a hierarchy of mitigation measures that is familiar to occupational health and safety advisers. At the top of the hierarchy, with the highest priority, are measures that reduce or remove the risk at source. This can include changing the design, changing the location and changing the process. These are measures taken by society, institutions or companies to remove the source of the impact. Much lower down in the hierarchy are measures that seek to protect the individual from the risk. These are measures taken to change individual behaviour. They include protection gear, such as hard hats and warning signs. A similar hierarchy can be used in HIA (see Table 8.1). In parallel with the hierarchy there are precautionary measures such as strengthening local medical facilities. Measures taken by society that remove the source of the hazard are always to be preferred to measures that depend on changing human behaviour. For example, dangers associated with cigarette smoking are best controlled by protecting people from passive smoking by making it illegal to smoke in public buildings. Adding warning labels to cigarette packets is much lower in the hierarchy and much less effective. Higher-level measures are even more important where an involuntary risk is created by a proposal, that is, a risk that local people did not choose for themselves.

We are equally interested in enhancing health and Table 8.2 provides an example of an enhancement hierarchy (ICMM, 2010).

Table 8.2 *Example health enhancement hierarchy*

Hierarchy	Explanation
Build in benefits for all	Design the proposal so features of the physical, social and economic environment that enhance or lead to a positive health impact for affected communities as a whole are included from the start, e.g. health promotion programmes, access to green space, hygienic and well-ventilated worker accommodation, training and development for employees, minimum incomes and social investment programme for affected communities.
Affirmative action for equity	Put in place measures to ensure that disadvantaged groups reap the benefits of the proposal, e.g. targeted health education and disease prevention programmes, policies or quotas that ensure employment of local people, profit sharing with local community.
Make healthy choices the easy choice	This involves adding something to the basic design or operational policies to encourage and reward health-promoting behaviour (e.g. physical activity). Examples include providing secure bike-parking facilities.
Proactive education and information	Utilize opportunities to provide information and education to enable people to make informed choices about issues like nutrition, safe sex and active travel.

8.2.1 Mitigation examples

8.2.1.1 Malaria

Table 8.3 illustrates the difference between societal and individual approaches for malaria control.

Table 8.3 *Societal versus individual recommendations for malaria control*

Societal	Individual
Improve housing design and location	Use impregnated bednets
Reduce poverty and overwork	Use repellents
Provide social marketing of bednets	Improve treatment-seeking behaviour
Provide rapid diagnosis and treatment	Increase knowledge about transmission
Prevent proposals from creating mosquito breeding sites	Prophylaxis
	Reduce risk-taking behaviour

8.2.1.2 HIV

Mitigation can involve actions at policy, strategy and local level. The mitigation hierarchy for HIV on large infrastructure construction projects could include those listed in Table 8.4.

Table 8.4 *Societal versus individual recommendations for HIV control*

Societal	Individual
Negotiate with government to overcome denial, agree culturally sensitive and effective means of control, and decriminalize adult sexual behaviour	Peer-to-peer health promotion for workforce by contractors
Ensure sexual health clauses are included in contracts, such as requiring condom distribution (the construction company will not implement such measures unless they have been budgeted and agreed)	Anonymous HIV testing and counselling
	Available condoms
Facilitate alternative means of livelihood for vulnerable men and women who might be drawn to commercial sex	

The oil and mining sectors, among others, have produced additional guidance (IPIECA/OGP, 2005; ICMM, 2008).

8.2.1.3 Obesity

In high-income urban communities, obesity is an increasing challenge and mitigation depends in part on physical activity and access to healthy foods. During the design of new communities, mitigation could include the suggestions in Table 8.5.

Table 8.5 *Societal versus individual recommendations for obesity control*

Societal	Individual
Ensure good public transport systems	Use of sports facilities
Make provision for walking and cycling	Healthy eating campaigns
Ensure local shopping centres are within 3 kilometres	Avoiding private car transport
Provide good quality green space	

8.3 SOME CRITERIA FOR APPRAISING RECOMMENDATIONS

There are no definitive sets of criteria for evaluating the recommendations made in an HIA. The following criteria are proposed. A good recommendation should:

- be supported by evidence and justification;
- be practical;
- aim to maximize health gain and minimize health loss;
- be socially acceptable (a degree of pragmatism may be inevitable);
- align with government policies and programmes;
- consider the cost of implementation;
- consider the opportunity cost (could the money be better spent elsewhere?);

- include preventative as well as curative measures;
- be prioritized as short, medium or long term;
- identify the drivers and barriers for change;
- identify a lead agency or individual to implement and fund it;
- be capable of being monitored and evaluated;
- improve social equity;
- be specific to the stage of the proposal;
- address a priority impact based on consequence of the effect and the probability that it will occur (see Chapter 7);
- affect a number of health determinants at once;
- be technically adequate;
- include a method of monitoring;
- have taken account of fossil fuel scarcity;
- be sustainable.

An exercise at the end of this chapter uses some of these criteria.

Recommendations are subdivided according to the stage of the proposal and the priority of the issue that they address. The best recommendations affect many different health determinants simultaneously. They should lead to improvements in social equity: in other words, they should ensure that the most vulnerable and marginalized are protected in addition to those who are powerful and vocal.

The technical adequacy of the intervention should be clear as well as its cost. The mechanism that will be used to monitor its effectiveness should be described.

The very best mitigation measures benefit the proposal as well as the community. For example, by substituting a process it may be possible to increase productivity while reducing pollution. Such 'win-win' mitigation measures are likely to gain the support of the decision-makers and are discussed in Chapter 7.

The need to consider the energy requirements of recommendations follows from the global debate on climate change and fuel scarcity discussed in Chapter 12.

Example of a recommendation for a housing project in a hot country

The architectural design should be modified to ensure appropriate ambient heat levels and reduce the risk of heat stress.

8.3.1 SMART

More generally, the criteria used for judging any management measure can be used for HIA recommendations. One set of criteria is the SMART concept (CDC, 2007; Wikipedia, 2009). See Table 8.6.

Table 8.6 *SMART recommendations*

Criteria	Explanation
Specific	Provide the 'who' (target population) and 'what' (action/activity) of the activities. A specific recommendation uses only one action verb.
Measurable	Focus on how much change is expected. A measurable recommendation provides a reference point from which a change can clearly be measured.
Achievable	Attainable given the proposal's current resources and constraints. An achievable recommendation is attainable within a given time.
Realistic	Address the scope of the impact and propose reasonable programmatic steps.
Timely	Provide a timeframe indicating when activities will be implemented and when the outcome will be measured.

8.3.2 Other tips

A recent guide provides additional tips for developing recommendations (Harris et al, 2007):

- highlight the positive impacts of the proposal before the negative impacts;
- keep the wording concise and action-oriented;
- focus on a small number of achievable recommendations;
- communicate clearly;
- develop the recommendations with input from stakeholders;
- justify the recommendations.

Recommendations are also discussed in the context of a general tool for reviewing HIAs (Fredsgaard et al, 2009). See Chapter 4 for more details. This tool assumes that the level of commitment of the proponent to the recommendations and mitigation methods is known, and can be stated in the HIA report. This represents a different view of the link between the report and the proponent. Recommendations are usually made, in my experience, before the proponent decides what commitment to make.

The recommendations are presented to the proponent/decision-makers and they are free to accept or reject them. They are more likely to accept them if they trust the assessor and have a good and close working relationship. They are also more likely to accept them if a consensus has been built over time by early and regular communication with all stakeholders. This is hard to achieve and requires specific skills in networking and communication. Ideally, the cost of implementation would be estimated before recommendations are presented to the decision-maker. In practice, this would require a deeper knowledge of the proposal than is usually available to the assessor.

Recommendations are not necessarily reversible (EC Health and Consumer Protection Directorate General, 2004; Mindell et al, 2004). Removing a negative impact does not necessarily produce a positive health effect.

Recommendations can offer different options. For example, a recommendation to reduce the adverse health effects of traffic-generated air pollution could offer the following options: reduce the road traffic, reduce emissions from road vehicles, increase healthier travel modes and/or develop local air pollutant 'alert' systems (EC Health and Consumer Protection Directorate General, 2004). From a public health perspective, a large-scale switch to healthier travel modes is probably the preferred option.

8.3.3 Create multiple barriers

Practical mitigation measures create barriers that reduce risk. Any one of these barriers may be sufficient. Nevertheless, any single barrier could fail. It is good management practice to construct multiple barriers in order to prevent the impact occurring. It is good practice to establish procedures for recovery, should the barriers all fail.

For example, the risk of obesity might be identified in a proposed new housing estate in the UK. The recommendations could include active travel. In order to promote active travel, cycleways might be provided together with secure cycle storage, riding and maintenance classes, and cycling clubs. At the same time, the location of food outlets may be planned so as to encourage healthy eating. A weight-monitoring programme could be established in collaboration with schools, with referrals to medical services.

8.4 INAPPROPRIATE RECOMMENDATIONS

It is very easy to propose inappropriate recommendations. Typically, these include recommendations to:

- change individual human health behaviour by 'health education' (the wrong end of the mitigation hierarchy);
- provide more hospitals for medical care;
- increase dependency of the community on the proposal.

Poor recommendations

Poor impact assessments often include poor recommendations such as building hospitals and changing people's behaviour.

Mitigation measures that are primarily associated with medical care are about 'repair or remedy'. But HIA should emphasize 'avoid or reduce at source' or 'abate on site'. An alternative to a general proposal for 'health education' might be to procure the services of a local specialist NGO to provide peer-to-peer education for several years. An alternative to providing more hospitals might be to pay for an extra doctor and ambulance for the trauma unit of the local hospital so that it can provide emergency services for the proposal's workforce.

Recommendations may be inappropriate because they are unsustainable. Low-income countries are littered with empty shells of buildings that were donated for one reason or another to local governments. The local governments did not have the resources to stock, staff or maintain them. Recommendations can be divided into those that require one-off capital payments, such as new infrastructure, and those that require recurrent expenditure, such as staff and maintenance. The latter require long-term commitments that may not be popular with proponents.

8.5 TIMING

Different HIA recommendations are likely to be required for each stage of the proposal: preparation, construction, operation and decommissioning. The most time-critical recommendations are the ones that need to be made and implemented at the beginning. Some may need to be implemented before the HIA is completed. While the HIA is being commissioned, developers may be negotiating construction contracts. The manner in which construction is managed will have health impacts. The relationship between the construction workforce and the local community will need to be regulated. For example, a sudden influx of migrant workers may create price inflation of food and rents. This will create responsibilities for the construction company and will therefore need to be specified in the construction contract.

During the construction stage there may be large increases in traffic, pollution and population. Some of these changes will be managed by other processes such as traffic management plans. There are likely to be gaps that the HIA should identify and develop recommendations to fill. For example, there may be a camp-follower community living in squalid conditions without water supplies and sanitation. The responsibility for managing that community will require careful negotiation. As they are informal, they fall outside the proponent's plans. During a European project, a travel management plan was created to ensure that local children would be protected from the movements of construction vehicles. The children cycled to school and the mitigation proposed was to bus them instead. However, any modal shift from active travel to motorized transport should be avoided because active travel is good for health (Active Travel Consortium, 2008, Transport Health and Environment Pan European Programme, 2009).

The health impacts of the operational stage may be much longer term. There may be more opportunities for systematic monitoring than are possible during the rush of construction. Social investment programmes are likely to come on stream and the interface with proposal impacts can be explored.

Decommissioning often takes place many years in the future. It would be impossible to identify the health impacts clearly or to make detailed recommendations for managing them. However, we can recommend that a new HIA will be required before decommissioning commences. We can also recommend that careful records are maintained from the first HIA onwards, and archived for use during that future assessment. For example, in the case of infrastructure there should be a record of everything significant that happens on site during the lifetime of the proposal. The UK is full of brownfield sites that are contaminated with unknown chemicals from earlier industrial projects, which have long since been abandoned and forgotten. We should not leave similar legacies.

8.6 CASE STUDY: RESETTLEMENT OF A FISHING VILLAGE

A fishing village had to be relocated. The existing fishing village had the following characteristics:

- The community was very poor.
- These were subsistence-level fishing folk who use dug-out canoes to access the inland waters and to take their produce to market.
- There were high levels of illiteracy.
- Houses were made of traditional materials including mud and straw.
- Water supplies were obtained from local ponds.
- There was open defecation.

Two main mitigation options were identified: sophisticated and simple.

8.6.1 Sophisticated mitigation option

The sophisticated option was to build a new village with all modern conveniences. This pleased the local political elite who were able to claim that they were bringing great benefits to the community. However, the community had no experience of flush toilets and these would soon become blocked. The new village would have port facilities and the fishing folk would be provided with motorized fishing launches. They would have no resources to pay for electricity and diesel, or to maintain the equipment.

8.6.2 Simple mitigation option

The simple option was to upgrade village standards by a small and maintainable amount. For example, VIP latrines provide a safe alternative to open defecation and are relatively easy to maintain. People like to build their own houses and need a rebuilding allowance to enable them to obtain supplies of raw materials of a higher quality. They can be helped to protect themselves from malaria by using treated bednets. Such nets cost about US$5 and this is often too expensive for village communities. The nets can be provided at a subsidized price. Micro-finance initiatives would provide an opportunity for the community to develop new businesses without burdening them with excessive debts and without encouraging them to default on loans.

All too often, proponents choose the sophisticated option in order to please the local political elite, even when they know that it is inappropriate and unsustainable.

8.7 CASE STUDY: HIA OF COMMONWEALTH GAMES IN GLASGOW

Glasgow was planning to host a major athletic event, the Commonwealth Games, requiring the development of considerable new infrastructure (Glasgow City Council, 2009). A strategic HIA was undertaken with inputs from a range of stakeholders and large-scale community consultations. It provides an example of good practice. Recommendations were allocated to 16 major themes: facilities, transport, civic pride, individual behaviour change, image of Glasgow, housing and public space, participation in cultural and sporting events, economy and employment, volunteering, community safety, antisocial behaviour and crime, community engagement, sports development legacy, environment, sustainable development, and carbon footprint. A large number of high-level recommendations were identified under each theme. For each recommendation, the potential health impact (positive or negative) was identified together with the agencies that would be primarily responsible. Some examples follow.

£2 billion of transport infrastructure was planned before the Games and one of the positive health impacts identified was a new demand for public transport. The recommendation was to ensure a long-term reliable, accessible, safe and low-cost public transport system. The responsibility would lie with three units within Glasgow City Council, as well as the Strathclyde passenger transport authority and Community Transport Glasgow.

There was concern that the new facilities built for the Games would not be sustainable in the long term and that this would be detrimental to public health. The community survey indicated that 95 per cent felt that access to affordable sports facilities was important. The recommendations were to ensure that the buildings were designed for mixed use and to develop an events strategy that would maximize their use. The community also felt that there should be access to affordable and healthy food within the facilities.

There was debate about sale and supply of alcohol in Games venues. A joint alcohol policy statement was produced that was intended to prevent the sale or supply of alcohol to drunk people. Implementation would be the responsibility of the Glasgow City Licensing Forum and the Working Group on Alcohol and Drugs. There were existing laws on the consumption of alcohol in public places and recommendations were made for strengthening their implementation.

An opportunity was identified for increasing the consumption of ethically sourced or local and seasonal produce as part of a carbon reduction strategy. One recommendation proposed an increase in the amount of public allotments and the development of a link between the Games and that the Market Garden Scheme.

8.8 MORE EXAMPLES OF MITIGATION IN LARGE-SCALE OVERSEAS PROPOSALS

When there is a need to build a road in association with a proposal, care should be taken to provide an apron, or pavement, on both sides of the road to accept non-motorized traffic. This will reduce road traffic injuries and assist the community to move goods and domestic animals.

Many proposals have a policy of sourcing their supplies as locally as possible. This brings secondary benefit to the local community in the form of employment and production opportunities. The increased wealth of the community may then promote increased health. A system of monitoring for local inflation could be included. However, if too many supplies are sourced locally then prices may rise and local communities may be deprived of basic necessities such as food.

8.9 MITIGATING PERCEPTION OF RISK

There is much that we still do not know about the mitigation of perceptions of risk. We do know that it is associated with an individual's and a community's sense of being in control of risks that affect their own lives. They need to understand the proposal and we need to understand their concerns. They need to feel empowered, not merely informed. They need to become familiar with the nature of the proposal. They need to feel that the proposal is bringing them some benefit and not merely some inconvenience.

In a recent US experiment, the local community were empowered to take independent air samples whenever and wherever they wished within their community. The company paid for the analysis of the samples by an independent laboratory and the results were communicated directly to the community by the laboratory. This reassured the community regarding the content of the air they were breathing. When major refurbishment was planned, the company took them on guided tours of a similar plant elsewhere, to enable them to envisage the changes.

A recent project in Pakistan included a social investment strategy that empowered local village women. The project supported a micro-finance (micro-credit) initiative (Grameen Bank, 2009; The Global Development Research Centre, 2009). The initiative enabled local women to develop their own business enterprises and hence their own status. In a subsequent interview, the women were very vocal and positive.

8.10 ENHANCING POSITIVE IMPACTS

Proposals clearly produce positive as well as negative health impacts. Some recommendations can enhance these positive impacts as well as enhancing the proposal. Here are examples for large international proposals:

- Instead of establishing a private clinic within the project boundary for the use of the workforce, it may be cost effective to improve the local public clinic for the benefit of both the workforce and the community.
- Instead of establishing separate water utilities for the proposal, it may be cost effective to improve the local public water utility for the benefit of both the proposal and the community.

- The construction phase is characterized by a boom and bust effect. Construction workers know that their period of employment is limited. Instead of merely laying them off at the end of construction, it may be possible to establish a system for assisting them to obtain new work. This will raise morale and ensure that they work more effectively during construction.
- In addition to providing comprehensive safety training on site, it may be practical to reach out to the domestic environment and improve safety there. This will ensure both that the workers require less time for managing domestic emergencies and also that the culture of safety is more firmly embedded in their minds.

Opportunities to enhance community health are based on an understanding of health needs, and health needs assessment, described in Chapters 1 and 4. Enhancement can include upgrading the skills of community members so that they can participate more fully in the cash economy. The proposal could provide new employment opportunities at both a skilled and unskilled level. Any action that reduces poverty is likely to enhance health.

The proposal may also enhance community health by upgrading access to services. For example, a good quality road might enable people to get to a clinic that they could not otherwise reach.

In addition to the direct employment opportunities on a project site, there may be many other employment opportunities associated with local supply chains. For example, perhaps the canteen could employ local truckers to move its supplies.

8.11 OFF-THE-SHELF RECOMMENDATIONS

It would be very convenient to have a catalogue of off-the-shelf recommendations for each sector. While these would, no doubt, need tailoring for specific proposals, they would provide a starting point. Creating such a catalogue is a project for the future. At present, there are small samples of recommendations scattered through many documents, including this book.

For example, the World Bank Group Environmental Health and Safety Guidelines ('EHS Guidelines') include small sets of recommendations catalogued by sector There are both general and sector-specific guidelines and each contains five to ten recommendations (IFC, 2007). The general guidelines include chapters on environmental issues, occupational health and safety, community health and safety, and construction and decommissioning. The section on community health and safety is subdivided into seven main categories and within each of these there are lists of f urther recommendations (see Table 8.7).

Another example is the set of actions for decision-makers provided in the healthy development measurement tool described in Chapter 11 (San Francisco DPH, 2006).

In Europe, standard mitigation measures for a range of proposals include the promotion of active travel. Components include high-quality intermodal transfer facilities, good information for travel planning, alternatives to private car ownership including car clubs, well-designed and continuous cycleways, frequent buses, dedicated bus lanes, travel advisers, secure cycle storage and well-lit walkways. See the London

Table 8.7   *World Bank Group EHS Guidelines: Summary of community health and safety section*

Main sections	Summary of contents
Water quality and availability	Prevention of adverse impacts and ensuring sustainable provision
Structural safety of project infrastructure	Safe design and construction; buffer strips; stability; building codes
Life and fire safety	New buildings standard
Traffic safety	Protecting those who are most vulnerable; best transport safety practices; maintenance; signage
Transport of hazardous materials	Compliance with standards
Disease prevention	STIs and labour mobility Vector-borne diseases
Emergency preparedness and response	Ensure that a plan is in place

Transport Strategy for more examples (Greater London Authority, 2010). The effectiveness of some of these measures is not yet confirmed.

8.12 EFFECTIVENESS OF RECOMMENDATIONS

More research and review of the effectiveness of standard recommendations is needed. There are a growing number of reviews in broad strategic areas. For example, a guide for transport initiatives considers the evidence for the health impacts of transport and transport interventions (Douglas et al, 2007). It identifies many gaps where recommendations are not supported by evidence, as well as cases where they are supported. The report also reviews the kinds of recommendations made in the HIAs of a number of strategic road transport proposals. For example, congestion charging has reduced traffic in London but there is little research evidence of the consequent impact on health. The charge was accompanied by substantial improvements in the public bus network and this was followed by increased use of buses. Lowering speed limits in residential areas can reduce personal injury rates, but can also increase the number of crashes in peripheral areas. The reviewers were unable to identify significant evidence that community severance caused by road building had health impacts, although this seems probable.

8.13 MONITORING

The recommendations made during an impact assessment always include monitoring. The objective of monitoring is to provide simple and rapid indicators to enable management to adjust mitigation measures. Monitoring provides an opportunity to keep the mitigation measures on track and to respond quickly and effectively to unforeseen impacts. The ideal indicators are simple to measure, sensitive and cover

Table 8.8 *Examples of simple health indicators*

Type	Examples
Water supply	Percentage of community with access
Outdoor and indoor air quality	Suspended particulate concentration
Poverty	Market inflation
Housing	Crowding

a wide range of health determinants and health outcomes. Indicators must respond rapidly so that monitoring information can be translated into management action.

8.13.1 Examples of indicators

Table 8.8 illustrates examples of simple and practical health indicators. The percentage of a community with access to safe water supplies provides an indicator of the risk of water-borne diseases. Access has to be carefully defined.

The density of PM_{10} particulates in the air provides an indicator of air quality. In many communities it is necessary to measure this both indoors and outdoors. Indoor air pollution can be much higher than outdoor air pollution where people use biomass for cooking and heating, or are cigarette smokers.

Inflation of the price of basic needs such as food, shelter, energy and water provides an indication of how poor people are being affected by a proposal. Is the local procurement policy functioning as intended?

What is the average household size? How many people sleep in each room? Levels of crowding may provide an indication of how well rental accommodation is being managed. Crowding is an important determinant of diseases like tuberculosis (TB).

8.14 INTEGRATION

The recommendations made in an HIA should be checked for consistency with the recommendations made in environmental and social impact assessments of the same proposal. Recommendations that mitigate or enhance several different kinds of impact at the same time are likely to be more cost effective. There is a risk of recommendations conflicting with each other. For example, a recommendation to enhance farm livelihoods by producing more milk can conflict with a recommendation to reduce milk fat consumption. A recommendation to make residential areas more permeable can conflict with measures to reduce crime. Recommendations to procure supplies locally can lead to inflation of local market prices.

8.15 DECISION-MAKING

HIA proceeds by analysing the health concerns and prioritizing them for each proposal stage and alternative. The next step is to submit the recommended mitigation measures to management for approval. There are often costs associated with the recommendations. The duty of management includes cost control. So managers will be looking for reasons why they do not have to accept the recommendations, or can reduce their cost.

One might think that when decision-makers are presented with the facts they will use logic to choose an optimum solution. In this model, the HIA report is produced, agreed and handed over to the decision-makers; the decision-makers then recognize the validity of the arguments raised, and take rational decisions to implement the recommendations. This does not accord with my observations.

A different view is that decisions are made irrationally through the slow build-up of consensus. In this model, the HIA report is one source of information among many, that may, or may not, influence decision-making. This is closer to what I observe and consistent with research in EIA (Cashmore et al, 2004).

If decision-making is irrational and based on the slow development of consensus, then it follows that an impact assessment report is but one component of the process. The following principles seem to apply:

- The final impact assessment report should contain no surprises. In other words, the decision-maker should be kept informed of what the report is going to contain, what recommendations are going to be made and why. Management should have been sufficiently well aware of what was taking place during assessment that they are not suddenly confronted with recommendations that alarm them and that they do not fully understand.
- The assessment process itself should be about consensus building. Even before the assessment proper starts, the decision-maker should be informed about what issues are likely to be examined and why they are likely to be important. Building consensus is by no means trivial or even resolved. It requires much additional research and debate.
- Persuasion is an art. The objective of an HIA is to enable decision-makers to modify their proposals so as to safeguard and enhance health. Ideally, decision-makers need to feel that their proposal is largely positive and worthwhile. They need to feel that on the whole they are doing a worthwhile job. Therefore, a good approach to impact assessment is 'affirmative impact assessment' in which all the good, healthy aspects of the existing proposal are acknowledged, maintained and strengthened. The critical elements are then more likely to be heard.

8.15.1 Example

A private developer was preparing outline plans for a new mixed residential community in the UK. The plans assumed that high priority should be given to private motorized transport. It was assumed that the new homeowners would need to keep two cars in their driveways and would access local shops, clinics, leisure centres and schools by car. New roads would be built and these would cut across existing footpaths. The roads would meander around the estate. The challenge was to build consensus that active travel should be a priority for safeguarding and enhancing the health of residents. This would lead to a radically different design and should not be brought to the attention of the proponent for the first time in the draft report.

8.16 MANAGEMENT PLAN

When the managers have accepted the recommendations, a plan can be drafted in order to ensure implementation. The management plan indicates who, when and what: who will be responsible for implementing each recommendation, the timing of implementation, the criteria for success and how this will be verified. Development of this plan is the responsibility of the commissioner and is generally not undertaken by the HIA consultant unless there is special budgetary provision.

The consultant can help the commissioner by including an outline management plan in the HIA report. Table 8.9 is an example for a housing regeneration programme (Birley and Pennington, 2009). The recommendations in the left hand column are described and justified in detail in the report. The summary table provides an opportunity for the commissioner to accept or reject each recommendation and then to allocate responsibility, budget and action date.

Table 8.9 *Example of outline management plan*

Recommendation	Acceptance/ Rejection	Who is responsible for implementation	Budget	Action date
Tenant control and involvement	Already implemented			
Identification of tenants and their special needs				
Communication of the proposal to the tenants				
Support for vulnerable tenants				
Links with social services				
Housing Health and Safety Rating System (see Chapter 11)				
Climate change adaptations				
Additional home improvements required as a result of disability				
Working to the lifetime homes standard				
Improvements to the external physical environment				
Phasing of works				
Implementation of recommendations				

8.17 EXERCISES

8.17.1 San Serriffe exercise continued

Refer back to the San Serriffe exercise in Chapters 2, 4 and 5. Suggest two recommendations for mitigating, safeguarding or enhancing the health of a specific stakeholder group during a specific proposal stage (such as construction or operation). Justify your recommendations with a paragraph of text. For example, you might decide that the risk of STIs during the construction stage should be mitigated by supplying condoms freely to all construction workers. You might provide opportunities for local people to seek forms of employment other than commercial sex through micro-finance initiatives.

8.17.2 An appraisal exercise

Here are two sample recommendations based on a UK transport HIA. Table 8.10 provides space for testing these recommendations against a set of criteria. This exercise is derived from existing training material (IMPACT, undated).

A Ensure that the proposal makes adequate provision for active transport as this has been demonstrated to promote physical and mental health. Build cycleways wherever possible.

B The health benefit–cost ratio of cycling is at least 5:1 and opportunities should be sought within the development proposal for promoting active travel. One barrier to cycle use identified by the community was fear of collisions with motorized vehicles. Separate cycleways should be constructed alongside all new roads and pathways and funded from planning gain. Residential areas should be permeable to walkers and cyclists. The local cycling campaign committee have expressed an interest in identifying priority cycleways and providing cycle training and maintenance (contact Mr X). The local council maintains automatic cycle counters on some throughways and these provide comparative annual counts of cycle usage that can be used for monitoring (the officer in charge is Ms Y).

Table 8.10 *Examples of criteria for appraising recommendations*

Appraisal criteria	Tick if met	
	A	B
Is practical		
Aims to maximize health gain and minimize health loss		
Is socially acceptable (be pragmatic)		
Considers the cost of implementation		
Considers the opportunity cost		
Includes preventative as well as curative measures		
Is prioritized as a short-, medium- or long-term objective		
Identifies the drivers and barriers to change		
Identifies a lead agency or individual		
Is capable of being monitored and evaluated		

8.18 REFERENCES

Active Travel Consortium (2008) 'Travel actively', www.travelactively.org.uk, accessed May 2010

Birley, M. and A. Pennington (2009) 'A rapid concurrent health impact assessment of the Liverpool Mutual Homes Housing Investment Programme', www.apho.org.uk/resource/item.aspx?RID=95106, accessed November 2010

Cashmore, M., R. Gwilliam, R. K. Morgan, D. Cobb and A. Bond (2004) 'The interminable issue of effectiveness: Substantive purposes, outcomes and research challenges in the advancement of environmental impact assessment theory', *Impact Assessment and Project Appraisal*, vol 22, no 4, pp95–310, www.ingentaconnect.com/content/beech/iapa/2004/00000022/00000004/art00005

CDC (2007) 'Program evaluation: SMART cards for SMART objectives', www.cdc.gov/healthy youth, accessed February 2011

Douglas, M., H. Thomson, R. Jepson, F. Hurley, M. Higgins, J. Muirie and D. Gorman (eds) (2007) 'Health impact assessment of transport initiatives: A guide', NHS Health Scotland, Edinburgh, www.healthscotland.com

EC (European Commission) Health and Consumer Protection Directorate General (2004) 'European policy health impact assessment: A guide', http://ec.europa.eu/health/index_en.htm, accessed July 2009

Fredsgaard, M. W., B. Cave and A. Bond (2009) 'A review package for health impact assessment reports of development projects', Ben Cave Associates Ltd, Leeds, www.hiagateway.org.uk, accessed September 2009

Glasgow City Council (Chief Executive's Office Corporate Policy Health Team, Glasgow Centre for Population Health, NHS Greater Glasgow and Clyde Public Health Resource Unit and Medical Research Council Social and Public Health Sciences Unit) (2009) '2014 Commonwealth Games health impact assessment report, planning for legacy', Glasgow City Council, Glasgow, www.glasgow.gov.uk

The Global Development Research Centre (2009) 'Microcredit and microfinance', www.gdrc.org/icm, accessed May 2010

Grameen Bank (2009) 'Bank for the poor: Grameen Bank', www.grameen-info.org, accessed May 2010

Greater London Authority (2010) 'Mayor's transport strategy', www.london.gov.uk/publication/mayors-transport-strategy, accessed May 2010

Harris, P., B. Harris-Roxas, E. Harris and L. Kemp (2007) *Health Impact Assessment: A Practical Guide*, Centre for Health Equity Training, Research and Evaluation (CHETRE), University of New South Wales, Sydney, www.health.nsw.gov.au

ICMM (2008) *Good Practice Guidance on HIV/AIDS, Tuberculosis and Malaria*, International Council on Mining and Metals, London, www.icmm.com

ICMM (2010) *Good Practice Guidance on Health Impact Assessment*, International Council on Mining and Metals, London, www.icmm.com/document/792

IFC (International Finance Corporation) (2007) *Environmental Health and Safety Guidelines*, www.ifc.org/ifcext/sustainability.nsf/content/ehsguidelines, accessed May 2010

IMPACT (undated) International Health Impact Assessment Consortium, www.liv.ac.uk/ihia, accessed May 2010

IPIECA/OGP (2005) 'HIV/AIDS management in the oil and gas industry'. IPIECA, London, www.ipieca.org/system/files/publications/hiv.pdf accessed February 2011

Mindell, J., A. Boaz, M. Joffe, S. Curtis and M. Birley (2004) 'Enhancing the evidence base for health impact assessment', *Journal of Epidemiology and Community Health*, vol 58, no 7, pp546–551, http://jech.bmjjournals.com/cgi/reprint/58/7/546.pdf

San Francisco DPH (Department of Public Health) (2006) 'The healthy development measurement tool', www.thehdmt.org, accessed April 2010

Transport Health and Environment Pan-European Programme (2009) 'The PEP toolbox, healthy transport', www.healthytransport.com, accessed May 2010

Wikipedia (2009) 'SMART criteria', http://en.wikipedia.org/wiki/SMART_%28project_manage ment%29, accessed December 2009

World Bank Institute (2007) Global Public-Private Partnerships in Infrastructure Web-portal, http://info.worldbank.org/etools/pppi-portal/index.htm, accessed May 2010

CHAPTER 9

Water resource development

- Health concerns associated with different types of water resource development are described.
- The association between water development and communicable disease transmission is explained.
- An earlier guide, published by PEEM, is reviewed.
- A detailed summary of the health effects of involuntary resettlement is provided.

9.1 INTRODUCTION

Fresh water is an increasingly scarce resource and water development proposals of all kinds are expected to grow in importance. However, the presence and absence of water has a profound affect on the transmission of communicable diseases. In warm climates, pathogens multiply fast and pose great challenges to human health. The prevalence of many diseases can be attributed to health determinants associated with water. Climate change will extend the range of these diseases to some extent.

Malaria in Africa

In Africa alone, about 1 million children die of malaria every year (WHO, 2010).

Water resource development proposals can be subdivided into those with and those without a primary health or safety objective, as summarized in Table 9.1. The focus of this chapter is the HIA of water resource development proposals that have no primary health objective.

In addition to communicable diseases, there are many other health concerns associated with such proposals because of infrastructure development, displacement and associated irrigated agriculture. Some of these concerns are associated with cooler climates as well. Table 9.2 provides a summary.

Table 9.1 *Water resource development categories*

Category	Examples of objectives
Primary health or safety objective	Provision of safe drinking water Safe disposal of wastewater Flood control Recreation
No primary health or safety objective	Generation of hydroelectricity Irrigation of agricultural crops Provision of transportation routes Water storage Freshwater fisheries Wetland conservation, creation or removal

Table 9.2 *Summary of health concerns associated with water resource development*

Health outcome categories	Example of health concerns
Communicable diseases	Water-associated diseases and others
Non-communicable diseases	Poisoning from minerals and algae
Nutrition	Food security, micronutrients, household entitlements
Injury	Drowning, traffic, violence
Mental illness	Depression, stress, substance abuse
Well-being	Resettlement, uncertainty, livelihood

There is a vast literature on the health impacts of water resource developments and the associated pathogens, vectors, modes of transmission, disease prevalence rates, medical interventions and preventative measures. One of the historical roots of HIA arose from concern about the impacts of dams and irrigation on health in warm climates. The following references are illustrative, others are listed in Chapter 3:

- publications of the joint WHO/FAO/UNEP Panel of Experts on Environmental Management (PEEM) for Vector Control (including Mills and Bradley, 1987; Tiffen, 1989; Birley, 1991; Bos et al, 2003);
- man-made lakes and human health (Ackerman et al, 1973; Stanley and Alpers, 1975);
- man-made lakes and man-made diseases (Hunter et al, 1982);
- parasitic diseases in water resources development (Hunter et al, 1993);
- environmental health engineering in the tropics (Cairncross and Feachem, 1993);
- human health and dams (WHO, 1999);
- dams and disease (Jobin, 1999);
- the future of large dams (Scudder, 2005);
- health impact assessment for sustainable water management (Fewtrell and Kay, 2008).

The most recent citation in this list is an edited book that provides specific information on the future use of rainwater harvesting, sustainable drainage and greywater reuse

Table 9.3 *Groups of communicable diseases directly associated with water*

Group	Example	Transmission
Water-related	Vector-borne diseases	Acquired from insect bites
Water contact	Schistosomiasis (bilharzia)	Acquired by water contact
Waterborne	Gastrointestinal infections	Prevented by clean water supply and sanitation
Water washed	Skin infections and scabies	Prevented by washing and bathing

technologies in the UK; the HIA of dams in Zimbabwe and Laos; and the consequences of flooding in the UK and raw wastewater irrigation in Pakistan (Fewtrell and Kay, 2008).

9.2 COMMUNICABLE DISEASES ASSOCIATED WITH WATER

Four groups of communicable diseases have been directly associated with water: these are summarized in Table 9.3.

9.2.1 Vector-borne diseases

Vector-borne diseases illustrate the level of technical detail that an HIA may need to deliver in order to safeguard human health. They can occur in both natural and disturbed ecosystems. Transmission is critically dependent on temperature because the development time of the parasite in the vector is temperature dependent. Typically, an average minimum temperature of 16–18°C is required.

Vectors include insects, ticks, snails and water fleas. Insects and ticks can transmit certain communicable diseases while they are acquiring a blood meal from a human host. Some aquatic snails can harbour parasites that are released into water bodies and later burrow through the skin of human hosts. One species of water flea can harbour a parasite that infects humans when it is accidentally consumed in drinking water.

Infectious agents responsible for vector-borne diseases are either parasites or arboviruses. Arboviruses are simply viruses transmitted via the bite of insects and ticks. Parasites include protozoa and nematodes. For example, the diseases dengue and yellow fever are caused by arboviruses; the diseases malaria and schistosomiasis are caused by parasites. Table 9.4 summarizes vector-borne diseases. Some of the diseases are zoonoses – they have an animal reservoir – and this adds an additional layer of complexity.

Water resource development proposals in warm climates often have a large positive or negative impact on some of these diseases. The most important diseases are usually malaria and schistosomiasis. The impact should be managed by altering the design or operation of the proposal as well as providing medical care. A large amount of technical knowledge is required in order to do so and this can be provided by specialists, such as medical entomologists. For example, it is inappropriate to rely solely on chemical methods of vector control or curative drug therapies. Integrated methods of control

Table 9.4 *Vector-borne diseases associated with water resource development*

Disease	Vector
Dengue, yellow fever, Japanese encephalitis, Rift Valley fever	Culicine mosquitoes such as *Aedes aegypti*
Dracunculiasis	Water fleas
Bancroftian and Brugian filariasis	Culicine and anopheline mosquitoes
Loiasis	Tabanid horsefly
Onchocerciasis (river blindness)	Simuliid blackfly
Leishmaniasis	Phlebotomine sandfly
Malaria	Anopheline mosquito
Schistosomiasis (bilharzia)	Water snail (bulinid, *Biomphalaria*, *Oncomelania*)
African trypanosomiasis (sleeping sickness)	Tsetse fly

should be used. There is a unique opportunity to incorporate these methods during planning (Birley, 1991).

9.2.2 Malaria and mosquitoes

Transmission of vector-borne diseases like malaria requires a suitable ecosystem and a vulnerable community. The ecosystem is subdivided between the aquatic system used by the larval stages and the non-aquatic system used by the adult stages. A suitable aquatic ecosystem includes characteristics of shading, depth, turbidity, vegetation, speed of water movement, organic content and seasonality. There are many different species of malaria vectors and each one requires a different combination of these characteristics. About 50 malaria mosquito species are important worldwide. Many species are now known to be species complexes that can only be distinguished using advanced technology. See Table 9.5 for examples.

Adult female mosquitoes transmit pathogens through their blood-feeding habits. The suitable ecosystem for the adult depends on temperature, humidity, seasonality, shade, wind-speed, and proximity to a suitable animal or human blood supply. Within each ecosystem there will usually be one or two mosquito vector species in a community of 10–20 species. Within the population of vector species, 0.1–1 per cent are actually infective. For some vector-borne diseases, such as malaria, a single bite from an infective vector is sufficient to cause disease in the human host. The number of infective bites per person per year tends to range from 1 to 50. At the lower end of the range there is considerable seasonal variation in disease rates. At the upper end of the range, the transmission environment is saturated. Saturation implies that the addition of extra mosquitoes makes no practical difference. Consequently, an intervention that increases or decreases mosquito density may or may not make a difference to disease prevalence rates. The practical implications have to be determined at local level and for specific proposals.

Vulnerable communities have intimate contact with the blood-feeding mosquito as a result of their work, habitation, location, behaviour, immunity and poverty. The primary determinant of clinical malaria may be poverty – as with many other diseases. Poverty tends to determine where and how people live and work and how well their

Table 9.5 *Examples of malaria mosquito species*

Malaria mosquito species	Location	Characteristics
Anopheles gambiae	Sub-Saharan Africa	Sunlit rain pools on soil. Adults prefer human blood and enter houses to feed and rest.
Anopheles balabacensis	Southeast Asia	Shaded pools in forest. Adults bite animals or humans, often feed and rest outdoors.

immune system functions. The association between malaria and poverty in Gambia is discussed in Chapter 2 (Clarke et al, 2001).

Further evidence is provided in studies of the prevalence rate of malaria on and off irrigation systems in Africa. In a number of such studies, a counter-intuitive result has been observed: there is less malaria in irrigation than in non-irrigation communities. For example, malaria incidence was compared in a rice growing, a sugar growing and a savannah community in the same region of Tanzania (Ijumba and Lindsay, 2001). People's wealth was also estimated, based on the percentage who could afford metal roofs. Although the density of malaria mosquitoes was much greater in the rice growing community, the incidence rate of malaria was much lower than elsewhere. A larger number of the rice growers were able to afford metal roofs for their houses, indicating that they were wealthier than the other two communities. There are two explanations for this observation:

- vector ecologists suggest that the mosquitoes must be genetically different;
- sociologists suggest that the communities must be different.

The second explanation is the most likely. Those who benefit financially from an irrigation project become richer, more able to protect themselves, have better nutrition and have better access to medical care. This is consistent with the evidence from health inequalities research more generally.

Malaria control relies partly on targeted pesticide use, insecticide-impregnated bednets, and both prophylactic and curative drugs. Resistance of mosquitoes to pesticides, and of parasites to drugs, is an ever evolving challenge that makes environmental methods more important.

9.2.3 Schistosomiasis and snails

Three principal forms of schistosomiasis and their snail hosts are listed in Table 9.6. The parasites live in the human host; eggs are passed in either faeces or urine. Eggs hatch in water and the larva enters the aquatic snail. Multiplication takes place and the infective stage of the parasite is released from the snail after three to eight weeks. Snails may survive and remain infective for up to six months. The free-swimming infective parasites are active in the water for 12–48 hours. They penetrate undamaged skin of a person who enters the water and infect the human body. The parasites are most active

Table 9.6 *Snails and schistosomiasis*

Snail genus	Parasite	Type of disease
Oncomelania	*S. japonicum*	Intestinal
Biomphalaria	*S. mansoni*	Intestinal
Bulinus	*S. haematobium*	Urinary

during the heat of the day. In warm climates both adults and children may enter water to swim, bathe, do laundry, wash dishes, collect water or bring livestock to drink.

Each of the three genera of snails contains many species and each species contains many varieties with different susceptibility to the parasite. A particular waterbody could contain different species of snails, only some of which are important.

A range of chemicals is available for snail control and there is a relatively cheap and effective curative drug. However, recolonization by snails is rapid and reinfection is frequent. Integrated methods of control should therefore be used.

Each species of snail prefers different water conditions. Some like stable, slowly flowing water while others like unstable, semi-stagnant water and colonize newly inundated water bodies. Other variables include light penetration, turbidity, partial shade, water velocity, pollution with excreta, gradient, slow changes in water level and firm mud substrate.

Figure 9.1 illustrates how an irrigation structure could be designed so as to prevent snail breeding. The design ensures that pooling does not occur.

Jobin (1999) provides a detailed discussion of schistosomiasis control in the context of HIA. His monumental work is too large to be summarized adequately here. He includes examples based on his own experience of dams and irrigation systems throughout the world. In addition, he provides a synthesis chapter that summarizes his lifetime experience. Some of his observations include:

- Control starts with the appropriate location of settlements. Locate settlements on reservoir shores exposed to the prevailing wind where wave action deters vector breeding. Locate settlements at least 1500 metres from the nearest minor canal.
- Sugarcane and banana plantations are often associated with schistosomiasis while cotton plantations are often associated with malaria. The causal linkage is demand for water and the large labour requirements of these plantation crops.
- Connect all health care centres by all-weather roads to the provincial hospital. Ensure health posts are adequately supplied with drugs and diagnostic laboratories.
- Provide communities with safe and adequate water supplies and prevent access to canals and drains.
- Ensure high water velocities in canals and irrigation ditches.
- Avoid night-storage systems (open reservoirs that provide temporary water stores).
- Maintain water distribution systems, remove weeds and dry the channels out regularly.

Figure 9.1 *Irrigation structures designed to avoid snail breeding in Zimbabwe*

Copyright Tropix.co.uk

9.2.3.1 Case study: Individual determinants of health

An irrigation project was planned in Zimbabwe, funded by an aid agency. Schistosomiasis was one of the priority health outcomes. About one-third of the community belonged to a Christian sect that believed that all fresh water should be used for religious immersion, that no health precautions were necessary and that schistosomiasis could be prevented by prayer. Here, an inappropriate set of beliefs and behaviours by the community increased the health risk (Konradsen et al, 1997). The project was not in a position to change these behaviours.

9.2.4 Summary of the PEEM method for assessing the vector-borne disease impacts of water resource development

Guidelines have been published by PEEM for assessing the vector-borne disease risks of a water resource development (Birley, 1991). The method consists of scoring three main categories of health determinants: community vulnerability; environmental receptivity; and the vigilance of health services. The meaning and content of these categories is explained below. They are all recognizable determinants of health used today, although designations and categories have subsequently changed and been extended.

The method is based on answering a series of 45 questions. The guidelines suggest how answers could be obtained, explain the technical terms used, and provide examples and summaries of facts and experience. The questions represent a semi-structured interview that can be used with key informants. Questions are grouped in two separate worksheets: according to method of analysis (see below); and according to key informant interview. Questions are also presented in a series of logical flow charts.

Key informants are identified as project planners; managers of similar projects in the region; ministry of health and specialist units; entomologist, vector control department or pest control officer; game or animal control officer; and proposal-affected community. The questions designed for the communities include the following. Have you been informed about the project and its implications for your home and livelihood? What diseases do you fear and how do you believe they are contracted? What health facilities are available to you?

Beliefs about malaria

In some rural communities of low-income countries relatively few people believe that malaria is transmitted by mosquitoes, compared to the numbers who believe that it is caused by supernatural means. They observe that mosquitoes are everywhere, but only some people have malaria.

9.2.4.1 Community vulnerability

The community vulnerability component of the method consists of the following 12 questions:

1   Which diseases are important in the region?
2   How prevalent are these diseases?
3   Is there any drug resistance?
4   Is there a human parasite reservoir?
5   How could the number of vulnerable people be changed by the project?
6   Which communities are likely to be affected?
7   Which communities are susceptible to specific diseases?
8   How will the health status of each community be changed by the project?
9   Does human behaviour favour contact with vectors or unsafe water?
10   Do people enter rural habitats for the project or for other work?
11   Do human activities at the project site present special problems?
12   Will the project change human behaviour?

The guide includes a table of indicators of well-being that may reduce a community's vulnerability to the challenge of vector-borne diseases; the table refers to educational and economic inequality, and to the existing health profile.

9.2.4.2 Environmental receptivity

The environmental receptivity component of the method consists of the following 19 questions.

1   Which vector species are important in the region?
2   Which pathogens do or can they transmit?
3   Is the vector abundant?
4   Does abundance vary seasonally?
5   Are the vectors more numerous in some places rather than others?
6   Are the vectors resistant to any insecticides?
7   Will the project affect the vector abundance?
8   Are the vectors abundant on similar projects in the region?
9   How will the project affect the number of vector breeding sites?
10   Could new species of vectors colonize the site from elsewhere?
11   Does the behaviour of the vector favour contact with the human community?
12   Do vector species associate with human communities?
13   Do the vectors inhabit undisturbed rural habitats?
14   Will the project affect vector behaviour?
15   Will settlement design affect vector abundance and contact?
16   Is there an animal reservoir of infection that could be affected by the project?
17   Will the animals invade the project site?
18   Could the reservoir population increase as a result of the project?
19   Could the reservoir population be eradicated?

Table 9.7 *The flight range of some vectors (kilometres)*

Vector	Local movement	Migration
Simuliid (blackfly)	4–10	400+
Anopheline (malaria mosquito)	1.5–2.0	50
Culicine (mosquito)	0.1–8.0	50
Tsetse (vector of sleeping sickness)	2.0–4.0	10
Phlebotomine (sandfly)	0.05–0.5	1

The guide includes a table that classifies the effect of human activity on the environment in relation to the breeding sites of vectors. It also includes an indication of the distance that vectors can travel, as this may affect the siting of projects (see Table 9.7).

9.2.4.3 Health service vigilance

The vigilance component of the method consists of the following 14 questions:

1 Is there effective, routine control of vectors in the project area?
2 Are animal reservoirs controlled?
3 Is pesticide applied effectively?
4 Is there insecticide resistance?
5 Are vector populations monitored effectively?
6 Are there effective curative measures for the disease?
7 Is curative medicine locally available and effectively used?
8 Are there effective prophylactic drugs and are they accessible?
9 Can district health services cope with additional project-related workloads?
10 Has vector control been incorporated in project design or operation?
11 Do any features of the design help to prevent vector breeding or contact?
12 Does the operating schedule ensure periodic destruction of breeding sites?
13 Can contact with unsafe water be avoided?
14 Can the project design be modified to reduce health hazards?

This guide includes a description of environmental designs that prevent vector breeding or contact. For example, Figure 9.2 illustrates an irrigation channel that is overgrown with vegetation and situated close to a human habitation. The community uses the water in the channel for washing and drinking. The channel provides breeding sites for mosquitoes and snails. By contrast, Figure 9.3 illustrates a well-managed irrigation channel on a sugar cane estate. The channel is concrete lined, free from vegetation and self-drying. Irrigation water is abstracted from the channel using siphons.

9.2.4.4 The scoring system

Table 9.8 is an example of the scoring system used in the PEEM Guidelines. The example refers to the construction phase of a commercial irrigation scheme somewhere in sub-Saharan Africa.

Figure 9.2 *Poorly managed irrigation in Kenya*

Copyright Tropix.co.uk

Figure 9.3 *Well-managed irrigation in Zambia*

Copyright Tropix.co.uk

Table 9.8 *Example of a summary scoring system*

Disease	Vulnerability	Receptivity	Vigilance	Health risk
Malaria	high	moderate	treatment only	high
Schistosomiasis	low	moderate	none	low
Onchocerciasis	low	none	none	none

The table was accompanied by the following statements:

- Malaria is expected to represent a health risk during the construction phase because susceptible people will be exposed to the vector and no preventative measures are planned.
- Schistosomiasis does not occur near the project site but a potential vector is present. The risk is moderate but will increase unless immigrants are screened on arrival for infection, or other preventative measures are instigated.
- Onchocerciasis occurs in the region but there is no vector at the project site and none is expected to become established during the construction stage.

9.2.5 Diarrhoea and malnutrition

According to WHO, some 1.4 million childhood deaths from diarrhoea are preventable each year (Prüss-Üstün et al, 2008). These are caused by ingestion of pathogens in drinking water or contaminated food, and from unclean hands. In many cases, repeated infection leads to malnutrition. Malnourished children are underweight and more vulnerable to infection. Some 860,000 preventable childhood deaths per year are attributed to malnutrition. WHO estimate that the three diseases with the largest contribution of DALYs lost are diarrhoeal diseases (39 per cent), consequences of malnutrition (21 per cent) and malaria (14 per cent). They further estimate that the biggest reductions can be achieved by improvements in hygiene (37 per cent), sanitation (32 per cent), water supply (25 per cent) and water quality (31 per cent). WHO has used these figures to estimate the costs and benefits of investing in water supply and sanitation. They estimate that an investment of US$11.3 billion per year would produce a saving of US$84 billion per year.

Many development proposals have impacts on drinking water and sanitation. The social investment associated with these proposals could target water supply and sanitation.

9.3 DRINKING WATER

Drinking water supply systems have upstream and downstream components. The upstream components include dams, reservoirs and protected catchments. Typical concerns of a drinking water supply manager are listed in Table 9.9. These are some of the issues that an HIA of a proposed drinking water supply reservoir should address (WHO, 2008).

Table 9.9 *Typical health concerns affecting drinking water supply catchment*

Attribute	Health concern
Meteorology and weather patterns	Flooding, rapid changes in source water quality
Seasonal variations	Changes in source water quality
Geology	Arsenic, fluoride, lead, uranium, radon, swallow-holes
Agriculture	Microbiological contamination, pesticides, nitrate
Forestry	Pesticides, fires
Industry	Chemical and microbiological contamination
Mining	Chemical contamination
Transport	Chemicals
Development	Run-off
Housing – septic tanks	Microbiological contamination
Abattoirs	Organic and microbiological contamination
Wildlife	Microbiological contamination
Recreational use	Microbiological contamination
Competing water uses	Sufficiency
Raw water storage	Algal blooms and toxins
Unconfined aquifer	Water quality subject to unexpected change
Well/borehole headworks not watertight	Surface water intrusion
Flooding	Quality and sufficiency of raw water

The downstream component is provision of domestic and industrial water supplies. Domestic supplies may be obtained from reticulated pipes, wells, boreholes, springs, ponds, streams, lakes and water sellers. There is a large literature that establishes minimum requirements for quality and access (for example, WHO, 1983; Cairncross et al, 1990). Many studies have sought to demonstrate the positive public health impacts of providing clean water supplies (see Section 9.2.5 above). This has proved elusive because there are many confounding factors; poor hygiene and sanitation can disguise the benefits of clean water. Clearly, drinking water supply is a basic necessity of life and this is the context in which it should be analysed in the HIA of many different kinds of proposals.

Figure 9.4 illustrates the stress of limited supply in poor urban communities. There is a standpipe and a queue of water containers. Supply is limited to a few hours each night. During the remaining periods, no water flows. This often creates a negative pressure so that contaminated water from septic tanks, sewerage pipes and surface pollutants is drawn into underground water pipes through cracks and bad joints. Water that has been treated at source may be contaminated when it is delivered at the standpipe. Domestic water is often stored in open containers in and around the home. Here it is subject to contamination by both inorganic and organic pollutants. In hot climates, such containers may provide breeding sites for *Aedes aegypti* and *Aedes albopictus,* the mosquito vectors responsible for dengue fever transmission. See Chapter 2 for a rural example of water supply stress.

There is a tension between the water needs of industry and agriculture and the domestic water needs of communities. An HIA provides an opportunity to ensure that industrial water use also supplies the domestic needs of surrounding communities. See Section 9.3.1 for an example.

Figure 9.4 *Urban domestic water supply stress*

Copyright Tropix.co.uk

9.3.1 Case study of Sakhalin integrated impact assessment

An example of drinking water supply issues is provided by an integrated impact assessment of the Sakhalin Energy project, discussed further in Chapter 4. Sakhalin Island is in the cold climate of Pacific Siberia (Birley, 2003; Sakhalin Energy Investment Company, 2003). The rural communities experience water stress in winter although they are surrounded by snow and ice. The water is not suitable for domestic use until it has been melted and this requires energy. Existing supplies were poorly maintained and shortages and contamination were common. The Sakhalin Energy proposal involved long, onshore oil and gas pipelines. There were various processing stations on the route and large quantities of water were required for both industrial and staff use. There were many small rural communities at varying distances from the pipeline and communication routes.

The energy project had large water requirements and was located among communities with unmet drinking water needs. On the whole, the proposal probably did not directly affect community access to potable water. On the other hand, it did provide a missed opportunity for improving that access.

9.3.2 Case study of arsenic poisoning in Bangladesh

During the 1980s, there was a programme of sinking relatively deep tube wells in Bangladesh in order to overcome the chronic domestic water pollution associated with surface flooding. There was no HIA of the programme. Unknown to the aid agencies, the sediments in which the groundwater was located contained compounds of arsenic. When the boreholes were drilled into this sediment a chemical reaction took place that released the arsenic in solution. The lack of appropriate assessment procedures has meant that many millions of Bangladeshis risk arsenic poisoning (Dhaka Community Hospital Trust and Disaster Forum, 1997; Anon, 1998). One unanswerable question is whether an HIA performed at the start of the programme would have identified the issue, or not. If not, then the assessors might have been held responsible for an epidemic of mass poisoning. At the time, the association between arsenic and that type of groundwater was unknown and so the appropriate tests were not made (British Geological Survey, undated).

9.4 DAMS

Dams, and especially large dams in developing countries, are the subject of much controversy as a result of the environmental, social and health impacts that they cause. In order to resolve the controversy, the World Commission on Dams (WCD) was set up by a consensus of governmental, non-governmental and international organizations (WCD, 2000). Scudder (2005) provides a detailed critique of large dams in a follow-up to the WCD and his work is drawn on extensively in the following discussion.

There are some 50,000 large dams in the world and 80 per cent of them are in China, the US, India, Spain and Japan. Some 40–80 million people have been involuntarily

---

Core values and strategic principles for safeguarding resettled people in dam projects

Core values

Equity, efficiency, participatory decision-making, sustainability and accountability.

Strategic principles

Gaining public acceptance; comprehensive options assessment; addressing existing dams; sustaining rivers and livelihoods; recognizing entitlements and sharing benefits; ensuring compliance; sharing rivers.

---

displaced by large dam projects and over 100 million have been indirectly affected. Many millions of poor people have been impoverished and their human rights have been ignored. In addition, dams have numerous health impacts and this was the subject of a WHO submission to the WCD (WHO, 1999).

In many circumstances large dams remain a necessary development option. The WCD proposed a set of five core values and seven strategic principles to ensure that new dam developments safeguard people and environments. Scudder believes that these principles were still being ignored by project authorities and governments. Budgets, political will and community consultation were frequently inadequate. Impact assessments were underfunded or recommendations remained unimplemented.

Dam projects affect many stakeholder communities. Three major categories are as follows:

1   the communities located above the dam who must be resettled;
2   the host population who must receive the resettlers;
3   the communities located below the dam who are affected by the changing water regime in the river or delta.

The resettlement community has been carefully studied and documented, perhaps because they are most clearly visible (Scudder, 2005). See Section 9.5.

For graphic descriptions of the downstream impacts of big dams, Scudder's book should be consulted directly. In summary, many millions of downstream populations have had their livelihood seriously disrupted by upstream dam building. Such populations are often dependent on natural flood regimes for flood recession agriculture, floodplain grazing, foraging and aquifer recharge. Scudder cites evidence that the economic productivity of flood recession agriculture is often underestimated. River deltas are also impacted by upstream dams and this affects biodiversity, ecosystem services, fisheries and the communities that depend upon them. Dam management systems should be designed to maintain 'environmental flows' for the benefit of downstream communities and ecosystems.

There are many positive and negative health impacts associated with dam projects. They are the result of changes in both the social and environmental determinants of health. A small selection of the negative impacts is illustrated in Table 9.10.

Table 9.10 *Examples of some of the negative health impacts of dams*

Health outcome categories	Examples of health outcomes	Examples of health determinants
Communicable diseases	Vector-borne diseases, STIs, zoonoses	Changing water regimes, migration, resettlement
Non-communicable diseases	Poisoning	Toxic algal blooms and minerals
Nutritional issues	Malnutrition	Resettlement and loss of food security
Injuries	Drowning, trauma	Changing social and physical environments
Mental illness, psychosocial issues and well-being	Stress, suicide, substance abuse	Resettlement

In a recent example from Laos, high rates of communicable disease and childhood malnutrition were present in the communities prior to the project. These included malaria, dengue, opistorchiasis, various STIs, TB and schistosomiasis (Krieger et al, 2008). The assessment concluded that resettlement design could affect malaria, and migrant labour could introduce schistosomiasis and elevate HIV rates.

9.5 RESETTLEMENT

Involuntary resettlement is associated with a range of infrastructure development projects including roads, pipelines, industries and urban development, as well as dams. Involuntary resettlement is a contentious issue and successfully resettling even a single household is difficult. Some financial institutions recognize this and I have observed that they try to screen out all loans that involve involuntary resettlement. Nevertheless, it is likely that more than 10 million people experience involuntary resettlement as a result of development every year. The World Bank has long had criteria for involuntary resettlement and these are sometimes cited as best international practice (World Bank, 2001). These criteria are designed to ensure that resettled people are at least as well off as they were before.

This discussion of resettlement is within the water chapter because resettlement has been studied and described in great detail by Scudder and others in the context of large dams (Scudder, 2005). The analysis applies equally to other forms of resettlement, especially where these are associated with poor people in rural areas of low-income

Underestimating number of resettlers

It has been estimated that the number of people who must be resettled is on average 40 per cent higher than the number identified at the planning stage.

Table 9.11 *Summary health impacts of resettlement*

Category	Determinants
Physiological	Communicable diseases associated with:
	• Higher living densities • Poor water supply and sanitation • Migration • Zoonoses • Environmental change
	Malnutrition associated with:
	• Loss of cultivation • Inadequate food security support provided by planning authorities • Reduced access to common property resources
Psychological	• Stress • Grieving for a lost home • Anxiety about the future • Gender inequalities
Socio-cultural	Temporary or permanent loss of:
	• Livelihood • Customs • Institutions • Symbols

countries. I try to draw out the general principles from Scudder's analysis but the original text is required reading for a proper understanding.

Table 9.11 summarizes Scudder's view of the health impacts of resettlement. The physiological health impacts of resettlement mentioned by Scudder include communicable diseases and malnutrition. These are discussed elsewhere in this chapter. The psychological impacts are associated with what has been lost and what uncertainties the future may hold.

Women are particularly vulnerable to and disadvantaged by involuntary resettlement. Their general welfare and their status within the household suffer. They and their children are vulnerable to physical abuse by male relatives whose self-esteem has been undermined. Women are more dependent than men on common property resources, kinship and other social ties, and on services and provisions that are essential for the welfare of their children. They have less access to wage labour.

Scudder describes socio-cultural impact as a threat to a community's cultural identity. Involuntary resettlement forces people to examine primary cultural questions such as the following: 'Who are we? Where are we?' This is a stressful process that takes time.

Examples of this phenomenon include:

• People who are relocated inland, or away from river banks, can no longer engage in cultural practices associated with water.

- Rituals associated with particular sites may be lost when those sites are submerged.
- Customs may be reduced because they are criticized by outsiders.
- The authority of leaders is frequently undermined.
- There are conflicts with the host community.

If asked, involuntary resettlers are likely to express a range of concerns starting with the level of compensation, the likelihood of actually receiving compensation, the kind of house they prefer to live in and the crops they prefer to grow. They would probably mention the loss to inundation of spiritual elements in their environment such as trees, mounds, bushes, rocks and pools; as well as shrines and graves. They may prefer living on the edges of rivers and find the resettlement sites inappropriate.

9.5.1 Phases of resettlement

Scudder identifies four phases of resettlement spanning two generations.

9.5.1.1 Pre-resettlement

The pre-resettlement phase can last many years and creates great uncertainty. Rumours of displacement abound, but there is little evidence on the ground. There is much denial. Once the certainty of future resettlement is realized and the time for removal draws near, people begin to cut back on livelihood support activities.

9.5.1.2 Early post-resettlement

The early post-resettlement phase lasts a minimum of two years and usually ten or more years. During this phase living standards fall and behaviour is conservative, or risk averse, as a coping strategy to the stresses of resettlement. Expenses rise and income generation is often 50 per cent or more lower than anticipated in the planning documents.

9.5.1.3 Late resettlement

During the third stage a new community has formed and economic development has restarted. Risk-taking increases, as do inequalities. The settlers may begin to feel 'at home': they paint traditional symbols, name geographical features, and incorporate resettlement into stories, songs and dances. They form co-operatives and are willing to improve schools and clinics. Community religious activities increase and shrines are constructed. This renaissance has only been observed in a minority of cases. More frequently there is social disintegration.

9.5.1.4 Incorporation

The second generation following resettlement become part of the wider socio-political economy. Services provided by project authorities may be handed over to normal line

---

Effect of dam-induced resettlement on living standards

7 per cent	improved
11 per cent	restored
82 per cent	declined

---

ministries. For example, the clinics provided for resettlers become the responsibility of the ministry of health.

9.5.2 Risk and resettlement

An alternative model of resettlement, attributed by Scudder to Cernea and the World Bank, focuses on the eight major categories of risks that resettlers face:

1 landlessness;
2 joblessness;
3 homelessness;
4 marginalization;
5 increased morbidity and mortality
6 food insecurity;
7 loss of access to common property; and
8 social dis-articulation.

This concept was promoted in the WCD report and distinguishes the voluntary risks taken by proposal managers from the involuntary risks imposed on communities who are proposal recipients. This is discussed in Chapter 7 under perception of risk.
Scudder analysed 50 case studies of dam resettlement projects and concluded that in the majority of cases the resettlers experienced an increase in the risk of landlessness, joblessness, food insecurity, marginalization and loss of common property resources. These are all determinants of health.

Managing risk

Distinguish voluntary risk takers and involuntary risk receivers.

9.5.2.1 Case study of the health impacts of the Mahaweli development project

Longitudinal studies of the Mahaweli development project in Sri Lanka identified the health issues highlighted in Table 9.12, among others (Scudder, 2005).

Table 9.12 *Example of health impacts of the Mahaweli development project*

Health issue	Note
Malnutrition	Malnutrition increased and 75 per cent of children under five were affected.
Non-communicable diseases	Snakebite increased and was associated with clearing new land.
Stress	Tensions between related households increased leading to the severing of relationships and charges of sorcery.
Suicide	Suicide increased and was the leading cause of mortality during the early resettlement phase, accounting for 70 per cent of hospital-reported deaths. The majority of cases were male and the favoured method was insecticide poisoning.
Services	During the initial stages there was a lack of health facilities and potable water supplies.

9.5.3 Resettler participation

Scudder suggests that one of the keys to successful resettlement is the involvement of the community in the planning and management of resettlement from an early stage. He recommends that the resettlement community should be provided with resources to employ their own independent consultants. In addition, the baseline study should clarify what the community know, value and do. Newly resettled communities should be expected to 'cling to the familiar' as a coping strategy. This will include replicating their original way of life until they feel secure enough to explore new opportunities. There should be a balance between continuity and change. In two-thirds of Scudder's case studies, the resettlement community had no involvement in the resettlement site selection. In many cases communities were not resettled in kinship groups, households or social units of their choice. Housing was often designed without reference to the occupiers and without taking account of their environmental and social needs. These are health determinants.

Success also depended on the resettlers becoming project beneficiaries. This could be accomplished in many ways. Joint-venture companies could be created between the proponent and the community with approximately equal shareholdings. The resettlers could be given responsibility for managing the resettlement funds. About half the world's large dams have significant irrigated agriculture components that can provide a substantial benefit to resettled communities. However, few projects had successfully integrated the majority of resettlers within the irrigated command area, or even tried to do so. There were also benefits for resettlers in using the drawdown zones of large reservoirs, reservoir fisheries and tourism opportunities. However, the nature reserves established around the edges of reservoirs frequently undermined customary land rights and communities were not consulted. In many cases appropriate job training was not provided to resettlers and attempts to establish new rural industries were often unsuccessful.

Services such as water supply and sanitation have frequently worsened following resettlement. In 66 per cent of the cases surveyed, Scudder found that water and sanitation facilities were defective. Resettlement sites were often situated over poor aquifers so that boreholes provided for community water supply were inadequate. In the Batang Ai resettlement site in Sarawak, I observed that potable water was piped from the river immediately downstream of the dam where the water was heavily contaminated with hydrogen sulphide associated with anaerobic conditions at the bottom of the reservoir.

The line ministries responsible for delivering dam or irrigation systems frequently ignored the impacts of their own projects on other ministries, such as health.

9.5.4 Resettlement funding

Scudder estimated that the resettlement costs sometimes exceeded 30 per cent of the total project cost. He suggests that the World Bank guidelines arbitrarily limit the cost of resettlement to direct economic and social impacts. The critical costs ignored included reintegrating and restarting disrupted economies, social institutions and educational systems and combating physiological, psychological and socio-cultural stress. Resettlement budgets were frequently raided to finance cost overruns. In the Batang Ai resettlement in Sarawak, I observed that cost overruns had led to the cancellation of clinics, schools and a marketplace.

Scudder is particularly critical of attitudes of proponents towards resettlement compensation. Cash compensation for tangible assets, on its own, had frequently led to impoverishment. Inexperienced subsistence agriculturalists may spend compensation monies on consumer goods. In the Batang Ai resettlement I observed a proliferation of colour televisions and excessive gambling on fighting cocks. There may be unequal entitlements within the household and there is an ever present risk of corruption. Many intangible social assets can get damaged, such as those associated with culture and identity, and these also need financial support.

9.5.5 Managing resettlement

Most major resettlement today is subject to the international standard created by the World Bank operational directive on involuntary resettlement (World Bank, 2001). Scudder regards this standard as inadequate because it leaves the majority worse off. Improved resettlement planning and implementation processes that eliminate the drop in living standards, multidimensional stress and conservative coping behaviour have yet to be documented. He suggests that the complexity associated with large-scale resettlement is so great that it exceeds the capacity of proposal managers to accomplish satisfactory outcomes. Involuntary resettlement should be avoided. Where it is unavoidable, a successful outcome is only possible if the necessary inputs and opportunities are there and the appropriate principles are followed. In reality, the funds and political will necessary for ensuring a successful outcome are rarely available. Scudder's conclusions are very briefly summarized in Table 9.13. The original text should be consulted for explanations and case studies.

Table 9.13 *Requirements for successful resettlement*

Project authorities	Single agency responsible for resettlement
	Sufficient staff capacity
	Adequacy of funding
	Political will of national government
	Creating new opportunities for resettlers
Project-affected people	Minimizing the number of people who must be resettled
	Resettlers' participation in project planning
	Resettlers' ability to compete with host and migrant communities

Whenever a resettlement is contemplated, the HIA report should contain very strong and detailed recommendations for managing the impacts. The work of the WCD and Scudder should be used as a reference.

9.6 MIGRANTS

Inward migration is a parallel issue to that of involuntary resettlement. Large proposals attract migrants during both the construction and the operational stages. Construction camps attract 'camp followers'. They set up informal settlements around the periphery of the project and supply goods and services to construction workers. These include alcohol, food and sex.

Migrants can have higher skills and more experience than the local community into which they move. For example, new reservoirs offer fishing opportunities. Fishing folk are attracted from coastal regions, deltas and other reservoirs. They bring with them fishing skills, equipment and access to markets. Migrants may set up informal villages at the margins of the reservoir. Poor sanitation and water supply promote communicable disease transmission. Reservoirs also provide communication routes into largely untouched hinterlands. These are attractive to logging companies, tourism operators and migrants. They make additional demands on already overloaded health services and are frequently overlooked during proposal development.

Migrants of this kind often remain when the construction phase is complete and slowly integrate into the host community.

9.7 WASTEWATER

Wastewater collection, treatment, disposal and recycling are urgent themes, especially in urban environments. Raw wastewater (sewage), concentrates a range of pathogens such as *Ascaris, Trichuris, Giardia, E. coli,* hookworm, tapeworm and various viruses. Each of these can survive a different length of time and requires different treatment in order to render wastewater safe to reuse. *Ascaris* eggs, for example, are latent and persistent – surviving over one year and sticking to the leaves of salad crops.

Figure 9.5 *Illegal abstraction of raw sewage in Syria for irrigating a tree crop*

Copyright Tropix.co.uk

In many cities, wastewater is used in peri-urban areas for agriculture and the produce is then consumed in the city (Birley and Lock, 1999). Such agricultural systems have important sustainability elements and there is great interest in promoting them in a safe way. Consequently, there is a need for HIAs of wastewater proposals.

There is a considerable literature on the safe reuse of wastewater in agriculture and aquaculture (for example, Mara and Cairncross, 1989; Cifuentes et al, 1991/1992; Wahaab, 1995; Ayres and Mara, 1996). Recent research challenges conventional wisdom regarding raw wastewater reuse in irrigation in Pakistan (Fewtrell and Kay, 2008). Some 80 per cent of all cities and towns in Pakistan use wastewater in agriculture. Farmers use raw wastewater in preference to regular irrigation water. It costs more than regular irrigation water, but saves on fertilizer costs. Much of the vegetable crop in Pakistan is thus irrigated with raw wastewater. It is generally thought that use of such water creates additional risks for both farmers and consumers. This research, however, suggests otherwise – at least for communities where the risk of such infections is already very high. For example, the main source of contamination of vegetables with *E. coli* was the marketplace, not the field. The prevalence rate of other infections was explained more by socio-economic differences than by use of wastewater.

There are different methods of treating sewage before reuse. Activated sludge treatment systems are capital and technology intensive and require relatively small amounts of land. Lagoon systems are relatively low cost and low tech but require much larger amounts of land.

In order to improve the safety of wastewater, WHO proposed a hierarchy of crop types. Tree crops, for example, are suitable for irrigation with untreated wastewater, providing that the crop is protected from contact with the ground. The lagoon system in Figure 9.6 is surrounded by olive trees. Partially treated water can also be used for irrigating ornamentals, parks and golf courses. Salad crops require pathogen-free irrigation water in order to prevent the external surfaces from becoming contaminated. But even when grown in such water, contamination can occur during transporting, marketing, storage and preparation.

Many assessments of wastewater treatment proposals are limited to the confines of the plant itself. However, it is common to find illegal abstraction of treated water just beyond the perimeter fence. Such water looks clean, because suspended solids have been removed, but it is usually still pathogenic. The assessment scope must include downstream use.

In many cities, domestic water supplies are provided to communities before wastewater removal systems are planned. Consequently, wastewater is often discharged directly onto streets and then percolates to the ground. See Figure 9.7. Such pools can provide ideal breeding sites for a mosquito called *Culex quinquefasciatus*. In contrast to most mosquitoes that require relatively clean water, this mosquito thrives in large numbers in heavily polluted water. In many coastal cities of the tropics it can transmit a pathogenic filarial nematode, giving rise to a disease called elephantiasis. The assessment of water supply systems should include inquiries into wastewater disposal.

Figure 9.6 *Sewerage lagoons and tree crops in Jordan*

Copyright Tropix.co.uk

9.7.1 Case study: Health impacts of a wastewater treatment plant

Some years ago, I was able to undertake a rapid HIA of a sewage treatment plant in Damascus, Syria. The plant was intended to gather mixed sewage discharged into drains leaving Damascus, treat the product, and produce irrigation water and dried sludge. The key stakeholders included the farmers who would receive the treated wastewater for irrigation and the dried sludge for fertilizer, and the consumers of the agricultural products. Additional details about this case are described elsewhere (Birley, 2004). Some of the key issues identified during the HIA were as follows:

- Industrial and domestic sewage were mixed so that potentially hazardous industrial chemicals were included.
- Planners had no access to the key books on the safe reuse of wastewater referred to above.
- The station for testing the quality of treated wastewater belonged to the treatment plant and not to the irrigation department who were the recipients of the product.

In 2006, the World Health Organization published an update to its guidelines for the safe reuse of wastewater, excreta and greywater in agriculture and aquaculture (WHO, 2006). The new guidelines build on the previous work and recognize the need for a

Figure 9.7 *Pools of sewage on the street of an Indian town*

Copyright Tropix.co.uk

series of critical control points to reduce the contamination of food crops between the field and the table. An edited volume discusses the latest evidence in depth (Drechsel, et al, 2010).

9.8 WETLANDS

Wetlands make a major contribution to human health and well-being through the provision of ecosystem services. These services include cleaning dirty water, producing edible biomass, fixing carbon and nitrogen, and regulating floods. Concerted efforts are required and are being made to conserve and manage the world's wetlands and these are outside the scope of this chapter. See the Ramsar Convention for more information (Ramsar, undated). Many wetlands have been destroyed by the building of dams and reservoirs and there are proposals to restore some of them by decommissioning dams or by seasonal opening of floodgates. The health impact of restoring seasonal floods has been a subject of assessment and debate (Salem-Murdock, 1996; Verhoef, 1996; Acreman, 2000; WRI, 2005).

Thirteen different kinds of wetlands are identified in a standard classification (WRI, 2005). There are both positive and negative health associations. Many natural wetland systems and some human-made wetland systems have positive health impacts in the form of contributions to livelihood, water supply, culture and sense of place. Health impacts are associated with the manipulation of wetlands through restoration of flooding, changes in irrigation regimes and increased abstraction.

9.9 EXERCISE

You have been appointed as HIA consultant to the donor agency that is funding a hydropower and irrigation project in the tropical republic of San Serriffe. The electricity produced by the project will be sold to neighbouring countries, which are industrializing.

The project is located in a woodland savannah environment (see Figure 9.8). There are high mountains in the north. The urban and rural communities are largely from different cultural backgrounds. The rural people are called Flong. About 12,000 will have to be resettled. The Flong are subsistence agriculturalists who grow maize, keep pigs, goats and cows, and gather other food from the woodlands. The capital city is Villa Pica. The country is ruled by life president, His Excellency General Pica. The town nearest to the project site is Woj. There is a productive fishing community based at Zapf.

A smaller hydropower scheme was completed ten years ago. The Flong community, 2000 people from 20 villages, was displaced and resettled. Some were resettled as the workforce on a commercial irrigated farm, in place of their subsistence agriculture. The baseline health report notes that the following communicable diseases are common: malaria, filariasis, schistosomiasis, sleeping sickness, dengue, HIV and diarrhoea. In addition, high levels of childhood malnutrition are reported.

Here is a partial list of project stakeholders:

- **Displaced**: 12,000 Flong subsistence farmers including dependents will be displaced from the reservoir and irrigation sites. Many will be offered a resettlement site on the planned irrigation scheme, the rest will move onto the slopes of Mount Flong, or drift into town.
- **Settlers**: 5700 cash crop farmers plus dependents will also be settled on the irrigation scheme from outside the area. These will include some Flong and some town dwellers.
- **Fishing folk**: 300 plus dependents will be attracted to the new reservoir from the coast.
- **Police**: about 20 plus dependents – to regulate activities around the reservoir.
- **Electricity workers**: 100 workers plus dependents to produce hydroelectricity and maintain the dam.
- **Irrigation management and agricultural extension**: ten workers plus dependents.
- **Existing health centre**: two nurses and a technician plus dependents.
- **Existing school**: seven teachers, four assistants plus dependents.
- **Construction workers:** 4000 males for two years including 100 international.
- **Camp followers**: 350 food sellers plus dependents; 14 professional sex workers plus dependents; 29 merchants plus dependents.
- **Seasonal labour for farmers:** 3700 plus dependents.

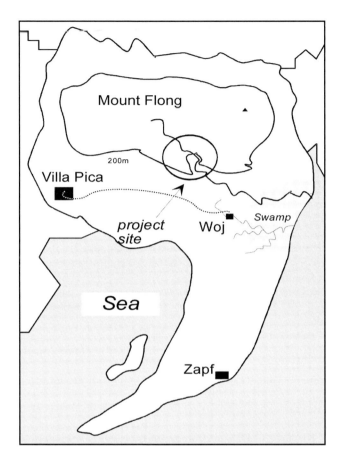

Figure 9.8 *Map of project site in San Serriffe*

- **Others**: loggers and poachers, an unknown number; places of worship, at least two; tourists and tour operators; rich city people making holiday homes; land speculators; administrators; ordinary traders; surveyors; and visiting consultants.

Answer the following questions about the project:

- What health concerns should be considered?
- What health determinants may change?
- How would you expect health outcomes to be distributed between project stages, geographical locations and stakeholder groups?
- What recommendations would you make to safeguard and enhance health, or mitigate risks?

9.10 REFERENCES

Ackerman, W., G. White and E. Worthington (eds) (1973) *Man-made Lakes, Their Problems and Environmental Effects*, American Geophysical Union, Washington, DC

Acreman, M. C. (2000) *Managed Flood Releases from Reservoirs: Issues and Guidance*, Centre for Ecology and Hydrology, UK, www.dams.org

Anon (1998) 'Arsenic in the water', the *Guardian*, London, pp1, 2, 3, 10, 11

Ayres, R. M. and D. D. Mara (1996) 'Analysis of wastewater for use in agriculture: A laboratory manual of parasitological and bacteriological techniques', World Health Organization, Geneva

Birley, M. H. (1991) 'Guidelines for forecasting the vector-borne disease implications of water resource development', World Health Organization, www.birleyhia.co.uk, accessed 2010

Birley, M. (2003) 'Health impact assessment, integration and critical appraisal', *Impact Assessment and Project Appraisal*, vol 21, no 4, pp313–321

Birley, M. (2004) 'Health impact assessment in developing countries', in J. Kemm, J. Parry and S. Palmer (eds) *Health Impact Assessment: Concepts, Theory, Techniques and Applications*, Oxford University Press, Oxford

Birley, M. H. and K. Lock (1999) *The Health Impacts of Peri-urban Natural Resource Development*, Liverpool School of Tropical Medicine, Liverpool, www.birleyhia.co.uk/Publications/periurbanhia.pdf

Bos, R., M. Birley, P. Furu and C. Engel (2003) *Health Opportunities in Development: A Course Manual on Developing Intersectoral Decision-making Skills in Support of Health Impact Assessment*, World Health Organization, Geneva

British Geological Survey (undated) 'Arsenic contamination of groundwater', www.bgs.ac.uk/arsenic/home.html, accessed April 2010

Cairncross, S. and R. Feachem (1993) *Environmental Health Engineering in the Tropics*, John Wiley & Sons

Cairncross, S., J. E. Hardoy and D. Satterthwaite (eds) (1990) *'The Poor Die Young': Housing and Health in Third World Cities*, Earthscan, London

Cifuentes, E., U. Blumenthal, G. Ruiz-Palacios and S. Bennett (1991/1992) 'Health impact evaluation of wastewater use in Mexico', *Public Health Review*, vol 19, pp243–250

Clarke, S. E., C. Bogh, R. C. Brown, M. Pinder, G. I. L. Walraven and S. W. Lindsay (2001) 'Do untreated bednets protect against malaria? ', *Transactions of the Royal Society of Tropical Medicine and Hygiene*, vol 95, pp457–462

Dhaka Community Hospital Trust and Disaster Forum (1997) 'Arsenic disaster in Bangladesh environment', Workshop on arsenic disaster in Bangladesh environment, Dhaka, Bangladesh

Drechsel, P., C. Scott, L. Raschid-Sally, M. Redwood and A. Bahri (eds) (2010) *Wastewater Irrigation and Health: Assessing and Mitigating Risk in Low-income Countries*, Earthscan, London

Fewtrell, L. and D. Kay (eds) (2008) *Health Impact Assessment for Sustainable Water Management*, IWA Publishing, London

Hunter, J. M., L. Rey and D. Scott (1982) 'Man-made lakes and man-made diseases: Towards a policy resolution', *Social Science and. Medicine*, vol 16, pp1127–1145

Hunter, J. M., L. Rey, K. Y. Chu, E. O. Adekolu-John and K. E. Mott (1993) 'Parasitic diseases in water resources development, the need for intersectoral negotiation', WHO, Geneva

Ijumba, J. N. and S. W. Lindsay (2001) 'Impact of irrigation on malaria in Africa: Paddies paradox', *Medical & Veterinary Entomology*, vol 15, pp1–11, www.ingentaconnect.com/content/bsc/mve/2001/00000015/00000001/art00001

Jobin, W. (1999) *Dams and Disease – Ecological Design and Health Impacts of Large Dams, Canals and Irrigation Systems*, E & FN Spon, London and New York

Konradsen, F., M. Chimbari, P. Furu, M. H Birley and N. O. Christensen (1997) 'The use of health impact assessments in water resource development: A case study from Zimbabwe', *Impact Assessment,* vol 15, pp55–72

Krieger, G., M. Balge, Soutsakhone Chantthapone, M. Tanner, B. Singer, L. Fewtrell, S. Kaul, P. Sananikhom, P. Odermatt and J. Utzinger (2008) 'Nam Theun 2 hydroelectric project, Lao PDR', in L. Fewtrell and D. Kay (eds) *Health Impact Assessment for Sustainable Water Management,* IWA Publishing, London

Mara, D. and S. Cairncross (1989) 'Guidelines for the safe use of wastewater and excreta in agriculture and aquaculture', WHO and UNEP, Geneva

Mills, A. J. and D. J. Bradley (1987) 'Methods to assess and evaluate cost-effectiveness in vector control programmes', in *Selected Working Papers Prepared for the 3rd, 4th, 5th and 6th Meeting of the WHO/FAO/UNEP PEEM.* WHO, Geneva

Prüss-Üstün, A., R. Bos, F. Gore and J. Bartram (2008) 'Safer water, better health: Costs, benefits and sustainability of interventions to protect and promote health', World Health Organization, Geneva

Ramsar (undated) *The Ramsar Convention on Wetlands,* www.ramsar.org, accessed April 2010

Sakhalin Energy Investment Company (2003) Health, social and environmental impact assessments, www.sakhalinenergy.com, accessed March 2003

Salem-Murdock, M. (1996) 'Social science inputs to water management and wetland conservation in the Senegal River Valley', in M. C. Acreman and G. E. Hollis (eds) *Water Management and Wetlands in Sub-Saharan Africa,* IUCN, Gland, Switzerland

Scudder, T. (2005) *The Future of Large Dams: Dealing with Social, Environmental, Institutional and Political Costs,* Earthscan, London

Stanley, N. and M. Alpers (1975) *Man-made Lakes and Human Health,* Academic Press, London

Tiffen, M. (1989) *Guidelines for the Incorporation of Health Safeguards into Irrigation Projects Through Intersectoral Cooperation,* WHO/FAO/UNEP

Verhoef, H. (1996) 'Health aspects of Sahelian floodplain development', in M. C. Acreman and G. E. Hollis (eds) *Water Management and Wetlands in Sub-Saharan Africa,* IUCN, Gland, Switzerland

Wahaab, R. (1995) 'Wastewater treatment and reuse: Environmental health and safety considerations', *International Journal of Environmental Health Research,* vol 5, no 1, pp35–46

WCD (World Commission on Dams) (2000) *Dams and Development: A New Framework for Decision-Making,* Earthscan, London

WHO (1983) 'Minimum Evaluation Procedure (MEP) for water supply and sanitation projects', World Health Organization

WHO (1999) 'Human health and dams, submission by the World Health Organization to the World Commission on Dams', www.who.int/docstore/water_sanitation_health/Documents/Dams/Damsfinal.htm#References, accessed November 2009

WHO (2006) 'Guidelines for the safe use of wastewater, excreta and greywater: 1 Policy and regulatory aspects; 2 Wastewater use in agriculture; 3 Wastewater and excreta use in agriculture; 4 Excreta and greywater use in agriculture', World Health Organization, www.who.int/water_sanitation_health/wastewater/gsuww/en/index.html

WHO (2008) 'Water safety plan manual: Step-by-step risk management for drinking-water suppliers' www.who.int/water_sanitation_health/publication_9789241562638/en/index.html accessed February 2011

WHO (2010) Media Centre fact sheets, www.who.int/mediacentre/factsheets/fs310/en/index.html, accessed January 2010

World Bank (2001) 'Safeguard Policies, Operational Policy 4.12: Involuntary resettlement', http://web.worldbank.org/wbsite/external/projects/extpolicies/extsafepol/0,,menupk:584441~pagepk:64168427~pipk:64168435~thesitepk:584435,00.html, accessed October 2006

WRI (World Resources Institute) (2005) 'Millennium Ecosystem Assessment, ecosystems and human well-being: Wetlands and water synthesis', www.maweb.org/documents/document.358. aspx.pdf, accessed October 2009

CHAPTER 10

Extractive industries

- Some special characteristics of the oil and gas, and mining and minerals sectors are discussed. These are often massive projects located in remote areas among poor communities and operated by transnational corporations. The host governments, for various reasons, may place a low priority on human rights issues.
- Personal experiences with the oil and gas sector are described.
- Approaches to social investment are explained.

10.1 INTRODUCTION

The extractive industries produce fossil fuels and minerals by drilling and mining. These are non-renewable, finite resources. Their sustainable development has economic, social and environmental components and these include human health and well-being (IIED and WBCSD, 2002; World Bank, 2004). Non-renewable resources are subject to increasing scarcity and the unit price varies in relation to supply and demand (Heinberg, 2007). As the price rises, more difficult deposits become financially attractive. Exploitation of such deposits is likely to involve greater impacts. The impacts are diverse, but many affect human health at the local, regional or global scale. More inaccessible deposits require greater energy and associated greenhouse gas (GHG) emissions that contribute to climate change. At the same time, consumption of fossil fuel products contributes to climate change. The associated public health consequences are discussed in Chapter 12.

The IFC (International Finance Corporation) Standards (IFC, 2006) and the Equator Principles (Equator Principles, 2006), described in an earlier chapter, grew out of the World Bank's Extractive Industry Review (World Bank, 2004). Many oil and gas projects are located in remote areas inhabited by marginalized and vulnerable communities. Environmental, social and health impacts can be severe, and strenuous efforts have been made to document and correct this. An edited book, based on papers presented at the 2005 IAIA conference, examines recent experience with pipeline construction (Goodland, 2005). The book documents examples from Africa, South America and Asia. Topics include the contamination of the Amazon rainforest in Ecuador, epidemics of introduced diseases in Peru, damage to fisheries in Sakhalin (Siberia) and human rights abuses in Asia. The health impacts of contamination in

Ecuador are also described by San Sebastián and Hurtig (2005). Opencast mining can have a significant effect on land use and fisheries, undermining food production. A recent book provides detailed examples from the Philippines (Goodland and Wicks, 2009). Other relevant publications include an analysis of the health impact of mining in northern Canada and the gaps associated with EIA (Noble and Bronson, 2005).

Extractive industries tend to be transnational corporations working in joint ventures with national governments. There is a debate about whether these can act rationally and compassionately, or whether they are inherently psychopathic entities (Achbar and Abbott, 2004). Public welfare is enhanced by regulation, transparency and public scrutiny of corporations and their host governments (Extractive Industries Transparency Initiative, undated). HIA contributes to this process. There is an increasing body of resources available for assessing health impacts (for example, Barron et al, 2010).

The extractive industries have trade associations: the International Association of Oil and Gas Producers (OGP), the International Petroleum Industry Environmental Conservation Association (IPIECA) and the International Council on Mining and Metals (ICMM). These publish a wide range of guidance on HIA (IPIECA/OGP, 2005a; ICMM, 2010a), HIV/AIDS (IPIECA/OGP, 2005b; ICMM, 2008) and much else.

10.2 OIL AND GAS

The first section of this chapter is about oil and gas (O&G) exploration and production. A number of O&G companies have made internal decisions, based on the business case discussed in Chapter 1, to include HIA in their impact assessment procedures. Initially, HIA was carried out in parallel with EIA, but it is increasingly being integrated and may be called ESHIA (environmental, social and health impact assessment). The trade association's guidance on HIA is accompanied by a set of background material and a pocket guide (IPIECA/OGP, 2005a). It provides a succinct introduction to HIA for the O&G sector.

Royal Dutch Shell was one of the first major oil companies to establish an enabling policy which supports HIA (Birley, 2003, 2005). They publicly stated their aims as follows:

- Conduct business as responsible corporate members of society.
- Contribute to sustainable development.
- Pay proper regard to health, safety and the environment.
- Express support for fundamental human rights in line with the legitimate role of business.
- Recognize that investment decisions are not exclusively economic.
- Take a constructive interest in societal matters that may not be directly related to the business.

In support of these aims they agreed an internal set of minimum health standards (Shell International, 2001). One of the standards requires an HIA in conjunction with any environmental and social impact assessments (ESIAs) for all new projects and major modifications, and prior to abandonment of existing projects where there is a potential impact on the health of the local community, the company and contract workers, or their families.

Shell had already established a health, safety and environment management system. The management system incorporated multiple feedback processes to ensure monitoring, corrective action and improvement. This seemed to be part of their general approach to risk management. Key components were procedures to identify, assess, control and recover from risks. Negative health impacts of projects on communities represent a risk and HIA is one tool for managing that risk.

In 2002, Shell established a time-limited post of senior health adviser on HIA in the corporate health service and they offered me the post. This opportunity provided me with insights into corporate thinking. Material for this chapter derives from experiences at Shell and subsequent consultancy work for other O&G corporations.

10.2.1 Social investment

Some O&G corporations have established social investment policies in addition to their impact assessment policies. Royal Dutch Shell's aims, above, reflect this. These policies recognize a corporate social responsibility to invest in the well-being of their host communities. See Chapter 1 for more details. One of the most advanced examples that I have observed of social investment policies was from Eni, an Italian corporation, which has a set of internal guides and principles making social investment a cornerstone of its new ventures. Its policy includes seeking opportunities for public–private partnerships with government and NGOs. By contrast, I have worked with an American corporation that explicitly excluded social investment.

10.2.2 Joint ventures

The modern O&G business is often based on joint venture (JV) agreements although there are other approaches such as production-sharing agreements. An O&G corporation will make contracts with a national government, investment companies and other corporations to create a national company for the purposes of exploration and/or production. Ownership of equity in the company is split so that the government retains a majority of the profit, while operational control lies with the corporation that has the technical expertise (Yergin, 1991). The JV company will have the technical expertise to deliver the proposal. It will be staffed by a mixture of expatriate and local staff. The expatriates will provide expertise that is missing locally. Local staff may be drawn from the public sector.

Corporations compete to obtain the licence from governments to set up JV companies. Finance for the venture is raised by each partner from a variety of sources including internal reserves and lending banks. Loans from lending banks are now subject to the Equator Principles discussed in Chapter 1 and these require health to be protected.

As HIA is not legislated by most national governments, at the time of writing, there is often a curious contradiction. The corporate partners must undertake HIA for internal or external reasons, but the government may not want an HIA because it could reduce national profits from the venture. In addition, the government may undertake to provide cleared land for new infrastructure. This land may be inhabited, as well as having important ecosystem functions. Key new opportunities are often located in regimes with poor human rights records; here, the government may exclude consideration of land clearance from the impact assessment. The corporation is often powerless to challenge this. In addition, the corporation may have a policy of public disclosure of all impact assessments once the final decision to proceed with development has been made. The government, however, may object to public disclosure because transparency is not consistent with their politics. If things go wrong, the government tends to shift the blame for their own omissions and commissions to the corporation.

Corporate structures can also prevent HIA policies from being implemented by JVs. The local company will have a degree of autonomy from the corporate parent. Managers drawn from national government may not wish to comply with the corporate policies of their international partner. They can resent the interference. As HIA is a relatively new requirement, many managers may have no idea what it is, or how to commission it. They may, for example, ask local physicians to produce an HIA report and then simply state that they have complied with corporate requirements. The local physicians will usually have no training in or experience of HIA.

Managers of new O&G projects are very skilled at overcoming the many obstacles that interfere with their primary aim, which is the profitable delivery of oil and gas. They are skilled at cost control. To them, HIA policy requirements from head office may simply look like another obstacle. It is unfamiliar and of no immediate benefit. Head office can manage their non-compliance in three ways:

1 Include the completion of good impact assessments in the annual performance review of managers; this affects their salary bonus and career advancement.
2 Tighten up the impact assessment process so that there are fewer loopholes.
3 Build understanding of the benefits of good impact assessment for reducing risk.

10.3 BUILDING UNDERSTANDING AND CAPACITY

I have tried a strategy of building understanding and capacity with a number of different O&G companies. The idea is as follows. An HIA training course is convened at the proposal initiation or scoping stage. The course is run in-country and attended by representatives from the JV, local consultants, public sector officials from public health, environmental health and environment, and other professional stakeholders. The intention is to transmit basic skills in commissioning, managing and conducting an HIA. Unexpected challenges associated with this approach have been described elsewhere (Birley, 2007). For example, local consultants named in proposals are often figureheads who have no intention of undertaking the work for which they have been commissioned. Instead they may employ junior research assistants to do the work on their behalf. At the same time, they are unsure about what they have contracted to do, but cannot reveal their ignorance. See also Chapter 4.

Such courses usually build enthusiasm among the government officers who attend. They often finish the course saying that HIA should be a requirement for all new proposals. As a result of this enthusiasm, they are more likely to support the HIA by supplying health baseline information, unpublished reports and surveys, and by suggesting key informants.

The attendees from the JV are usually health, safety and environment (HSE) managers, occupational health physicians and hygienists. They advise the project managers but usually do not have executive roles. In order to ensure a good HIA, these attendees need to be equipped with persuasive arguments and to be persuaded themselves. Their own careers depend on being effective and responsible members of a team, but they are unlikely to 'rock the boat'. They regard the additional skills acquired by attending the course as helpful to their career; they may require a certificate of attendance, or a formal examination and certificate of achievement.

10.4 SCREENING

There is usually great interest in the screening stage, because it provides a loophole for managers to state that an HIA is not needed. At its simplest, the screening stage becomes a 'tick-box exercise' with the HSE manager ticking the appropriate boxes to ensure that no HIA is required. The person completing the checklist may have no public health training. They can respond in the negative to a question like 'Could this proposal have potential health impacts on the local community?'. Mechanisms for ensuring an appropriate screening process are described in Chapter 4.

10.5 POTENTIAL HEALTH IMPACTS OF OIL AND GAS SECTOR PROPOSALS

There are many different scenarios associated with O&G proposals. For example, some proposals take place entirely offshore such as the middle of the North Sea or the Gulf of Mexico. If there are no human communities affected and the proposal is located far from fishing grounds and shipping routes, then an HIA may be unnecessary. However, proposals usually have onshore support infrastructure, such as ports and construction facilities, and these may have project-affected communities. The onshore facilities may be well established so that no new impacts are expected. The Gulf of Mexico oil spill of 2010 provides an example of the impacts of catastrophic events. If a major spill occurs, ecosystems are seriously damaged, food supplies are lost and human livelihoods are disrupted over thousands of kilometres. All of these have consequences for public health. Should HIA include consideration of catastrophic engineering failures, or is this better managed by other processes? Up till now, I have assumed that catastrophe management was part of health risk assessment. This needs reviewing.

Another broad scenario consists of prospecting and finding oil or gas in a remote rural region of a poor country. There may be a scattering of villages, towns and transportation routes. Typical community health issues are illustrated in Table 10.1. There may be a gap of 10–40 years between prospecting and producing. Operation may then continue for 30+ years, perhaps with major expansion. Finally, the project will be

Table 10.1 *Overview of community health issues by proposal stage*

Stage	Typical health outcomes	Typical health determinants
Prospecting (exploring)	Communicable diseases	Commercial sex workers and male exploration crews
Feasibility studies	Anxiety, expectation	Uncertainty
Construction	Communicable diseases, malnutrition, injury, psychosocial disorder	Loss of food security, communal violence, loss of livelihood
Operation (production)	Communicable diseases, non-communicable diseases	Inhalation of particulates or toxic gasses
Decommissioning	Communicable diseases, malnutrition, injury, psychosocial disorder	Loss of livelihood

Table 10.2 *Examples of determinants and outcomes for an industrial plant in a developing economy*

Determinant Category	Subcategory	Examples of health outcomes
Resources	Land take	Malnutrition associated with loss of wild foods
	Settlements and labour camps	HIV/AIDS, malaria, gastrointestinal infections; alcoholism, violence, injury; vector-borne disease
Discharges	Air	Dust-induced lung disease; noise induced deafness
	Water	Poisoning; waterborne diseases
Machinery	Traffic	Injury; asthma; heart disease; psychosocial disorders
	Pipelines	Zoonoses; explosions
Incidents	Petroleum refinery explosion	Burns; poisoning

decommissioned some 40, or more, years in the future. Each of these stages has impacts and is discussed in more detail below.

O&G proposals often include the construction of large industrial plants. Table 10.2 summarizes some of the main components and the community health linkages.

10.5.1 Exploration

Oil and gas exploration, in common with other extractive industry, frequently takes place in poor, remote regions. This is because more accessible reserves have already been exploited. Exploration usually requires seismic surveys and test drilling. Engineers set off underground explosions and listen for echoes using a network of sensitive microphones.

Seismic survey crews move in straight lines and are based in temporary camps. Contact with local communities is brief but intense. In many cases, no reserves of oil and gas are identified but local expectations nevertheless begin to rise. When promising reserves are found, test rigs staffed by small crews may be set up for longer periods. Traffic increases and includes larger trucks, four-wheel drives and helicopters. All supplies are likely to be obtained from elsewhere. The results are taken away for analysis and the crews and materials are withdrawn. Negotiations and feasibility studies begin that may last a decade.

10.5.2 Construction

The construction stage may last two years and, for a large proposal, require 10,000 workers. Most of the workers will be male, skilled and from other countries and regions. There may be extreme inequality between the local community and the workforce. Construction camps require resources such as food and water and rented accommodation. Local procurement is likely to be limited but may lead to price inflation. Wages paid to the construction workers can be much higher than the standard wages paid in the local community. Nurses and doctors from local hospitals may seek work in the project clinic. School teachers may seek work as secretaries and translators. Policeman may seek work as drivers and security guards. Many local young women and men may become commercial sex workers as a result of unequal opportunities, unequal status, inflation and poverty. The rapid change in the region can create a 'boom town' atmosphere. Alcohol and substance abuse increase and those who cannot benefit from the new opportunities may become mentally unwell. The influx of different cultural groups can cause social tensions leading to communal and domestic violence. Suicide rates may increase. At the end of the construction stage, most of the migrant workers leave, rents fall, jobs are lost and procurement of goods and services tend to fall to much lower levels. This is a boom and bust cycle.

Table 10.3 summarizes some of the health issues associated with construction in poor regions.

Table 10.3 *Example health issues for large construction projects in poor regions*

Outcomes	Examples	Determinants
Communicable diseases	Malaria, TB, STIs, diarrhoea	Commercial sex, crowding, creating mosquito breeding sites, poor water supply
Non-communicable diseases	Chronic obstructive lung disease	Indoor and outdoor air pollution
Injury	Trauma from road traffic or communal violence	Pedestrian and domestic animal use of roads
Nutritional	Wasting and stunting associated with lack of food	Food price inflation, loss of earnings
Mental health/psychosocial well-being	Substance abuse, depression, suicide	Loss of control, rapid change, loss of community structure

Table 10.4 *Some of the summary recommendations for the construction stage of a European, land-based, gas proposal*

Impact	Rating	Mitigation	Timing
Cycle route safety	Moderate	Transport plan	Detailed design
Overweight/obesity	Moderate	Encourage active transport	Detailed design
Pumping well blowout	Minor because barriers already designed	Recovery barriers Emergency response plan See EIA	Detailed design
Transportation and storage of chemicals	Minor	Transport plan Storage standards Recovery barriers Emergency response plan	Detailed design
Employment opportunity	Positive	Local employment policy	Before invitation to tender

By contrast, Table 10.4 is an example of the summary recommendations associated with gas development in an economically developed, European country. The proposal was for a major refurbishment of an existing facility consisting of new pumping stations connected to pipelines on existing routes. There was concern that the construction stage would affect the health and safety of the entire transport-using community, especially children. The company had a separate procedure for making transport plans. The proposed health and safety solution was to bus children to school. Obesity was a major public health concern in the country and obstacles to active travel had to be avoided. The HIA provided advice that modified the transport plan to safeguard active travel.

Some concern was also expressed about risks associated with drilling, such as blowouts, and with the transportation and storage of chemicals. However, there were already engineering designs to prevent these. An emergency response plan was negotiated with the local health authorities and similar agencies. There was also an opportunity for a positive health impact by establishing a local employment policy.

10.5.3 Operation

The staffing requirements of the operation stage are much smaller than the construction phase and have the following elements:

- Small numbers of skilled, non-local operations staff are permanently on site, including a community liaison officer and medical doctor. They will be accompanied by their families.
- There are some permanent positions for skilled local staff such as nurses, secretaries, cooks and drivers.

- There are some permanent positions for unskilled local labour as guards, gardeners, cooks' assistants and cleaners.
- There is continuing use of construction camps for maintenance and refurbishment work, occupied by a series of different companies with their own teams.
- Streams of professional staff visit the site for periods of a few days or weeks.

Continual arrivals and departures create traffic. On remote sites a landing strip may have been constructed for passenger aircraft, or a jetty for supply vessels. Roads and bridges may also have been constructed.

There will usually be a permanent medical clinic staffed by a medical doctor, nurses and a safety officer. The clinic will be equipped with advanced technologies and well-stocked medical supplies. There will be a plan for stabilizing and evacuating seriously injured or sick personnel. This will include arrangements with national or international tertiary hospitals and appropriate transportation such as ambulances, helicopters and other aircraft.

Permanent housing is usually constructed for families of professional staff. This is often spacious and air-conditioned, with domestic staff. The site may include a school, sports centre, swimming pool, golf course and restaurant. It will be gated and secured. There may also be recreational opportunities in the local community such as bars, restaurants and dance clubs.

Catering will be subcontracted to international catering companies who will be responsible for food procurement and food safety. They may fly in frozen meat and fresh vegetables from all over the world.

A permanent water supply system will have been constructed. Water will probably be sourced from deep aquifers and used for both the production process and domestic supply. Solid waste management systems will have been established for sewage, domestic waste and production waste. Special systems will be in place for hazardous waste disposal such as incineration and lined landfill. Routine air and water emissions will have been designed, on a new plant, to comply with international best practice.

Semi-permanent communities of camp followers will have been established during the construction stage and may continue and grow during the operation stage.

In addition, there will probably be some system of social investment or philanthropy in place. This is likely to include support by the project to local schools and clinics.

Typical health impacts during the operation stage include the following. The list is illustrative rather than comprehensive:

- The clinic is not designed to supply services to the general community. On the other hand, local government clinics may be comparatively poorly equipped. A procedure will be required for managing medical emergencies arising in the community. For example, the government clinic may call on the project clinic for humanitarian assistance to manage an obstetric emergency.
- Sexual interactions between the workforce (mobile men with money) and poor local people will continue. Some expatriate staff will form semi-permanent liaisons and move their partners into their accommodation. It is common, in certain countries, to see overweight, middle-aged white men dining in company restaurants with pretty, young, local women.

- Traffic movements increase the risk that local people or their livestock are injured.
- Food price inflation may occur as the catering company tries to procure goods in the local markets, leading to under-nutrition in the community.
- There may be a risk of communal violence between migrants and locals or between locals and project staff. Theft is likely to increase.
- The peripheral community may perceive an increase in disease that is attributed to the operation, such as birth abnormalities, cancers, asthma and hypertension.

10.5.4 Decommissioning

It is challenging to make detailed statements about a decommissioning stage that will occur many years in the future, as discussed in Chapter 4.

Contaminated land and plant are issues during the decommissioning of large infrastructure projects. The issues arise because of poor waste management practices in the past and lack of detailed archives of the chemicals and materials used on site. During the life of a large project, management and ownership may change hands several times and records are easily lost. An obvious recommendation is to establish a robust archival system from the beginning.

When a proposal reaches the end of its life, operations staff may lose their jobs or be redeployed to other regions and the local economy may suffer as procurement of goods and services ceases. Secondary benefits of the proposal may also cease including support for schools, clinics, recreation centres and road maintenance. Towns that grew up alongside the project may become ghost towns. Recommendations include providing advance notice, assisting with relocation, seeking other support for community facilities, and employing local labour for the clearing and removals and site rehabilitation. Some of these features can be built in from the beginning. For example, public–private partnerships can be created for supporting the secondary benefits.

A new HIA should be undertaken before the decommissioning commences and all previous HIAs associated with the proposal should be part of the archived documentation available.

10.6 INFORMAL SETTLEMENTS

Infrastructure proposals in poor countries can attract large numbers of temporary migrants, or camp followers. They arrive during construction and may remain when construction is over. They live in temporary and unsanitary housing around the periphery of the site and sell goods and services to the workforce. Over time, the settlements may enlarge as families become established.

Settlements of this kind can be seen on satellite imagery, if you know where to look. Google Earth software provides convenient and free access to such imagery. Figure 10.1 is an image from the Bonny Island O&G terminal, Nigeria, coordinates: 4°24'22.52" N, 7°10'43.43" E, viewed from an elevation of about 9 kilometres. It identifies where some of the informal, squatter, settlements are located, on the edge of a saltwater lagoon. The lagoon is used by the community for waste disposal. The resettlement village is called New Finima.

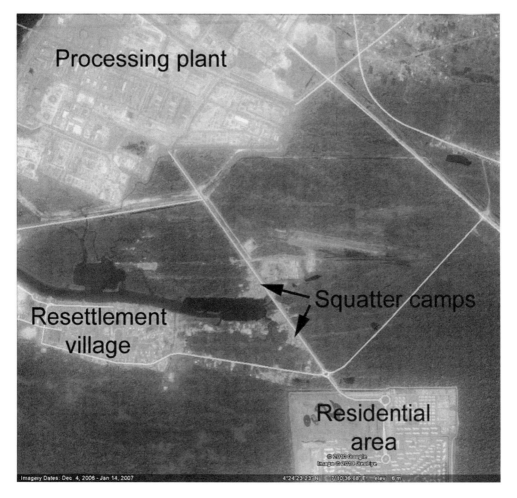

Figure 10.1 *View from Google Earth of part of Bonny Island, Nigeria*

Copyright 2010 Europa Technologies, 2010 Google Image, 2010 GeoEye; used under their fair use policy

10.7 CASE STUDY FROM THE UNITED ARAB EMIRATES

An HIA was commissioned in Dubai, part of the United Arab Emirates (UAE) in 2007. The proposal was to construct infrastructure associated with the O&G sector and the client was a transnational corporation. I had previously worked in Dubai in 1984 and photographed some of the conditions in labour camps at that time. I was interested to determine whether conditions had changed.

The population of UAE is small and most of the land is desert. Much new infrastructure can be located reasonably far from human communities, either in the desert or on offshore land reclamations. Many of the health impacts of new infrastructure may be small or easy to manage. However, there is one glaring issue: construction labour. The management of health and safety and other 'inside the fence' issues is not strictly part of HIA, as mentioned in the introduction. However, in the Gulf there are special conditions associated with the procurement and treatment of labour.

The UAE has experienced rapid economic development and growth during the past several decades, but it lags in the development of its civil society: in 2007, the country did not hold elections for any public office, and political participation was limited to the ruling families. The government had not signed most international human rights and labour rights treaties, such as International Labour Organization (ILO) core conventions No. 87 on Freedom of Association and Protection of the Right to Organise (1958) and No. 98 on Right to Organise and Collective Bargaining Convention (1949) (ILO, 2009). Trade unions did not exist and strikes and lock-outs were expressly banned. Reforms have been promised but there was much doubt as to their efficacy. Migrant workers were vulnerable to serious human rights violations, there were obstacles to monitoring and reporting, and NGOs were actively discouraged (Human Rights Watch, 2007b; Mafiwasta, 2008).

There were about 2.8 million migrant workers in the UAE, some 90 per cent of the working population, and 20 per cent were employed in construction. Almost all construction workers were male, many of them illiterate and from impoverished rural communities in South Asia (Human Rights Watch, 2006). Some 5938 construction companies operated in Dubai alone and the majority were small companies employing fewer than 20 workers.

The recruitment of migrant workers was carried out either by a recruitment agent (who must be a UAE national) or by a UAE company and required a licence from the ministry of labour. Workers must have a work permit, supplied by the employer. Laws did exist to regulate maximum working hours, breaks, annual leave and overtime. Laws had also been introduced to ensure workers were not required to work in extreme summertime temperatures. Employers had to meet the costs of treatment for work-related injuries, and in the event of a work-related death, the members of a deceased worker's family were entitled to compensation. Workers were entitled to severance pay, and repatriation costs on completion of a contract. A worker could abandon his work if the employer failed to honour either contractual or legal obligations. At the time, there was substantial doubt about the implementation and enforcement of these rules (Human Rights Watch, 2007a; Mafiwasta, 2008).

The labour supplier held pools of available labour in construction camps and then hired them out on short or long contracts. Standards of accommodation were frequently poor including shelter, crowding, water supply, sanitation, food and recreational facilities. Large numbers of men were separated from their families for long periods of time

There were persistent, credible reports of abuses committed by employers, especially in small firms and against low-skilled workers. A main factor was said to be the immigration sponsorship laws that granted employers extraordinary control over the affairs of migrant workers. Reported abuses committed against migrant workers included non-payment of wages, extended working hours without overtime compensation, unsafe working environments resulting in death and injury, squalid living conditions in labour camps, and withholding of passports and travel documents (Human Rights Watch, 2006). For example, in 2005 some 800 workers, who were

Figure 10.2 *Cramped, crowded and hot: Labour camp interior, Dubai 1984*

Copyright Tropix.co.uk

Figure 10.3 *Labour camp exterior, Dubai 1984*

Copyright Tropix.co.uk

part of 6000 employed by a development and construction company in Abu Dhabi, protested against non-payment of wages for more than five months. The minister of labour ordered immediate payment. In 2006, some 1300 companies had permits suspended for late payment and the total sum owed to poor migrant workers was said to be US$52million.

In order to be recruited from their home country to work in UAE, workers often had to incur several years of indebtedness to recruitment agencies for fees that UAE law said only employers should pay. This is indentured labour and it is specifically prohibited, for example, by the IFC Performance Standards (IFC, 2006) and the ILO Conventions (ILO, 2009). Women domestic workers may often be confined to their places of work, at particular risk of abuse including unpaid wages, long working hours, and physical or sexual harassment (Human Rights Watch, 2007b). Observers have suggested that the economic boom in the UAE has been built and operated over several decades on the gross exploitation of labour that approximates slavery.

Since 2006, the UAE claimed to have enacted a number of reforms, in response to widespread international criticism (Embassy of the UAE in Washington, 2009). These included improving working and living conditions; combating human trafficking of children; establishing a labour court to handle workers' complaints; negotiating bilateral labour agreements with the governments of nations supplying large numbers of labourers to the UAE; regulating contracts of foreign labourers; setting fixed working hours for domestic employees; requiring the labour ministry to create a mechanism to prevent delays in wage payments; and introducing health insurance for all categories of workers. However, one campaigning NGO (Mafiwasta) claimed that UAE had rejected all recommendations relating to the introduction of trade unions, the right to strike, the right to freedom of association or the right to collective bargaining (2008). The UAE government has disputed Human Rights Watch findings (UAE Interact, 2007).

There have been regular waves of heat-related illness among workers during the summer months, when effective temperatures climb above 50°C (Human Rights Watch, 2006). For example, as many as 5000 construction workers per month were admitted to accident and emergency at Rashid hospital, during the summer months of 2004, suffering from heat-related symptoms (Anon, 2005). There appears to have been a common practice used over many decades in the Gulf to disguise heat-related occupational health issues. An afternoon break for workers has been directed by the minister, but a number of companies were said to have not complied. One small study found that 80 per cent of workers had been instructed to take a break, 64 per cent had only one break during the working day, 87 per cent had the break in an area without air conditioning or fans and 59 per cent had no water available on site (Barss, unpublished). More recently, compliance appeared to have increased but resting workers were frequently not provided with proper shelter or cooling facilities during their enforced two- to three-hour midday break (Egbert, 2007).

In the absence of a free press, and in the light of NGO comments and documented historical experience, the HIA analysis concluded that procurement and safety of labour required special attention. The labour conditions were reflected in the HIA recommendations: labour should only be procured from a reputable supplier; the reputation of the supplier should be independently verified; contractors should be screened for best practice; the screening should include working conditions, accommodation, health insurance, payment of workforce and indenture practices. In addition, standards set and applied by the main construction company should be propagated to all tiers of subcontractors and regularly monitored. A partnership with a reputable specialist NGO should be established to independently and continuously verify that best practice is implemented. I do not know whether or not these recommendations were put into practice.

10.8 MINING

The mining and minerals sector appears to have recognized the business case for HIA later than the O&G sector. For example, during 2009 the International Council on Mining and Metals (ICMM) commissioned an HIA guide (ICMM, 2010a, 2010b). The equivalent trade organization for the O&G sector, IPIECA, had published their guide in 2005. ICMM represents many of the world's leading mining and metals companies as well as regional, national and commodity associations. ICMM policy is a commitment to the responsible production of the minerals and metals that society needs. Its vision is for a respected mining and metals industry that is widely recognized as essential for society and as a key contributor to sustainable development. ICMM has also published good practice guidance on HIV/AIDS, TB and malaria (ICMM, 2008), occupational health risk assessment and much else. A number of mining proposals have commissioned HIAs independently of the ICMM guidance and there is a Canadian online course on the HIA of mining projects (Université Laval, 2009). Typical health impacts of mining include tuberculosis, dust-induced lung diseases, STIs and traumatic injury.

There are some scientific studies of the health impacts of existing mines. For example:

- Mines in Zaire (now Republic of Congo) created large bodies of standing water that were ideal transmission sites for malaria and schistosomiasis (van Ee and Polderman, 1984; Polderman et al, 1985).
- The Ok Tedi gold mine opened in a remote mountain region of Papua New Guinea during the early 1980s. The local community only had their first contact with the outside world in 1963, so they were vulnerable in many ways to the rapid changes that came about through the creation of a mining town in their midst. A series of papers by Lourie and colleagues monitored the effect of the mining project on local nutrition, heavy metal contamination, obstetrics and stress (Taufa et al, 1986; Ulijaszek et al, 1987; Hyndeman, 1988). Birth rate increased while infant mortality and birth spacing fell. Infant survival and anthropometric measurements improved for communities living closer to the mine. Social determinants of health were affected in complex ways. Suicides occurred and traffic injury became common.
- The health impacts of gold mining in South Africa under the apartheid regime received considerable attention (Moodie, 1989). Migrant males from neighbouring countries living in crowded, all-male institutions could acquire STIs and TB and take these diseases back to their home villages.

There are various reviews, for example:

- The Mining, Minerals and Sustainable Development (MMSD) programme commissioned reviews of worker and community health (Brehaut, 2001; Stephens and Ahern, 2001; IIED and WBCSD, 2002).
- Large numbers of NGOs provide information and comment on the Internet of the environmental, social and health impacts of mining, and scrutinize individual mining projects (for example, Mines and Communities, 2008).
- An earlier publication on the HIA of development projects provides additional examples (Birley, 1995).
- One review is entitled 'Resource impact – curse or blessing?' (Stevens, 2003).

The following two case studies illustrate HIA of mining proposals in two contrasting settings and using contrasting methods.

10.8.1 Case study from Wales, UK

An HIA in Wales examined the impact of a proposed extension to an opencast coal mine, at the request of local communities who were opposed to the proposal (Golby and Lester, 2005). Local people felt that health concerns had not been considered as part of the planning process. The proposal was located in one of the most deprived regions in Europe. There was a general need to improve the determinants of health in the indoor and outdoor physical environment, as well as social and community conditions. The method used in the HIA was based on focus group discussions with the peripheral community. The community was concerned about a wide range of physical diseases as well as stress, anxiety and depression. The community's experience of the existing opencast mine included nuisance dust; this affected the appearance of homes, ability to walk outdoors, cleaning, and sitting in gardens. A range of medical conditions were also attributed to the dust, and some studies of asthma rates in local schoolchildren supported an association with proximity of schools to the mine. The study also considered transport, plant on site, physical activity, social capital, severance, local economy, noise and vibration, safety, light pollution, loss of amenity, visual impact, property values, heritage and climate change. Impacts were framed within the context of local policies and standards towards human rights, equity and development objectives. The report argued that mine extension would produce impacts that were in direct contrast to those standards. For example, there was a policy to encourage schoolchildren to play outdoors, but the dust and particulates from the mine would discourage them from doing so.

The HIA report quoted a large number of comments from focus groups. The following is typical (quoted by permission):

> *When I heard* [about the proposed extensions] *I thought 'They can't do this to us again, they just can't'. I just felt completely hopelessly in despair. I just couldn't go through all that again* [campaigning].

The study concluded that likely negative impacts on health and well-being of an extension to the existing opencast mine would be far in excess of the positive health impacts. Plans for the extension were postponed. In 2007 the Welsh Government made HIA compulsory for all opencast coal proposals (WHIASU, undated).

10.8.2 Case study of a mining HIA from the Democratic Republic of the Congo

The second case study, by contrast, is the Moto gold mine in one of the lowest-income countries in the world (Winkler et al, 2010). The proposal was a major extension to an existing gold mine and would consist of both open pits and underground mines, a processing plant, power generation, local and international roads, water supply and treatment, workforce housing, management facilities and related services. Substantial resettlement of local villages would be required. In-migration was expected and the project would become a major employer during the construction and operational stages. The proponent intended to adhere to the performance standards of the international financial institutions. The published paper reports a rapid HIA of the project.

Baseline conditions in the local communities indicated that a number of communicable diseases were very common and the health system was extremely weak. There were little reliable government public health data. There were some 12,000 people in 2300 households located in 20 villages that might require resettlement. Some 65 per cent of the population were under 15 years old. Five different affected communities were identified.

Key informant interviews were carried out with three medical doctors and one community health representative. There were also focus group discussions with men and women from the local communities. The analysis was based on the 12 environmental health areas discussed in Chapter 3. The impacts were prioritized using the risk assessment matrix (RAM) discussed in Chapter 7.

The study concluded that 9 of the 12 environmental health areas would be aggravated by the project; many of the residual impacts would be significant; and 8 of the environmental health areas could be improved, relative to the baseline, as a result of additional mitigation measures. Table 10.5 is an extract from the published paper and summarizes part of the analysis. For example, there was some indication that HIV rates were relatively high in the community, although baseline data were poor. Commercial sex workers were common in the vicinity of the gold mines, condom usage rates were low and the number of patients seen in medical clinics with STIs was high and second only to malaria. The study concluded that the significance of STIs was very high and that even with mitigation it would still be high.

Many of the proposed mitigation measures represented social investments that were required to improve the health of the local community independently of the impacts of the project. Issues specific to the proposal included housing price inflation, overcrowding and transport-related injuries. The reported study was regarded as a rapid HIA and one of the recommendations was that a comprehensive HIA should be conducted as part of the revised feasibility studies.

10.9 SOME CHARACTERISTICS OF MINING PROPOSALS

Mining proposals normally require large amounts of labour for both construction and operation. They tend to be surrounded by artisanal and small-scale mines (ASMs) because mineral-rich rocks lie close to the surface. ASMs are characterized by very poor health and safety conditions, environmental pollution and degradation, abject poverty and exploitation, child labour, gender inequality and dangerous technology (D'Souza, 2009). There are often uneasy relationships between the formal mining sector and the ASM sector leading to physical conflict. The ASM sector is growing and there may be 20–30 million ASM workforce worldwide and 80–120 million dependents. They may be responsible for 10–15 per cent of the global mineral produce. Those conducting HIAs of mining proposals need to understand ASM, analyse its impacts and understand the options for management.

The mining and O&G sectors have much in common. However, each has evolved within its own culture and so there are also differences. One difference I have observed is the approach towards social investment. There may be a more widespread assumption in the mining sector that social investment should be a normal part of proposal development. This is probably because of a greater reliance on local labour.

For example, a very large mining project is under development in Mongolia at the time of writing (Ivanhoe Mines, 2005). The proponent issued a call for consultancy services on the design of a community health, safety and security programme (Oyu Tolgoi LCC, 2009). Numerous other studies had already been completed on water, housing, macro-economic benefit, influx and socio-economic baseline. Many of

Table 10.5 *Reproduction of part of summary table for the Moto gold mine project*

Environmental health areas	Significance		
	Baseline	Without mitigation, aggravation compared to baseline	Residual, improvement compared to baseline
Communicable diseases	xxx	xxxx	xx
Malaria	xxxx	xxxx	xx
Arboviruses	xxx	xxx	xx
Soil-, water- and waste-related diseases	xxx	xxxx	xx
STIs including HIV/AIDS	xxxx	xxxx	xxx
Zoonotic viral haemorrhagic fever	xxx	xxx	x
Malnutrition	xx	xxx	xx
Non-communicable diseases	xxx	xxxx	xxx
Accident/injuries	x	xxxx	xx
Exposure to potentially hazardous materials, noise and malodours	xxx	xxx	x
Social determinants of health, lifestyle	xx	xxx	xx
Cultural health practices	xxx	xxx	+
Health system issues			
Infrastructure and capacity	xxx	xxxx	+
Maternal health	xx	xx	x
Child care	xxx	xxx	xx
Programme management and delivery systems	xx	xxx	+

Significance: x low, xx medium, xxx high, xxxx very high, + potential for positive effect
Reprinted from Winkler et al, 2010 with permission from publisher

these studies had identified health issues. These included communicable and non-communicable diseases, STIs and HIV/AIDS, nutritional change or deficiencies, dust and environmental health hazards, inadequate hygiene and sanitation, alcohol and drug abuse, domestic violence, human trafficking, and urban, industrial and occupational health and safety. They also identified significant opportunities for awareness raising, education, prevention, and improvement of individual and institutional capacities. A plan was formulated to develop and implement an effective health, safety and security programme. The programme would function to minimize health threats to both the project and the surrounding communities, while maximizing health gains. The intention was to integrate the programme with other public sector, non-government and donor-funded projects. Participatory methods and stakeholder engagement were emphasized. The objectives included the following:

● Minimize and mitigate, to the greatest extent possible, health, safety and security impacts directly and indirectly stimulated by company operations.

- Measurably improve health, safety and security service delivery, capacity and indicators in the target area.
- Complement and build upon existing company policies, standards, knowledge and best practices pertaining to employees and contractors, and extend such standards to local stakeholders where appropriate and feasible.

Associated with this, there appears to be a blurring of the distinction between the health impacts of the proposal itself and the health benefits of the social investment. This is anecdotal and based on a limited number of discussions that took place during the preparation of the ICMM guidance, together with examples such as a diamond mine in northern Canada (Kwiatkowski and Ooi, 2003). Staff of corporations in which social investment is confused with project impact may become confused by the HIA process. This could be avoided by making a clear distinction between the health impacts of the proposal, the existing health needs of the community and the health impacts of any proposed social investment programmes. In addition, the differential impacts must be kept in mind. For example, employment opportunities may benefit men while disbenefits may fall more on women.

10.10 REFERENCES

Achbar, M. and J. Abbott (2004) 'The Corporation', Zeitgeist Films, Canada: 145 minutes, http://en.wikipedia.org/wiki/the_corporation

Anon (2005) 'Many victims of heatstroke are not being accurately diagnosed by A&E hospital staff', *Construction Week*, vol 83 (online journal) www.constructionweekonline.com

Barron, T., M. Orenstein and A.-L. Tamburrini (2010) *Health Effects Assessment Tool (HEAT): An Innovative Guide for HIA in Resource Development Projects*, Habitat Health Impact Consulting & Environmental Resources Management, http://apho.org.uk/resource/view.aspx?RID=83805

Barss, P. (unpublished) 'Knowledge, attitude and practice of construction workers toward prevention of heat related illnesses in al Ain district, student project No 211 (2005)'

Birley, M. H. (1995) *The Health Impact Assessment of Development Projects*, HMSO, London, www.birleyhia.co.uk, accessed February 2011

Birley, M. (2003) 'Health impact assessment, integration and critical appraisal', *Impact Assessment and Project Appraisal*, vol 21, no 4, pp313–321

Birley, M. (2005) 'Health impact assessment in multinationals: A case study of the Royal Dutch/Shell Group', *Environmental Impact Assessment Review*, vol 25, no 7–8, pp702–713, www.sciencedirect.com/science/article/b6v9g-4gvgt8v-1/2/01966b5af4f9ae9ecd390e4dd382a5a3

Birley, M. (2007) 'A fault analysis for health impact assessment: Procurement, competence, expectations, and jurisdictions', *Impact Assessment and Project Appraisal*, vol 25, no 4, pp281–289, www.ingentaconnect.com/content/beech/iapa

Brehaut, H. (2001) *The Community Health Dimension of Sustainable Development in Developing Countries*, IIED, London, www.iied.org/sustainable-markets/key-issues/business-and-sustainable-development/mmsd-working-papers, accessed 2010

D'Souza, K. P. (2009) 'Artisanal and small-scale mining, the poor relation', Environmental and Social Responsibility on Mining. European Bank for Reconstruction and Development, London, www.ebrd.com accessed 2010, now removed

Egbert, C. (2007) 'Is it really a break?', *Construction Week*, www.constructionweekonline.com/article-1094-is_it_really_a_break

Embassy of the UAE in Washington (2009) 'Initiatives to combat human trafficking', www.uae-embassy.org/uae/human-rights/human-trafficking?id=63, accessed September 2009

Equator Principles (2006) 'The Equator Principles', www.equator-principles.com, accessed October 2009

Extractive Industries Transparency Initiative (undated) www.eiti.org, accessed January 2010

Golby, A. and C. Lester (2005) *Health Impact Assessment of the Proposed Extension to Margam Opencast Mine*, Welsh Health Impact Assessment Support Unit and National Public Health Service for Wales on behalf of the Margam Opencast and Health Steering Group, Cardiff, www.wales.nhs.uk/sites3/Documents/522/Kenfig%20Hill%20Final%20-%20Dec%2005.pdf

Goodland, R (ed.) (2005) *Oil and Gas Pipelines, Social and Environmental Impact Assessment, State of the Art*, McLean, Virginia, www.goodlandrobert.com/PipelinesBK.pdf

Goodland, R. and C. Wicks (2009) *Philippines: Mining or Food?* Working Group on Mining in the Philippines, London, www.piplinks.org/system/files/Mining+or+Food+Abbreviated.pdf

Heinberg, R. (2007) *Peak Everything*, New Society Publishers, Vancouver, http://richardheinberg.com/

Human Rights Watch (2006) *Building Towers, Cheating Workers: Exploitation of Migrant Construction Workers in the United Arab Emirates*, E1808, www.hrw.org/reports/2006/uae1106, accessed November 2007

Human Rights Watch (2007a) *Exported and Exposed: Abuses against Sri Lankan Domestic Workers in Saudi Arabia, Kuwait, Lebanon, and the United Arab Emirates*', www.hrw.org/reports/2007/srilanka1107/index.htm, accessed November 2007

Human Rights Watch (2007b) *United Arab Emirates (UAE)*, http://hrw.org/english/docs/2006/01/18/uae12233.htm, accessed November 2007

Hyndeman, D. (1988) 'Ok Tedi: New Guinea's disaster mine', *The Ecologist*, vol 18, no 1, pp24–29

ICMM (2008) *Good Practice Guidance on HIV/AIDS, Tuberculosis and Malaria*, International Council on Mining and Metals, London, www.icmm.com

ICMM (2010a) *Good Practice Guidance on Health Impact Assessment*, International Council on Mining and Metals, London, www.icmm.com/document/792

ICMM (2010b) International Council on Mining and Metals, www.icmm.com, accessed January 2010

IFC (International Finance Corporation) (2006) *Policy and Performance Standards on Social & Environmental Sustainability*, www.ifc.org/ifcext/enviro.nsf/Content/EnvSocStandards, accessed April 2008

IIED and WBCSD (International Institute for Environment and Development and World Business Council for Sustainable Development) (2002) 'Breaking new ground: Mining, minerals and sustainable development', www.iied.org/sustainable-markets/key-issues/business-and-sustainable-development/mmsd-introduction, accessed December 2009

ILO (International Labour Organization) (2009) 'ILOLEX, database on international labour standards', www.ilo.org, accessed February 2011

IPIECA/OGP (2005a) 'A guide to health impact assessments in the oil and gas industry', International Petroleum Industry Environmental Conservation Association, International Association of Oil and Gas Producers, London, www.ipieca.org

IPIECA/OGP (2005b) 'HIV/AIDS management in the oil and gas industry', IPIECA, London, www.ipieca.org/system/files/publications/hiv.pdf, accessed February 2011

Ivanhoe Mines (2005) 'Oyu Tolgoi Project, Mongolia Integrated Development Plan', www.ivanhoemines.com, accessed February 2011

Kwiatkowski, R. E. and M. Ooi (2003) 'Integrated environmental impact assessment: A Canadian example', *Bulletin of the World Health Organization*, vol 81, pp434–438, www.scielosp.org/scielo.php?script=sci_arttext&pid=S0042-96862003000600013&nrm=iso

Mafiwasta (2008) 'For worker's rights in the United Arab Emirates', www.mafiwasta.com, accessed September 2009

Mines and Communities (2008) 'MAC: Mines and Communities', www.minesandcommunities. org, accessed October 2010

Moodie, T. D. (1989) 'Migrancy and male sexuality in South African gold mines', *Journal of South African Studies*, vol 14, no 2, pp228–256, www.jstor.org/pss/2636630, accessed February 2011

Noble, B. and J. Bronson (2005) 'Integrating human health into environmental impact assessment: Case studies of Canada's northern mining resource sector', *Arctic*, vol 58, no 4, pp395–405, http://pubs.aina.ucalgary.ca/arctic/Arctic58-4-395.pdf

Oyu Tolgoi LCC (2009) 'Oyu Tolgoi – Community Health, Safety & Security Program Design Consultancy Solicitation of Proposals', www.ivanhoemines.com (now removed)

Polderman, A. M., K. Mpamila, J. P. Manshande, B. Gryseels and O. van Schayk (1985) 'Historical, geological and ecological aspects of transmission of intestinal schistosomiasis in Maniema, Kivu Province, Zaire', *Annales de la Societe Belge de Medecine Tropicale*, vol 65, no 3, pp251–261

San Sebastián, M. and A. Hurtig (2005) 'Oil exploitation and health in the Amazon basin of Ecuador: The popular epidemiology process', *Social Science & Medicine*, vol 60, no 4, pp799–807

Shell International (2001) 'Minimum health management standards', Shell International, Den Haag

Stephens, C. and M. Ahern (2001) 'Worker and community health, impacts related to mining operations internationally: A rapid review of the literature', IIED, London, www.iied.org/sustainable-markets/key-issues/business-and-sustainable-development/mmsd-working-papers, accessed 2010

Stevens, P. (2003) 'Resource impact – curse or blessing? A literature survey', Centre for Energy, Petroleum and Mineral Law and Policy, University of Dundee, and IPIECA, Dundee

Taufa, T., J. Lourie, V. Mea, A. Sinha, J. Cattani and W. Anderson (1986) 'Some obstetrical aspects of the rapidly changing Wopkaimin society', *Papua New Guinea Medical Journal*, vol 29, pp301–307

UAE Interact (2007) 'Statement from the UAE Government on the Human Rights Watch Domestic Labour Report', http://uaeinteract.com/docs/Statement_from_the_UAE_Government_on_the_Human_Rights_Watch_Domestic_Labour_Report/27589.htm, accessed September 2009

Ulijaszek, S., D. Hyndman, J. Lourie and A. Pumuye. (1987) 'Mining, modernisation and dietary change among the Wopkaimin of Papua New Guinea', *Ecology of Food and Nutrition*, vol 20, pp143–156

Université Laval (2009) Health Impact Assessment (HIA) of Mining Projects, online course, http://132.203.105.207/eis/index.php?id=10&L=1, accessed January 2010

van Ee, J. H. and A. M. Polderman (1984) 'Physiological performance and work capacity of tin mine labourers infested with schistosomiasis in Zaire', *Tropical Geographical Medicine*, vol 36, no 3, pp259–266

WHIASU (undated) Wales Health Impact Assessment Support Unit, www.wales.nhs.uk/sites3/home.cfm?OrgID=522, accessed May 2010

Winkler, M. S., M. J. Divall, G. R. Krieger, M. Z. Balge, B. H. Singer and J. Utzinger (2010) 'Assessing health impacts in complex eco-epidemiological settings in the humid tropics: Advancing tools and methods', *Environmental Impact Assessment Review*, vol 30, no 1, pp52–61, www.sciencedirect.com/science/article/B6V9G-4WFPPC7-1/2/a9621176a138c680fd2e77e594b806d1

World Bank (2004) *Extractive Industries Review*, www.ifc.org/eir, accessed January 2010

Yergin, D. (1991) *The Prize: The Epic Quest for Oil, Money and Power*, Free Press, New York

CHAPTER 11

Housing and spatial planning

- Some initiatives for promoting a healthy urban environment are described.
- The spatial planning system in England, at the time of writing, is summarized.
- Examples are provided of HIAs in England for mixed residential development and social housing refurbishment.

11.1 INTRODUCTION

Housing and spatial planning are vital components of a healthy environment in both urban and rural areas. As approximately 50 per cent of the world population now live in cities there is an urgent need to ensure that urban environments are developed in a healthy manner. There are a number of initiatives to address this issue at both global and local levels. Examples include the WHO Healthy Cities movement and the healthy development measurement tool used in San Francisco. After summarizing some elements of these, the majority of this chapter focuses on housing and spatial planning in England.

11.1.1 WHO Healthy Cities

The WHO Healthy Cities programme is currently in its fifth five-year phase (Ashton, 1992; WHO Regional Office for Europe, 2009a). In the European region alone, over 1200 cities and towns were participating, from over 30 countries. These cities and towns were committed to health and sustainable development and their designation is reviewed every five years. Cities were expected to show explicit political commitment, leadership, institutional change and intersectoral partnerships.

The Healthy Cities movement includes a set of principles and values. These are equity, participation and empowerment; working in partnership; solidarity and friendship; and sustainable development. According to the Zagreb declaration (WHO Regional Office for Europe, 2009b): a healthy city is a city for all its citizens: inclusive, supportive, sensitive and responsive to their diverse needs and expectations; a healthy city provides conditions and opportunities that encourage, enable and support healthy lifestyles for people of all social groups and ages; a healthy city offers a physical and built environment that encourages, enables and supports health, recreation and well-being, safety, social interaction, accessibility and mobility, a sense of pride and cultural identity; and is responsive to the needs of all its citizens.

HIA was one of the core themes of Phase 4 (2003–2008) of the WHO Healthy Cities network. The aim was to integrate HIA into decision-making for all plans. Consequently, the WHO Healthy Cities and Urban Governance programme included a project on 'promoting and supporting integrated approaches for health and sustainable development at the local level' (PHASE). The project developed an HIA toolkit for European cities and towns (WHO Regional Office for Europe, 2005, 2009a). The toolkit included an online HIA training module that was based on the training course developed at IMPACT in the University of Liverpool.

The aim of Phase 5 (2009–2013) was health and health equity in all local policies. This built on the recommendations of the Commission on Social Determinants of Health (CSDH, 2008). Core themes included caring and supportive environments, healthy living and healthy urban design.

11.1.2 Healthy development measurement tool

The Department of Public Health in San Francisco created a healthy development measurement tool (San Francisco DPH, 2006). The tool assisted with the inclusion of health needs in urban development plans. The tool was developed by the programme on health equity and sustainability in the Department of Public Health. The programme was committed to assessing urban environmental conditions and responding to health inequities and environmental policy gaps using HIA. There has been increasing interest in using the tool in other cities and in other states in the US. A review of HIA in the US suggested that many of the existing examples had been undertaken in this context (Dannenberg et al, 2008).

The tool consisted of three core components. There was a community health indicator system that included over 100 indicators of social, environmental and economic conditions. This could be used to evaluate community health profiles and baseline conditions. There was a health development checklist that could be used to assess whether urban plans achieved community health objectives. There was also a list of potential actions that could be taken by decision-makers to achieve the health development targets in the checklist.

The tool was based on six elements that comprised a healthy city: environmental stewardship, sustainable and safe transportation, public infrastructure, social cohesion, adequate and healthy housing, and a healthy economy. Each of these elements was organized into a set of community health objectives. Each objective included a set of resources such as community health indicators, healthy development targets, policies and design strategies, and health-based rationales. There was a comprehensive geographical information system available to support the tool.

The tool had been used to assist local area plans under development by the San Francisco planning department. These are described on the website. In one example of the application of this tool, a residential development project was evaluated in order to determine the proportion of households that would be within half a mile of a full-service grocery store. The development plan did not include any such specification and the one existing grocery store was located in a relatively inaccessible location.

The assessment therefore concluded that the new development would have a negative health impact. It recommended that financial and political support should be obtained for a suitable grocery store. It further recommended that if the store was located off site, then transportation plans should be developed to enable direct and easy access by pedestrians, bicycles and public transport.

Other tools for helping American planners incorporate health concerns have been described by Forsyth et al (2010).

11.1.3 England

The housing and spatial planning sector in England provides an example that is in marked contrast to the extractive industries and water development sectors, discussed in Chapters 9 and 10. The setting is wholly a representative democracy in a post-industrialized economy. There is a huge body of planning law and an ocean of policy. In this setting, policy analysis is a significant component of an HIA. As a result of British devolution, there are differences in approach between England, Wales, Scotland and Northern Ireland. These differences are outside the scope of this chapter.

In England there are both statutory and non-statutory requirements, guidance and influential, government-funded, reports and reviews. White Papers, issued by government, lay out policy, or propose action, on topics of current concern. Green Papers are issued as consultation documents to propose strategies that can be implemented in other legislation. National policy has been translated into regional and local policies, strategies and frameworks. At the time of writing, much of this was subject to change.

English policy development has applied to housing, spatial planning, energy, transport, open spaces, the built environment, sustainability, public health, social services and much else. There have been a large number of policy planning guidance documents and many have made implicit reference to the determinants of health.

In the public health domain there are detailed data down to almost neighbourhood level that includes many health indicators as well as the physical, social and economic environment. The data have enabled detailed comparisons to be made of both physical and mental health indicators between areas and between socio-economic groups.

There has been a debate in England about whether HIA was an effective tool for safeguarding and promoting health within the spatial planning context. This chapter seeks to illustrate this debate both in general and through specific case studies.

11.2 SPATIAL PLANNING IN ENGLAND

Planning regulates the use of land in urban and rural areas and thus has great potential to influence health (Fredsgaard et al, 2009). The traditional focus is on ensuring that aspects of the physical environment such as air, water and noise do not harm health. There are increasing calls for planning decisions to contribute to the wider public health issues; for example reducing obesity, improving mental health and well-being, and addressing climate change (Butland et al, 2007; Royal Commission on Environmental

Pollution, 2007). There is a long history of association between planning and health in the UK (Cullingworth and Nadin, 2006). During the 19th century, public health legislation was directed at the creation of adequate sanitary conditions, as discussed in Chapter 3. This included the control of street widths and the design of buildings. Later, the association became more tenuous as town and country planning developed into a discipline in its own right. At present, there is increased interest in explicitly including health issues in spatial planning once again. For example, the Royal Town Planning Institute (RTPI) published guidelines on delivering healthy communities (RTPI, 2009).

A series of government policy papers between 1999 and 2008 set the basis for collaboration between local authorities and the health sector in England. The policy papers drew on a series of key research reports including the Acheson report into health inequalities (Acheson et al, 1998), the Wanless report on priorities for public health (Wanless et al, 2002), and the Royal Commission on Environmental Pollution in the Urban Environment (Royal Commission on Environmental Pollution, 2007). These influenced a series of planning policy statements (PPSs) (Department for Communities and Local Government, 2009). PPSs explained statutory provisions and provided guidance to local authorities on planning policy and the operation of the planning system. They also explained the relationship between planning policies and other policies that had an important bearing on issues of development and land use. Local authorities had to take their contents into account in preparing their development proposals. Planning policy is slightly different in the devolved countries of Scotland and Wales and this is summarized in the RTPI report (2009). In Wales, there has been substantial integration of HIA and planning, especially in relation to waste management, transport and local development plans (WHIASU, undated). There was no PPS that explicitly addressed planning and health and none was planned at the time of writing. However, the National Health Service had become a statutory consultee of local authority spatial plans, if not of individual planning permissions.

A new evidence-based strategy for reducing health inequalities in England was proposed in 2010 that had many links to spatial planning (Marmot, 2010). It identified six policy objectives. One of these was to create and develop healthy and sustainable places and communities. The strategy stated that physical and social characteristics of communities, and the degree to which they enable and promote healthy behaviours, all make a contribution to social inequalities in health. Their recommendations for reducing health inequalities included active travel, public transport, energy-efficient houses, availability of green space, healthy eating and reduced carbon-based pollution. The report concluded that community capital should be improved across the social gradient.

Priority recommendations were to improve:

- active travel;
- availability of good quality open and green spaces;
- the food environment in local areas;
- energy efficiency of housing.

The report also recommended:

- full integration of the planning, transport, housing, environmental and health systems to address the social determinants of health in each locality;
- support for locally developed and evidence-based community regeneration programmes that removed barriers to community participation and action, and reduced social isolation.

There have been a number of proposals that may lead to a specific legal requirement for HIA. For example, the 2009 Select Committee on Health Inequalities (House of Commons, 2009) stated the following:

> *340. The built environment has a crucial impact on health and on health inequalities and affects every aspect of our lives. We are concerned that it does not encourage good health.*

> *342. In our view, health must be a primary consideration in every planning decision that is taken, and to ensure that this happens, we recommend that*

> - *in collaboration with the Department of Health, Department of Local Government should publish a Planning Policy Statement on health; this Statement should require the planning system to create a built environment that encourages a healthy lifestyle, including giving local authorities the powers to control the numbers of fast food outlets.*
> - *Primary Care Trusts[1] (PCTs) should be made statutory consultees for local planning decisions; PCTs, for their part, need to ensure they have the knowledge of cost effectiveness of alternative policies and resources to make an informed contribution to such decisions.*

11.2.1 Local development frameworks and HIA

The UK government required each local authority to produce a local development framework (LDF) (UK Government Planning Portal, 2010). This consisted of a folder of documents that outlined spatial planning strategy for the local area. Community involvement and consultation was essential. The LDF was subject to a sustainability appraisal, which incorporated the requirements of the SEA Directive, to ensure economic, environmental and social effects of any new proposals were in line with sustainable development targets. The documents included a core strategy that set out the general spatial vision and objectives.

In some cases, the inclusion of HIA in LDFs was covered through the proper integration of health in sustainability appraisal/SEA. Some local authorities had explicit requirements. Brighton and Hove, for example, established a Healthy Cities objective as part of their core strategy. This stated that the local planning authority would support programmes and strategies that aimed to reduce health inequalities and promote healthier lifestyles through the following (Brighton and Hove City Council, 2009):

- carrying out HIA on all planning policy documents;
- requiring HIA on all strategic developments in the city;
- requiring developers to demonstrate how they maximize health benefits;
- ensuring that any development requiring an EIA should incorporate an HIA.

The inclusion of HIA built on earlier work in the local authority that enabled them to join the WHO Healthy Cities movement in 2004 (WHO Regional Office for Europe, 2009a).

Plymouth was another example of a city that had adopted HIA. In this case, it had been made a component of health care provision: to improve the health of the city by requiring all major development proposals to be subject to HIA (Andrew Platt, pers. comm.).

Following a change of government in 2010, major changes were under way in the policy and planning system that makes some of the previous statements historic. At the time of writing, it was not clear what those changes would be.

11.3 CASE STUDY: MIXED RESIDENTIAL DEVELOPMENT

The following case study is based on an unpublished HIA that was prepared in collaboration with Ben Cave Associates for a private developer. For reasons of confidentiality, full details of the specific project have been removed.

Because of the acute housing shortage in England, some regions were designated priority areas for large-scale house building projects. Private developers were encouraged to put forward proposals for building new communities. The proposals had to receive outline planning permission before they could proceed. Outline planning permission was granted subject to review by a number of statutory consultees. At the time of writing, the health authority, represented by the local primary care trust (PCT), was not a statutory consultee. However, in some areas and for some proposals the PCT was included. This was because of interest and influence by key individuals within the PCT or strategic health authority, or because the development had direct implications for the delivery of health care by the PCT. In this case study, the PCT was included for both these reasons. The project was located in the Milton Keynes and South Midlands subregion, which already had experience of developing healthy sustainable communities (Cave and Molyneux, 2004; Cave et al, 2004). Consequently, the private developer commissioned an HIA report for inclusion in the application for outline planning permission.

The project was located on a greenfield site at the edge of a small market town. The project would lead to an increase in the population of the town of some 75 per cent and this would require additional health care provision. The developer would be required to allocate land and contribute to the construction of a new secondary school, a new health centre and a bypass around the town.

The project included land for 3000 new homes of mixed type and tenure, light industrial use, a local complex comprising a shopping mall and health centre, a neighbourhood centre with local convenience retail, hotel and conference venue, primary

Table 11.1 *Documents submitted in the outline planning application*

Planning statement	Support planning statement
Design and access statement	Regeneration assessment
Travel assessment and green travel plan	Air-quality assessment
Contaminated land investigation report	Sustainability appraisal
Refuse storage plan	Energy impact statement
Flood risk assessment	Nature conservation and ecological assessment
Environmental impact assessment in three volumes	Historical and archaeological assessment
	Listed building and conservation area appraisal
Retail impact assessment	Sewage assessment
Secured by design plans	Utilities statement
Noise report	Affordable housing statement
Disabled access report	Planning obligations draft statement
Travel plan	
Tree survey	

school, secondary school, extensive areas of public open space and strategic landscaping, surface water and flood management works, drains and services, and a bypass.

The outline planning application required the submission of a suite of documents, in addition to the HIA (see Table 11.1). None of these documents was available for review during the preparation of the HIA, although there are many areas of overlap.

Figure 11.1 indicates the general layout. Two major roads intersected at the northern end of the town and caused traffic congestion. The proposed bypass, to the south, was very welcome to the community. Many of the existing and future inhabitants would need to commute to work in larger towns and cities nearby. There was no rail link and the existing bus service was poor. There was no bus station but simply a series of roadside stops. The majority of the community in and around the town enjoyed above average wealth and health. There was a low unemployment rate and few pockets of deprivation. A prestigious school in the old town attracted pupils from a wide area. Car ownership rates were high and there was little existing provision for walking or cycling.

In addition to the HIA, we were asked to produce a proposal for a health centre and this was conceived as a health and wellness centre that would incorporate a range of services under one roof.

When preparing the HIA, we needed to consider justification for any recommendations that would not suit the developer; and to seek out issues that the PCT would expect to see addressed when reviewing the proposal.

The PCT was likely to use a checklist entitled *Building in Health: A Checklist and Guide to Developing Healthy Sustainable Communities* (Ballantyne, 2006). This provided a series of questions for a local authority to ask when reviewing a planning proposal. The checklist has 34 questions in 7 major categories:

- governance;
- social and cultural;
- environment;
- housing;
- transport;

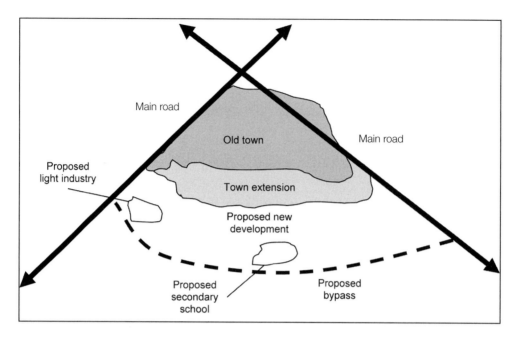

Figure 11.1 *Proposal for a mixed residential development on the edge of a small market town*

- economy;
- services.

Typical questions follow:

- Are large developments (2000+ dwellings) planned as balanced communities with a range of housing types and tenures? Developments should not have the effect of segregating areas or excluding certain groups.
- Are well-designed places available where people and voluntary groups can gather, for example shared places of worship, community centres, sports facilities, community spaces? Is there community involvement in the design and management of such places?
- Are developments designed to minimize opportunities for crime, and maximize opportunities for community control and defence of the local area? Is community involvement an integral part of this approach?

However, the project plans available to us as HIA consultants were at an early stage and did not contain enough detail to enable such questions to be answered. In preparing the HIA, an unorthodox approach was tried. The developer was provided with a set of healthy design principles and these were individually justified by reference to government regulations and planning guidance. A typical healthy design principle is illustrated in Table 11.2.

As the principles were of general interest we subsequently extracted them from the report, edited and published them on the Internet (Birley and Birley, 2007).

Table 11.2 *Example of a healthy design principle*

Category	Governance
Subcategory	Public involvement
Relevance	A high level of public engagement is associated with improved population health status and reduced demand for health care resources
Healthy design principle	Ensure proposals are developed with the active involvement of all those likely to be affected
Policy guidance justification	*Planning Policy Statement 1: Delivering Sustainable Development* (Department for Communities and Local Government, 2004)
Compliance	At outline planning stage
Cross-reference/ evidence	Statement of community involvement

We identified a series of compliance stages at which the healthy design principles should be applied: outline proposal; 'reserved matters'; implementation by the developer; and future.

The outline planning stage is the point where the local authority accepts in principle the outline plan, or master plan, submitted by the developer in preference to any competitors. The outline plan contains very little detail but should address issues that are of concern to the planning authority and its consultees. The impact assessment report was prepared as part of the documentation for this submission.

The reserved matters stage is when detailed plans are drawn, after submission and acceptance of the general concept in the outline plan. Ideally, a second iteration of the HIA would take place during the preparation of the detailed plan. In practice, no second iteration was scheduled so the HIA report provided a reference document for this and subsequent stages.

When all plans are agreed, construction or implementation can start. The original developer may still own the project and be responsible for construction. Alternatively, it may have sold the project to a new developer. In either case, there are health issues that should be addressed during construction. Examples include construction traffic management and dust control.

The fourth stage, future, takes place when construction is complete, or nearly complete. Compliance would be the responsibility of a third party such as a new developer or the local authority. At this stage, residents will have bought the new homes and occupied them and the developer will have partly handed control of the project site back to the local authority. Health issues that will require management include waste collection and 'snagging'. Snagging refers to any deficiencies in the quality of homes that are identified by the occupants and which require correction by the developer.

11.3.1 Active travel example

Active travel is an example of one of the health issues addressed in the HIA. It refers to walking, cycling and the use of public transport. Transport influences a range of health determinants including physical exercise, body mass, social isolation, fear of crime, climate change and air pollution (UNECE, undated). One of the healthy design principles was to maximize active travel.

There was a regional requirement to promote public transport with a target of a 20 per cent modal shift away from single occupancy private car commuting. There were several opportunities for promoting the modal shift, including:

- the provision of employment opportunities close to the new housing;
- enhanced high-speed bus routes through the new development to the main employment centres and railway stations in towns within a 10–20 kilometre radius.

At the time of the HIA, the primary focus of planning discussions and transport modelling within the project team seemed to be adding provision for private car transport, for instance design of intersections and bypass width. No information about modal shift was available and we did not have access to the draft travel plans. The sketches of residential street layouts seemed to assume that roadways would be needed for all journeys, however small. There had been minimal focus on the provision of foot bridges or underpasses where the bypass intersected public footpaths and bridleways. Little attention seemed to have been paid to footpaths or cycleways to the shopping centres, health centre and schools. There was no plan for a central bus station, no clarity about the effectiveness of new bus links and no evidence of interest in public transport.

The outcome of this HIA cannot be evaluated as the global financial crisis halted demand for new housing and the development proposal was not implemented. However, we were provided with a complete set of the submission documents after our own report was finalized and our formal involvement with the project had ended. This included both employment and residential travel plans that contained substantial detail of proposed active travel provision and cited the associated health gains, but without cross-reference to the HIA. The detail included:

- a survey of existing footpaths and cycleways that identified where they were poorly maintained and lit, and lacked dropped kerbs and road crossings;
- existing bus routes, frequencies, travel times and destinations;
- proposed new foot and cycleways;
- plans for high-speed bus links with intermodal transfer at mainline railway stations;
- appointing a travel plan coordinator;
- plans for car-share schemes;
- travel welcome packs for new residents, which would include maps and safe routes to schools and other key destinations, and vouchers for a year's free public transport;
- secure bicycle parking;
- recommendation to publicize the health benefits of cycling and walking;
- broadband access and travel-plan website;
- publicizing public transport availability;
- ensuring bus stops were to be within easy access distances from all homes.

Figure 11.2 *Regeneration of high-rise apartments*

Copyright Tropix.co.uk

A design and access statement had been prepared as one of the submission documents in accordance with government guidelines. The purpose of this statement was to explain 'the design principles and concepts that have been applied to the proposed development and how issues relating to access to the development have been dealt with'. The statement referred to principles of healthy living and the *Manual for Streets*, which establishes a hierarchy in which pedestrians receive priority (Department for Transport, 2007). This statement provided design principles for the project and summarized the conclusions of many of the reports. It made no reference to the IIIA.

One of the learning points for the HIA team was that we had not developed a close enough working relationship with the transport planning team and others during the course of our assessment. We had not been able to cite their work and they had not cited ours. There was a lack of integration and understanding on both sides.

11.4 CASE STUDY: SOCIAL HOUSING REFURBISHMENT

The following case study is based on an HIA of a housing regeneration project that was undertaken in collaboration with IMPACT (Birley and Pennington, 2009).

In England there has been a transfer of council-owned housing stock to registered social landlords (RSLs) and there is evidence that it has been successful in meeting tenants' needs (Pawson et al, 2009). Many of the homes had been poorly maintained as a result of council budget constraints and operating procedures. Substantial refurbishment was necessary.

Following the Housing Act (HM Government, 2004), the government set a standard for Decent Homes (Department for Communities and Local Government, 2006) and introduced a Housing Health and Safety Rating System (HHSRS) which identified 29 physical hazards in the home and quantified the risk (Office of the Deputy Prime Minister, 2003, 2006a, 2006b; HM Government, 2005). The most serious hazards had to be rectified.

Improving social housing was part of government policy for promoting health and tackling health inequalities (Department of Health, 2003), planning for an ageing population (Department for Communities and Local Government, 2008) and planning for climate change (Department for Environment, Food and Rural Affairs, 2008).

11.4.1 Summary of the proposal

The client (an RSL) had assumed responsibility for some 15,000 housing units in a very deprived area. The residents had agreed, by ballot, that the change of ownership could take place and inducements had included a substantial government grant to be invested in housing refurbishment. The RSL had agreed to include resident representatives on the management board and undertake neighbourhood regeneration. They had also agreed to undertake an HIA of their regeneration proposal. They were committed to:

- carry out a baseline HIA in liaison with local health agencies;
- ensure that the health of the community was known when shaping service delivery;
- target resources to maximize health impacts.

The home refurbishments were planned to take place over five years and included windows, doors, kitchens, bathrooms, insulation, heating, wiring, piping and ventilation. The work on each home was undertaken in phases so that windows and doors would be renovated before kitchens and bathrooms, or heating and insulation. Each phase could last 1–15 days and the phases would be separated by several months. The work was ongoing and, consequently, the HIA was concurrent. Residents would remain in their homes during the work, in most cases.

All refurbishment work was to be undertaken by contractors and subcontractors. The contractors had their own community liaison teams, responsible for informing tenants about proposed work and ensuring that any special needs of tenants were identified and managed.

There were assumed to be sufficient funds to ensure that all necessary improvements would be made to all properties during the programme. The refurbishments were intended to exceed the Decent Homes standard. The refurbishment would include an affordable warmth and energy efficiency strategy, and this would contribute to the reduction of fuel poverty. For example, there was an intention to double the energy efficiency of the homes by installing modern boilers, insulation, double glazing, temperature controls and showers.

11.4.2 The association between housing regeneration and health

There have been a number of reviews of the general association between housing and health (Taske et al, 2005; Care Services Improvement Partnership, 2006). Evidence varies from robust to anecdotal and there are complex interactions between determinants. The determinants are a mixture of social and environmental factors, such as:

- poor housing and local environments;
- limited social networks;
- income, poverty and worklessness;
- poor local transport and access to services;
- low educational attainment;
- drug and alcohol misuse.

The robust evidence indicated the following:

- Multiple housing deprivation leads to greater risk of disability or severe ill health in later life (Dedman et al, 2001).
- Improvements in mental health are reported consistently following housing refurbishment. The degree of mental health enhancement may be linked to the extent of the housing improvements and housing satisfaction.
- Housing improvements that ensure the provision of affordable warmth may have the greatest potential to reduce the adverse effects of poor housing. Optimal temperature is an essential component of domestic heating provision and may also affect damp and allergen growth. Energy efficiency improvements have led to enhancement in general health and respiratory health among asthmatic children. The elderly and very young are particularly at risk from both low and high indoor temperatures. Sudden increases in air pollutants are also most detrimental to the health of the elderly and asthmatics.
- When housing improvements are accompanied by increased rents there is a risk of negative health impacts.
- Regeneration can sometimes create divisions within the local area.
- Insulating existing houses leads to a significantly warmer, drier indoor environment and results in enhanced self-rated health, self-reported wheezing, days off school and work, and visits to general practitioners, as well as fewer hospital admissions for respiratory conditions (Howden-Chapman et al, 2007).

The process of estate regeneration can impact on health as a result of design and planning, refurbishment or new build operations.

Design includes the interior of homes, communal space and the public realm. Estate design changes many of the determinants of health. Excellent guidance is available from the Healthy Urban Development Unit, Commission for Architecture and the Built Environment and elsewhere (Barton et al, 2003; CABE, 2009; HUDU, undated).

Housing refurbishment can create indoor air pollution as a result of the use of formaldehyde and volatile organic compounds (VOCs). There is strong evidence for a relationship between exposure to recent indoor painting (within the previous 12 months) and clinically verified asthma, bronchial hyper-responsiveness and nocturnal breathlessness (Wieslander et al, 1996; Rumchev et al, 2004; Arif and Shah, 2007).

There were a number of studies of the health impact of housing renovation and some of these had been based on a quantitative HHSRS method (Gilbertson et al, 2006; Chartered Institute of Environmental Health, 2008). Others had been based on social survey methods (Barnes, 2003). One important conclusion from these studies was that refurbishment is a stressful process for the residents that can have negative impacts on their well-being as a result of disorientation, invasion and disruption. On the other hand, living in a refurbished home had a number of positive impacts including reducing heart and respiratory disease, reducing the number of accidents in the home and giving greater security and mental well-being.

11.4.3 Profile of the community

At the time of the transfer, some 25 per cent of the homes did not have central heating and 60 per cent failed the Decent Homes standard, compared with 40 per cent nationally. Many properties retained their original windows and doors, kitchens and bathrooms were basic, homes needed rewiring and there was a general lack of environmental enhancement and security measures.

The health profile of the population in which the housing was located can be summarized as follows:

- 26 per cent were claiming some form of social support (benefits);
- 25 per cent had limiting long-term illness;
- 14 per cent had 'not good' self-reported health;
- 11 per cent were aged 70+;
- 6 per cent of 65+ adults had been referred to mental health services;
- 1.3 per cent were claiming disability allowance.

A number of these rates were two or more times the English average.

The data and reports available suggested that 25–50 per cent of residents had a vulnerability that would require special care during home refurbishment. These included poor coping, literacy or planning skills. It is significant that there was insufficient information to be more precise. On average, 42 per cent of social renting households in England contain a member with a serious medical condition or disability and rates are increasing over time (Hills, 2007). The rate in social housing is about twice that in other forms of tenure. This is not surprising, as poorer socio-economic groups are concentrated in deprived areas, use social housing and tend to have higher levels of disability due to poorer health, more accidents and more mental health problems.

11.4.4 Analysis of refurbishment

The refurbishment programme represents a transition for the resident that had four main stages:

1 Before the works – when the resident is living in a poor physical environment that may be cold, damp and unsafe. This may be accentuated by worklessness and anxieties such as poverty and fear of crime. The resident may be disabled and awaiting housing adaptation. This is a time of anticipation and delay. The resident may be offered choices about home improvements.
2 During the works – when the resident is disrupted and there are new, temporary, environmental hazards including dust and solvents.
3 Early post-works – when the resident first experiences the improved home environment and the associated sense of improved well-being and euphoria. If the works themselves have had a major disruptive effect, then the stress may continue into this phase. There will also be raised concentrations of volatile organic compounds and some snagging and defects await correction.
4 Late post-works – when the resident has ceased to feel a sense of euphoria because the improvements have become part of everyday life. The determinants of some chronic medical conditions have been permanently improved and the home environment is safer and warmer. The physical and social environment outside the home and the social situation of the resident may remain just as before. Rises in rent and energy costs may have eroded any financial improvements.

In order to manage the health impacts associated with these stages, it is necessary to know the identity of the occupants of each of the housing units. In order to do so, at least two databases are required: a property database and a tenant database. The property database identifies the location, size and state of repair of each home, as well as name of the main tenant. The second database identifies the number of people living in the home and any special needs such as disability or vulnerability. A special need could arise either because the residents represent a hazard to the workforce or because the works, and the workforce, represent a hazard to the residents.

At the time of the HIA, the RSL had not had an opportunity to develop the second database. Consequently, the identity of the occupants was not known. Such a database would identify special needs such as the following:

- people with chronic illnesses;
- registered disabled (physical or learning);
- elderly;
- families with newborn infants;
- people with significant planning, organizing or coping skills deficits;
- people with significant mental health challenges;
- violent offenders;
- registered drug users;
- asthmatics and those with severe allergenic reactions;
- other certified medical needs.

11.4.5 Recommendations

A total of 12 main recommendations were made as an outcome of this HIA. The main recommendation was that the RSL should know who its residents were and whether they had any special needs. It should maintain a register and should transmit this information systematically to its contractors. The register should flag which residents required special care from the contractors. Communication of the register to officers, contractors and especially front-line staff, would require appropriate confidentiality management.

The register should indicate the following:

- residents requiring respite care during works;
- residents requiring priority home improvements because of medical needs;
- residents requiring multiple services;
- residents who were potentially dangerous to the workforce.

An associated management system would be required that ran in parallel with the normal refurbishment programme. This system would need to deliver:

- one-off home improvements to individuals with severe medical conditions whose homes need adaptation and repair as a medical priority;
- respite care to vulnerable residents during refurbishment work;
- hypoallergenic products to sensitive individuals, such as paints and kitchen work surfaces with low VOC emissions;
- multiple services to those residents that required them.

During the course of the HIA, workshops were held with the residents' representatives and with a sample of the residents themselves. The recommendations that were developed were discussed with the residents' representatives before and after submission to the client.

For each of these recommendations, we suggested to the client that they should create a health management plan. The plan should indicate whether the recommendation was accepted or rejected. For each recommendation that was accepted, it should identify a champion within the client team who would take responsibility for implementation, as well as a budget and a deadline. See Table 8.9.

11.5 NOTE

1   PCTs were local health departments with public health functions that were replaced by a new public health service in 2010.

11.6 REFERENCES

Acheson, D., D. Barker, J. Chambers, H. Graham, M. Marmot and M. Whitehead (1998) 'Independent inquiry into inequalities in health report', www.official-documents.co.uk/document/doh/ih/ih.htm, accessed 30 August 2000

Arif, A. and S. Shah (2007) 'Association between personal exposure to volatile organic compounds and asthma among US adult population', *International Archives of Occupational and Environmental Health*, vol 80, no 8, pp711–719

Ashton, J. (ed.) (1992) *Healthy Cities*, Open University Press, Milton Keynes

Ballantyne, R. (2006) *Building in Health: A Checklist and Guide to Developing Healthy Sustainable Communities*, www.mksm.nhs.uk/buildinginhealthannouncement.aspx, accessed December 2006

Barnes, R. (2003) *Housing and Health Uncovered*. Shepherds Bush Housing Association, London, www.housinglin.org.uk/_library/resources/housing/housing_advice/housing__health_ uncovered.pdf

Barton, H., M. Grant and R. Guise (2003) *Shaping Neighbourhoods: A Guide for Health, Sustainability and Vitality*, Spon Press, London and New York

Birley, M. H. and V. J. Birley (2007) 'Healthy design principles for use in the health impact assessment of mixed residential developments', www.birleyhia.co.uk/publications/healthy% 20design%20principles%20v7.pdf, accessed July 2008

Birley, M. and A. Pennington (2009) 'A rapid concurrent health impact assessment of the Liverpool Mutual Homes Housing Investment Programme', www.apho.org.uk/resource/ item.aspx?RID=95106, accessed November 2010

Brighton and Hove City Council (2009) A–Z of services, Community, www.brighton-hove.gov.uk/ index.cfm, accessed August 2009

Butland, B., S. Jebb, P. Kopelman, K. McPherson, S. Thomas, J. Mardell and V. Parry (2007) *Tackling Obesities: Future Choices*, www.foresight.gov.uk, accessed September 2009

CABE (2009) *Future Health: Sustainable Places for Health and Well-being*, Commission for Architecture and the Built Environment, London, www.cabe.org.uk/publications/future-health

Care Services Improvement Partnership (2006) 'Good housing and good health? A review and recommendations for housing and health practitioners', http://networks.csip.org. uk/_library/Resources/Housing/Housing_advice/Good_housing_and_good_health.pdf, accessed February 2009

Cave, B. and P. Molyneux (2004) 'Healthy sustainable communities, key elements of spatial planning', www.mksm.nhs.uk, accessed September 2007

Cave, B., P. Molyneux and A. Coutts (2004) 'Healthy sustainable communities: What works?', www.mksm.nhs.uk, accessed September, 2007

Chartered Institute of Environmental Health (2008) 'Good housing leads to good health: A toolkit for environmental health practitioners', Chartered Institute of Environmental Health, London, www.cieh.org

CSDH (2008) 'Closing the gap in a generation: Health equity through action on the social determinants of health, Final Report of the Commission on Social Determinants of Health', www.who.int/social_determinants/thecommission/finalreport/en/index.html, accessed July 2009

Cullingworth, B. and V. Nadin (2006) *Town and Country Planning in the UK*, 14th edition, Routledge, London

Dannenberg, A. L., R. Bhatia, B. L. Cole, S. K. Heaton, J. D. Feldman and C. D. Rutt (2008) 'Use of health impact assessment in the U.S: 27 case studies, 1999–2007', *American Journal of Preventive Medicine*, vol 34, no 3, pp241–256, www.sciencedirect.com/science/article/B6VHT-4RSS76V-C/2/094f349ebbf8eb230e003d37cf2548a2

Dedman, D., D. Gunnell, G. DaveySmith and S. Frankel (2001) 'Childhood housing conditions and later mortality in the Boyd Orr cohort', *Journal of Epidemiology and Community Health*, vol 55, no 1, pp 10-15

Department for Communities and Local Government (2004) *Planning Policy Statement 1: Delivering Sustainable Development*, www.communities.gov.uk/planningandbuilding/planningsystem/ planningpolicy/planningpolicystatements/pps1, accessed March 2011

Department for Communities and Local Government (2006) *A Decent Home: The Definition and Guidance for Implementation*, www.communities.gov.uk/publications/housing/decenthome, accessed March 2009

Department for Communities and Local Government (2008) *Lifetime Homes, Lifetime Neighbourhoods: A National Strategy for Housing in an Ageing Society*, London, www.communities. gov.uk/documents/housing/pdf/lifetimehomes.pdf

Department for Communities and Local Government (2009) Planning Policy Statements, www.planningportal.gov.uk/planning/planningpolicyandlegislation/previousenglishpolicy, accessed March 2011

Department for Environment, Food and Rural Affairs (2008) Climate Change Act, www.defra. gov.uk/ENVIRONMENT/climatechange/uk/legislation, accessed March 2009

Department for Transport (2007) *Manual for Streets*, www.dft.gov.uk/pgr/sustainable/manfor streets, accessed June 2007

Department of Health (2003) *Tackling Health Inequalities: A Programme for Action*, www.dh.gov. uk/en/publicationsandstatistics/publications/publicationspolicyandguidance/dh_4008268, accessed March 2009

Forsyth, A., C. S. Slotterback and K. J. Krizek (2010) 'Health impact assessment in planning: Development of the design for health HIA tools', *Environmental Impact Assessment Review*, vol 30, no 1, pp42–51, www.sciencedirect.com/science/article/b6v9g-4wcsys3-1/2/3049d07c3f32 e89b6cb56717628034a4

Fredsgaard, M. W., B. Cave and A. Bond (2009) 'A review package for health impact assessment reports of development projects'. Ben Cave Associates Ltd, Leeds, www.hiagateway.org.uk, accessed September 2009

Gilbertson, J., G. Green and D. Ormandy (2006) 'Sheffield Decent Homes health impact assessment', Sheffield Hallam University, Sheffield, www2.warwick.ac.uk/fac/soc/law/ research/centres/shhru/sdh_hia_report.pdf

Hills, J. (2007) *Ends and Means: The Future Roles of Social Housing in England*, CASE/LSE, London, www.communities.gov.uk/archived/publications/corporate/annual-report07?view=Standard

HM Government (2004) Housing Act 2004, www.opsi.gov.uk/ACTS/acts2004/ukpga_20040034_ en_1, accessed March 2009

HM Government (2005) 'Statutory Instrument 2005 No. 3208, The Housing Health and Safety Rating System (England) Regulations 2005', Office of the Deputy Prime Minister, London, www.opsi.gov.uk/si/si2005/20053208.htm

House of Commons (2009) *Health Committee – Third Report – Health Inequalities*, www.publications. parliament.uk/pa/cm200809/cmselect/cmhealth/286/28602.htm, accessed July 2009

Howden-Chapman, P., A. Matheson, J. Crane, H. Viggers, M. Cunningham, T. Blakely, C. Cunningham, A. Woodward, K. Saville-Smith, D. O'Dea, M. Kennedy, M. Baker, N. Waipara, R. Chapman and G. Davie (2007) 'Effect of insulating existing houses on health inequality: Cluster randomised study in the community', *British Medical Journal*, vol 334, no 7591, p460, www.bmj.com/cgi/content/abstract/334/7591/460

HUDU (undated) NHS London Healthy Urban Development Unit, www.healthyurban development.nhs.uk/index.html, accessed February 2010

Marmot, M. (2010) *Fair Society, Healthy Lives: A Strategic Review of Health Inequalities in England Post-2010*, Global Health Equity Group, UCL Research Department of Epidemiology and Public Health, www.ucl.ac.uk/gheg/marmotreview, accessed March 2010

Office of the Deputy Prime Minister (2003) *Statistical Evidence to Support the Housing Health and Safety Rating System, Volume II – Summary of Results*, London, www.communities.gov.uk/ documents/housing/pdf/138580.pdf_

Office of the Deputy Prime Minister (2006a) *Housing Health and Safety Rating System, Enforcement Guidance*, www.communities.gov.uk/publications/housing/hhsrsoperatingguidance

Office of the Deputy Prime Minister (2006b) *Housing Health and Safety Rating System, Operating Guidance*, www.communities.gov.uk/publications/housing/hhsrsoperatingguidance

Pawson, H., E. Davidson, J. Morgan, R. Smith and R. Edwards (2009) *The Impacts of Housing Stock Transfers in Urban Britain*, www.jrf.org.uk/publications/impacts-housing-stock-transfers-urban-britain, accessed March 2009

Royal Commission on Environmental Pollution (2007) 'Twenty sixth report. The urban environment', www.rcep.org.uk, accessed February 2011

RTPI (Royal Town Planning Institute) (2009) 'RTPI Good Practice Note 5: Delivering healthy communities', www.rtpi.org.uk/item/1795/23/5/3, accessed April 2009

Rumchev, K., J. Spickett, M. Bulsara, M. Phillips and S. Stick (2004) 'Association of domestic exposure to volatile organic compounds with asthma in young children', *Thorax*, vol 59, pp746–751

San Francisco DPH (Department of Public Health) (2006) 'The healthy development measurement tool', www.thehdmt.org, accessed April 2010

Taske, N., L. Taylor, C. Mulvihill and N. Doyle (2005) *Housing and Public Health: A Review of Reviews of Interventions for Improving Health Evidence Briefing*, National Institute for Health and Clinical Excellence

UK Government Planning Portal (2010) Welcome page, www.planningportal.gov.uk, accessed March 2010

UNECE (United Nations Economic Commission for Europe) (undated) 'The PEP, Transport, Health and Environment pan-European Programme', www.unece.org/thepep/en/welcome.htm, accessed May 2010

Wanless, D., M. Beck, J. Black, I. Blue, S. Brindle, C. Bucht, S. Dunn, M. Fairweather, Y. Ghazi-Tabatabai, D. Innes, L. Lewis, V. Patel and N. York (2002) *Securing Our Future Health: Taking a Long-Term View. Final Report*, www.hm-treasury.gov.uk./Consultations_and_Legislation/wanless/consult_wanless_final.cfm, accessed September 2007

WHIASU (undated) Wales Health Impact Assessment Support Unit, www.wales.nhs.uk/sites3/home.cfm?OrgID=522, accessed May 2010

WHO (World Health Organization) Regional Office for Europe (2005) 'Health impact assessment', www.euro.who.int/en/what-we-do/health-topics/environmental-health/health-impact-assessment_10, accessed February 2011

WHO Regional Office for Europe (2009a) 'Healthy Cities and urban governance', www.euro.who.int/healthy-cities, accessed August 2009

WHO Regional Office for Europe (2009b) 'Zagreb declaration for healthy cities: Health and health equity in all local policies', www.euro.who.int/__data/assets/pdf_file/0015/101076/e92343.pdf, accessed February 2011

Wieslander, G., D. Norbäck, E. Björnsson, C. Janson and G. Boman (1996) 'Asthma and the indoor environment: The significance of emission of formaldehyde and volatile organic compounds from newly painted indoor surfaces', *International Archives of Occupational and Environmental Health*, vol 69, no 2, pp115–124 www.metapress.com/content/6y4q8y2yv4akrqc9

CHAPTER 12

Current and future challenges

- Ethical challenges to HIA associated with human rights, involuntary resettlement and lack of democratic process are outlined.
- The practical meaning of spiritual well-being is discussed.
- Some of the challenges to establishing the effectiveness of HIA are explained.
- The importance of cumulative impacts is explained at local, national and global levels.
- Climate change and energy scarcity represent two of the most significant new public health challenges of our time. Some of the implications for HIA are outlined.
- The conclusions provide an overview of some of the themes explored in this book, such as balance, diversity, holism and uncertainty.

12.1 INTRODUCTION

The intention of this chapter is to highlight some of the diverse range of challenges that remain to be addressed in the rapidly developing field of HIA. It is not intended to be a complete or systematic list. The main challenges addressed are ethics, spiritual well-being, effectiveness and cumulative impacts.

12.2 ETHICS

The ethical values proposed for conducting HIA were defined in Chapter 1 as participation in decision-making, equity, sustainable development, use of evidence, adopting a comprehensive approach to health and respecting human rights.

In the UK, the National Institute for Health and Clinical Excellence (NICE) has issued guidance on social value judgements (NICE, 2008). Its guidance refers to seven principles for the conduct of public life. These are selflessness, integrity, objectivity, accountability, openness, honesty and leadership. NICE also refers to four moral principles and these are listed in Table 12.1.

Table 12.1 *Moral principles*

Principle	Explanation
Respect for autonomy	The right of individuals to make informed choices about their own health
Non-maleficence	The obligation not to inflict harm – 'first do no harm'
Beneficence	Seeking to provide benefits
Distributive justice	Acting in a fair and appropriate manner

There is a distinction between utilitarian and egalitarian distributive justice. The utilitarian approach seeks to provide the greatest good for the greatest number. This may allow the interests of the majority to override the interests of the minority; it may not help in eradicating inequality. The egalitarian approach involves distributing service fairly to all individuals. Everyone may get an adequate service but no one may get the best service. There are difficulties about both these forms of distributive justice, and procedural justice is also required.

Procedural justice provides accountability for decisions that are made. It is based on publicity, relevance, challenge, revision and regulation. The decision-making process should be made public, the grounds for making decisions should be reasonable, there should be a process for challenging and revising decisions, and there should be a voluntary or public mechanism for regulating the decision-making process to ensure that it possesses the other characteristics.

These ethical principles represent an ideal, where HIA is conducted in democratic societies that respect human rights using publicly accountable agencies. But even in this case, HIAs may not be free of bias. See Chapter 4 for further discussion of bias. Bias may arise because HIA specialists are commissioned by agencies or corporations whose primary aim is to promote and implement a proposal. The specialist owes a duty to both the client and the profession. They must inform the client of the health impacts of the proposal and recommend mitigation measures, to the best of their ability. They have no control over the mitigation actions that the client chooses to implement and little control over the scope of the task that they are assigned.

The principles and values summarized here and in earlier chapters are challenged when assessing a proposal in regimes that are non-democratic, have poor human rights records, extreme inequalities and high levels of corruption. There is an issue of moral and cultural relativism (the principle that a person's beliefs and activities should be understood in terms of his or her own culture). Those who work in such regimes have to accept compromises that they might not accept in their own countries. They must decide whether it is better to engage with the process and secure what improvements they can, or to disengage and not work there. HIA is, in my view, fundamentally about engagement. For those who are new to working in poor countries, the dilemma often becomes apparent when the first disabled, hungry child knocks on the car window. The nature of the ethical challenge may not become obvious until the HIA is under way and confidentiality agreements have been signed. In such cases, reports have to be phrased with great diplomacy and the opportunities for safeguarding human well-being are limited. Here are examples.

12.2.1 Involuntary resettlement

Large infrastructure proposals often require some involuntary resettlement of individual households, settlements or towns. It is probably impossible to resettle a single person against their will without damaging their health and well-being. Safeguard policies have been adopted by the World Bank (World Bank, 2001). These policies require that the livelihoods of resettled communities are fully protected and that they should be at least as well off afterwards as before. Nevertheless the challenges are so great that some lending institutions screen out all projects that require involuntary resettlement. In other cases, considerable resources are expended on new communities, services, training and economic activities. The challenges and consequences have been extensively documented in the case of large dams and these are described in Chapter 9.

12.2.2 Land clearance

In sectors such as the extractive industries sector there are additional challenges. These highly valued proposals are frequently joint ventures between national governments and transnational corporations, as discussed in Chapter 10. For example, governments may offer corporations land that has already been cleared and designated for industrial development. The means by which the land was cleared may not be revealed or it may be stated, without evidence, that there were only a few people and that they were properly compensated, or that there was no one living there. In these cases, resettlement issues are excluded from the scope of the impact assessment by the government. As the corporation must respect national sovereignty it is powerless to disagree. The corporate decision-makers know that if anything goes wrong they will be the first to be blamed and they also know that proceeding with a proposal under such conditions may violate their own code of practice as well as international best practice. The impact assessor is powerless to investigate the resettlement process or to change the scope of the assessment. The members of the displaced community are no longer there and cannot be questioned. For example, they may have migrated to a large city. When governments have poor human rights or corruption records, the displaced communities may have no redress. In my experience there are always people living there, although government officers may not see them.

12.2.3 Community engagement

In some nations community representatives are unelected or hereditary members of political elites. Gender, ethnic and other inequalities are commonly accepted as the norm. Under these circumstances there is no possibility of engaging with the community in a similar manner to that understood in high-income democracies. The health concerns of the proposal-affected community cannot be properly expressed. However, there may be local NGOs that are active in promoting change and empowering those who are disenfranchised. There may also be a literature that documents surveys of similar issues and identifies priority needs. The development proposal itself can become a positive agent of change. The IA process may provide an opportunity to speak for those who cannot speak for themselves. Recommendations might include a social investment programme that contributes positively to change. Examples include micro-finance initiatives or other empowerment for village women.

12.3 SPIRITUAL WELL-BEING

The WHO definition of health from 1946 refers to physical, social and mental well-being. There have been various debates about including spiritual well-being. For example, WHO debated an amendment to the constitution in 1999 that read 'Health is a dynamic state of complete physical, mental, spiritual and social well-being and not merely the absence of disease and infirmity' (WHO, 1999). Instead of approving this amendment, the Board decided to keep the matter under review. It is a matter of opinion as to whether spiritual well-being is part of or distinct from mental well-being.

Ancient sources from different cultures have associated health with spiritual well-being. For example, Pericles defined health in 340 BCE as follows: 'Health is more than the absence of sickness! It is a state of spiritual, mental and physical well-being, which enables a person to face any crisis in life.' The Hindu, Chinese and Buddhist systems of medicine also refer to a balance of body, mind and spirit (Koenig et al, 2001).

There are many definitions of spirituality. One of the most broad and succinct follows (quoted with permission, Royal College of Psychiatrists, 2006):

> *Spirituality is identified with experiencing a deep-seated sense of meaning and purpose in life, together with a sense of belonging. It is about acceptance, integration and wholeness.*

In addition:

> *The spiritual dimension tries to be in harmony with the universe, strives for answers about the infinite, and comes especially into focus in times of emotional stress, physical and mental illness, loss, bereavement and death.*

Spirituality involves a dimension of human experience in which psychiatrists are increasingly interested because of its potential benefits to mental health. Spiritual health provides meaning and purpose to people's lives, and these are health protective factors as well as outcomes.

Spirituality can be distinguished from religion (Koenig et al, 2001). Many people of faith may also experience spirituality, but not all do so. Similarly, many people who experience spirituality are not members of organized religions and do not have a faith. They may be members of post-secular cultures. Spirituality may be regarded as unifying and inclusive. In this section, all religions and faiths are given equal weight and the focus is the spiritual content and practice. Spirituality can be observed in the behaviour or comments of individuals and communities and can take many and diverse forms. The forms may include buildings, locations, artefacts and rituals as well as natural objects and plants.

One role of HIA may be to conserve and protect the expression of spirituality within the community. However, in some cases spiritual or religious practices are harmful to individuals or communities and in these cases preservation seems unwarranted. A useful test is whether a practice infringes the international declarations of human rights and the rights of the child (Council of Europe, 1950; Office of High Commissioner for Human Rights, 1989). Typical practices that could fall into this category include discrimination against women or other people, child abuse or mutilation, promoting violence and interfering with the right of people to control their own bodies.

A substantial review of spiritual well-being is beyond the scope of this text, but much needed. During the 2008 HIA conference in Liverpool there were a number of keynote addresses on spirituality and HIA (e.g. Signal, 2008). These referred to practices in Thailand, Australia and New Zealand, among others. Thailand, in particular, has a strong tradition of including spirituality in health and consequently in HIA (Chuengsatiansup, 2003). Subsequently, a discussion group was established to encourage appropriate publications.

The New Zealand Guidelines on HIA include the 'Te Whare Tapa Wha' model of health, which emphasizes spiritual health along with physical and mental health, as well as family and community well-being (Public Health Advisory Committee, 2005). In this model, cultural and spiritual participation are regarded as determinants of health.

During a visit to a dam project in Sarawak in the 1980s I was able to observe direct evidence of the effect that a project can have on the spiritual and mental well-being of a community (Birley, 2004). An upland rice growing community of Iban people was displaced by the construction of a reservoir. In this community there was a strong association between rice growing and spirituality. Rice was believed to have a soul that should be treated respectfully and propitiated (Malone, undated). The project deprived the community of the opportunity of growing rice, causing great unhappiness and migration.

Another example is the Kariba dam, on the Zimbabwe/Zambia border. When forcibly removed to make way for the building of the dam, the Tonga people lost everything including their shrines. Nyaminyami is the Valley Tonga river spirit. His promise is that, one day, the Tonga will return to their original homes on the banks of the Zambezi (Mulonga.net, 2005). The unfortunate consequences of Tonga resettlement have been described in detail (Scudder, 2005).

Proposals that are governed by World Bank standards are expected to comply with operational policy on the protection of indigenous peoples (World Bank, 2005). These require borrowers to pay particular attention to the ties that indigenous people have to the natural environment, including their cultural and spiritual values. These standards should go some way to preventing the kind of abuse outlined above, as well as providing authority for including spiritual issues in impact assessments.

Within European culture, examples of spiritual well-being that are not connected with faith groups can be found in many contexts. For example, the Findhorn Foundation is located on the banks of the Moray Firth in Scotland. Residents and visitors to the Foundation express a close affinity to the land, which they hold in sacred trust. They find spiritual meaning and well-being in many features (Findhorn Foundation, undated). Their village is said to have the lowest eco-footprint in the UK. All action is accompanied by mindfulness; before building they 'listen to the land'. See also Figure 12.1. Members of that community would be deeply distressed if they were asked to

Figure 12.1 *The spiritual practice of walking a spiral, photo from the island of Iona*

Copyright Tropix.co.uk

vacate their land to make way for development in a similar manner to the Iban or Tonga communities.

There is growing evidence that green space is beneficial to both physical and mental health (Sustainable Development Commission, 2008; APHO, undated). But many people also attribute a sense of spiritual well-being to contact with the natural world. This is part of a current dialogue in psychotherapy, referred to as ecopsychology.

Similar issues apply to the formal faith communities. Each has their places of worship, cemeteries, festivals, customs and community centres. Ways of safeguarding their spiritual well-being need careful and sympathetic consideration during new development or regeneration planning. For example, in a multicultural society, what provision should be made for spirituality when a new town is planned? What are the needs of a Muslim or Buddhist household when social housing is refurbished? How should the buildings in Hong Kong be aligned in order to conserve harmony?

12.4 EFFECTIVENESS OF HIA

As HIA becomes a respectable academic research topic, sources of research funding may emerge that enable effectiveness studies to be devised and implemented in a wide range of different settings. To date, there are relatively few (some examples: Quigley and Taylor, 2003; Dannenberg et al, 2008).

One of the practical challenges is the 'counterfactual argument', explained in Chapter 5. An HIA is intended to change the design or operation of proposals. Therefore the outcomes of the proposal are different to the outcomes that would have occurred if there had been no HIA. How, then, can one determine whether the HIA changed the outcome? One could argue that the HIA only changes the outcome if the recommendations made are actually implemented. But this is not the case: the observer changes what they observe. It is likely that the presence of an HIA process changes the way decisions are made, because decision-makers are more aware that health is being considered. As stated before, no one likes to be criticized for doing a bad job.

Another practical challenge is that the team responsible for carrying out an HIA is frequently not present after the proposal has been implemented and is not in a position to evaluate effectiveness. For example, a consultant may be employed to produce an HIA report. After the report is completed, the consultant has no further involvement. The proponent should be in a position to evaluate effectiveness, but often has other priorities. In some cases, a steering group has overseen the HIA and the consequent implementation of recommendations. However, the steering group is bound by confidentiality agreements and may be unable, or unwilling, to publish its observations.

12.4.1 EC effectiveness study

HIA effectiveness was investigated during an EC-funded project from 2004 to 2007, involving 21 research teams in 19 countries (Wismar et al, 2007). A set of 17 case studies were developed from an initial list of 470 documented HIAs. The output was a multiple author book containing case studies and conclusions. The case studies covered transport, urban planning, agriculture, environment, industry (workplace), infrastructure and nutrition (health promotion). The case studies were all European democracies and public sector proposals at policy and project level. The majority of the case studies were in the transport or housing/urban planning sectors and 65 per cent were prospective.

Effectiveness was defined as the capacity to influence the decision-making process and to be taken into account adequately by the decision-makers. There was no attempt to undertake health outcome evaluation because of the complex methodological difficulties in doing so. Effectiveness was tested using a standardized key informant interview and literature search. Four types of effectiveness were distinguished (see Table 12.2). The study noted that effectiveness has other properties and that more than one type of effectiveness can occur at the same time (see Table 12.3). The study identified several factors that contributed to effective HIAs (see Table 12.4).

The study reached the following conclusions, among others:

- Broadly, HIA can be employed effectively in all sectors, in all European countries, and at all levels.
- Each sector has its own primary objectives and these cannot be subsumed to secondary health objectives. In some cases there may be win-win situations. In other cases there may be trade-offs. HIA is a decision-support instrument that brings about

Table 12.2 *Types of effectiveness*

Type of effectiveness	Description	Example
Direct	A decision is dropped or modified as a result of the HIA	Decision not to allow 24-hour construction work
General	The assessment was considered adequately by the decision-makers but did not result in modifications to the proposed decision	Creating stronger health consciousness in decision-makers
Opportunistic	The HIA is conducted because it is assumed that it will support the proposed decision	A participatory approach helped resolve a long-lasting conflict
Ineffectiveness	Decision-makers do not take account of the assessment	

Table 12.3 *Some dimensions of effectiveness*

Dimension of effectiveness	Example
Magnitude	Blocking* an inappropriate policy Introducing additional mitigation measures Causing major redesign
Equity	Assisting most economically deprived
Community	Addressing community concerns
Organizational	Promoting intersectoral working Investing in staff skills that could be utilized in future Understanding other sectors' agendas

*The authors were considering the possibility of public policies being blocked. In general, HIAs are not used to block programmes or projects, but to modify them.

Table 12.4 *Some contributing factors for effectiveness*

Contributing factor	Comment/example
Timing	Neither too early nor too late
Involvement of appropriate institutions and support units	National repositories of public health information
Public health culture in a given country	Whether the decision-makers had a medical concept of health or a broader determinant-based concept of health
Political leadership, public support	Clarification of who bears the cost and if necessary provides funding
Communication	The quality of communication between the various parties involved

intersectoral, social and political compromises and contributes to consistency in decision-making.

- There is uneven development of HIA between countries.
- The supporting resources are incompletely developed or missing, including sustainable and adequate financing and databases.
- While HIA may be seen as a barrier to implementing proposals, it can help with avoiding conflict and extra costs.
- The majority of the case studies were intermediate between desktop and comprehensive HIAs.
- The majority of the case studies did not have an evaluation component.

12.4.2 Chad–Cameroon pipeline case study

One published example of effectiveness concerns the ESHIA of the Chad–Cameroon oil pipeline and associated works (Jobin, 2003). The project was owned by a consortium of oil companies, including Exxon, cost US$3.5 billion and involved 1000 kilometres of pipeline. Part of the loan was provided by the World Bank group. Two communicable diseases – malaria and HIV – were known to be important in the region. However, there were no recent health data from the project area. Trucking of materials required the construction of bridges and associated works. A particularly important stopping point was on the border between the two countries where an informal settlement of about 10,000 people included a conglomeration of brothels. HIV rates were estimated to be 55 per cent among commercial sex workers in that settlement.

An external panel of experts was recruited, as a steering group, to assist the countries evaluate the assessment and guide the management plan. Jobin was a member of that panel. There were two other groups of external monitors. One of these was responsible for monitoring corruption and ensuring the equitable use of funds and profits. Jobin's paper identified the following weaknesses, among others:

- Public meetings required armed guards for government and oil company officials and it is likely that their presence inhibited free discussion.
- The assessment proposed monitoring HIV transmission and, although various committees supported this, the proposal was not accepted by Cameroon.
- Greenhouse gas emissions were explicitly excluded from the assessment, against the advice of the panel.
- The consortium refused to allow any primary field data on health to be collected.
- There was a lack of continuity in the monitoring process. For example, the contracts of the panel were allowed to lapse randomly. In addition, the chairperson of the intergovernmental committee was replaced four times in three years and that committee lacked a health specialist.

The assessment concluded that HIV/AIDS transmission was the top priority health risk. The panel decided that the conventional methods of control recommended in the assessment were insufficient and proposed several additions, some of which were accepted and implemented. One important measure proposed was to control the distance that truckers had to travel. Instead of requiring them to travel the whole 1000 kilometres, there would be a system of relays. This would enable the truckers to sleep at home every night. This proposal was accepted by the consortium but rejected by the trucking contractors. Extraordinary measures were also proposed to counsel, diagnose and treat the workforce and camp followers for HIV infection before they dispersed at the end of the construction project. However this was rejected by the national governments and the consortium.

Jobin concluded that, in this project, decisions were based largely on cost and profit considerations and only gave passing attention to environmental and social concerns. Little or no decision-making power was transmitted to the affected population. Any concerns raised by the panel that might delay or obstruct completion faced powerful opposition. The appointment of the international panel by the project proponents gave the erroneous impression that the assessment and management plan were supervised and approved. Finally, he concluded that the World Bank group seemed unwilling to implement their own policies when these conflicted with the views of the consortium.

12.5 ANALYSIS OF POTENTIAL CATASTROPHES

To what extent should an HIA analyse potentially catastrophic failures of a proposal? Catastrophic failures are severe events that cause large-scale loss of life or livelihood but are very rare. A classic example is the deep sea, BP oil well leak that occurred n the Gulf of Mexico during 2010. Other examples include the Bhopal disaster, described in Chapter 1, the Bangladesh arsenic story, described in Chapter 9 and the Chernobyl nuclear disaster. Such failures are often the consequence of poor engineering practices. They represent conditions that are outside the design or operation of a proposal. The analysis and management are usually part of the health risk assessment. Normal engineering practice is to build in multiple safeguards to prevent catastrophes and then include an emergency plan in case all the safeguards fail. What role should HIA play in this process?

12.6 CUMULATIVE IMPACTS

Cumulative impacts are the consequence of multiple policies, programmes or projects and may be considered outside the scope of any single HIA. They are very important and can occur locally, nationally and globally. Cumulative impacts are included in SEAs, but unfortunately not all proposals are embedded within SEA processes.

12.6.1 Local cumulative impacts

Figure 12.2 is a simple example of a local cumulative impact. Imagine that three similar industries (A, B, C) are being established next to each other, inside a designated industrial zone. Project A is receiving an HIA. The other projects belong to separate and competitive companies. All three companies make similar products, produce airborne

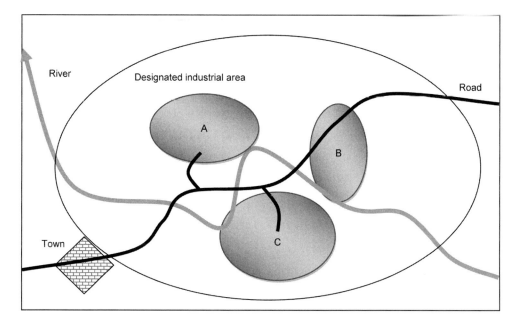

Figure 12.2 *Schema for local cumulative impacts for three projects in an industrial zone*

and waterborne emissions and send motorized transport along the local roads. The emissions from each industry are within statutory limits but the cumulative emissions from all three industries exceed statutory limits. A town is located downstream and the road to the industrial zone passes through it.

Responsibility for the total emissions and the road loading should be the responsibility of the industrial zone management company. They should have undertaken a strategic assessment before permitting each development. In many countries, there is no functioning industrial zone management entity and no strategic assessment of the industrial zone. Consequently, the town is polluted, noisy and damaged by traffic.

12.6.2 National cumulative impacts

National cumulative impacts are most apparent in the extractive industry sector, described in Chapter 10. The challenge is referred to as the paradox of plenty, 'Dutch disease' or the curse of oil. Empirical observation by economists, amongst others, suggests that developing a nation's natural resources, such as oil and gas, does not necessarily improve the national economy. The reason is that all other sectors of the economy suffer, as resources are pulled into the one important sector. For example:

● Nurses leave the public sector to work in company clinics.
● Policemen become company drivers and security guards.
● Teachers become company secretaries.
● Inflation reduces the value of agricultural production.

The consequences affect all aspects of life including human health. In general, nations with economies that depend on abundant natural capital tend to display the following effects (Gylfason, 2001):

- less trade and foreign investment;
- more corruption;
- less education;
- less domestic investment;
- lower life expectancy;
- more malnutrition.

These consequences can occur in both high- and low-income economies. Mitigation depends on high-level engagement with the government and with international agencies. It may be outside the scope of the proposal but not outside the scope of the HIA.

12.6.3 Global cumulative impacts

There are a number of global cumulative impacts of development (HM Government, 2008). These include population growth, water scarcity, non-renewable resources and climate change. For example, fossil fuel policies and projects have global cumulative impacts on climate change and energy scarcity. These twin global cumulative driving forces have implications for human health and the practice of HIA. A brief summary of some of the main issues is provided in the following sections.

12.7 CLIMATE CHANGE

One of the most pressing global cumulative impacts of fossil fuel production is arguably the increasing concentration of greenhouse gases (GHGs) in the atmosphere. This section explains the theoretical consequences for human health and considers the implications for HIA.

There is no longer any reasonable doubt about the contribution of fossil fuel usage to climate change or the effect of climate change on human survival (IPCC, 2007). Yet GHG emissions continue to rise and this has been referred to as a new kind of mutually assured destruction (terminology that was originally associated with nuclear weapons). At the time of writing, the best estimates of the current annual human health impact of climate change are as follows (Global Humanitarian Forum, 2009):

- 300,000 people dying;
- 325 million seriously affected;
- 500 million people at extreme risk;
- 4 billion people vulnerable;
- economic losses of US$125 billion.

Human health is affected as a result of extreme heat and cold, wind, vector distribution, water scarcity, agricultural failure, floods and much more. The future health impacts, and the requirements for managing them, have been discussed extensively (for example, Department of Health, 2008; Costello et al, 2009). It has implications for all HIAs (Haines et al, 2009).

The figures above represent averages based on projected trends and carry a significant margin of error. The real numbers could be lower or higher. The figures may represent the visible 'tip of the iceberg'. In any case, the figures should be sufficiently alarming to ensure that HIAs include this issue and the precautionary principle is adopted in all policies. The health and well-being of a large number of people will depend on climate stabilization and adaptation.

There is debate about the acceptable maximum GHG concentrations in the atmosphere. Elevated concentrations correspond to elevated risk, although the relationship need not be linear. One group of climate scientists have set the acceptable maximum at 350 parts per million (ppm) of CO_2 equivalent (Hansen et al, 2008). This can be regarded as a threshold of 'unacceptable risk'. Current concentrations are above this figure. Politicians appear to be proposing 450ppm as the target and accepting a higher level of associated risk. Normal industrial decisions are based upon risk levels of catastrophic failure of $1:10^6$ or lower. By contrast, politicians appear to accept much higher risks of climate collapse. Political decisions appear to be increasingly out of step with climate scientists. This can be illustrated by recent quotations (Milmo, 2007):

- 'We now have a choice between a future with a damaged world or a severely damaged world.'
- 'We are all used to talking about these impacts coming in the lifetimes of our children and grandchildren. Now we know that it's us.'

The Intergovernmental Panel on Climate Change (IPCC) is a body tasked to evaluate the risk of climate change caused by human activity. It produces regular assessments of the causes and consequences and is generally regarded as authoritative (IPCC, 2007). Its assessments are based on peer-reviewed and published scientific literature and its members are drawn from governments. As a result of processes such as time delays and intergovernmental consensus building, the assessments are believed to be overly conservative. For example, the summer Arctic Sea ice melt is proceeding much faster than predicted (Stroeve et al, 2007). Arctic ice melt may be an example of a tipping point – a trigger for catastrophic runaway climate change. The white ice sheet reflects some of the sun's heat and this helps to stabilize the climate. When the ice melts, the dark colour of the ocean is exposed and this absorbs heat causing the atmosphere to warm and more ice to melt.

There are two main management responses to climate change: mitigation and adaptation. Mitigation refers to actions that will stabilize the climate below an acceptable temperature threshold. The actions consist of reductions in GHG emissions or geo-engineering. Geo-engineering proponents advocate the use of untried technologies in order to remove GHGs from the atmosphere or reflect back more of the sun's heat. Adaptation refers to accepting major changes in the climate and building suitable defences on a local, national or global scale. These defences would have to ensure against challenges such as sea level rises, flooding, drought, widespread failure of food systems and extreme weather events. Successful adaptation is likely to be a function of wealth and geography. The majority of the world population may only adapt by migrating, leaving behind the very young and old and the most vulnerable. Economic analysis suggests that early action, in the form of mitigation, is much cheaper than later action in the form of adaptation (Stern, 2006).

At the time of writing, there is some consensus that average global GHG emissions should be reduced to 70–90 per cent of current levels by about 2050 in order to stabilize the climate. This figure appears to be too conservative when the damped oscillatory nature of the climate system is considered. In any case, major reductions are required within the lifetime of any proposal that is under consideration today. These reductions affect the choice of proposal, the energy by which it will be powered, the associated technology and the impact assessment. Proposals that create net GHG emissions have negative health impacts on a global scale.

Analysing net emissions is complex and specialist and not the role of HIA practitioners. However, we need to ensure that the work is done and to understand the issues involved. For example, fossil fuels can be arranged in a hierarchy according to the inherent carbon content. This is highest for coal and especially high for brown coals such as lignite. It is lowest for natural gas. The fuels can also be arranged in a hierarchy according to the GHG cost of production and use. A recent study examined the GHG cost of delivering one barrel of oil to an American refinery (Kristin and Skone, 2009). Nigerian crude has very high associated emissions because of the uncontrolled flaring of produced gas from many of the oilfields. Canadian oil sands are next in line because of the high energy cost of extracting the oil from the sand. When a project is sourcing oil for its operations, careful choice of supply can substantially reduce life-cycle emissions: oil refined from Saudi crude has 11 per cent of the emissions associated with Nigerian crude. In response to data of this kind, Early Day Motion 1250 was introduced into the British parliament in 2009 which asked the government to require all UK-listed companies in the oil, gas and power sectors to report on their total carbon liabilities (UK Government, 2009).

Table 12.5 illustrates GHG emissions associated with production and use as reported by several large corporations to the Carbon Disclosure Project (2008). Reporting is voluntary and incomplete as not all corporations are willing to comply. The table separates the direct GHG emissions associated with production and two kinds of indirect emissions: emissions associated with produced electricity and other emissions. In the case of fossil fuel projects, the other emissions are principally associated with use of the product. Shell, for example, creates about eight units of indirect emission for each unit of direct emission.

At the time of writing, many of the large oil companies were only reporting their global operation and not the lifetime emissions predicted for each new project. One possible recommendation for the impact assessment of a proposal would be full prior public reporting of both direct and indirect lifetime emissions. This information would enable public bodies to make rational decisions on the merits and demerits of the proposal, bearing in mind the global health consequences.

Table 12.5 *Examples of annual carbon disclosure reported in 2008 (1000 tonnes)*

Sector	Corporation	Direct GHG emissions	Electrically indirect GHG emissions	Other indirect GHG emissions, including product use
Oil and gas	Shell	92,000	13,000	743,180
	BP	63,460	10,670	521,000
	Chevron	63,759	?	?
	Eni	67,556	4070	303,000
	Exxon	141,000	?	?
	Tullow	234	?	?
Mining	Newmont	2886	983	0
Manufacture	Siemens	1550	2410	499

12.7.1 Co-benefits

There are significant co-benefits at local level from proposals that include reductions in GHG emissions (Davis, 1997; Bollen et al, 2009; Edwards and Roberts, 2009; *The Lancet,* 2009). Table 12.6 provides examples.

Measures to reduce emissions of greenhouse gases to 50 per cent of 2005 levels, by 2050, could reduce the number of premature deaths from the chronic exposure to air pollution by 20–40 per cent (Bollen et al, 2009). Compared with a normal population distribution of body mass index (BMI), a population with 40 per cent obese requires 19 per cent more food energy for its total energy expenditure (Edwards and Roberts, 2009). Greenhouse gas emissions from food production and car travel due to increases in adiposity in a population of 1 billion are estimated to be 0.4–1.0GT of carbon dioxide equivalents per year.

In 2009, *The Lancet* published a series of papers on the health co-benefits of actions to reduce GHG emissions (*The Lancet,* 2009). Four main sectors were analysed: household energy, urban land transport, electricity generation, and food and agriculture.

Table 12.6 *Examples of co-benefits*

Intervention	Effect
Reducing GHG by 50%	Reduce premature death from air pollution by 20–40%
Reduce car transport	Reduce morbidity from traffic injury and lack of physical activity
Reduce animal consumption	Reduce diseases associated with consumption of animal fats
Reduce obesity	Save 0.4–1 billion tonnes CO_2 per year per billion people associated with transporting overweight people
Reduce oil reliance	Reduce injury from oil wars
Contract and converge*	Reduce global inequalities

*This strategy was first proposed in 1995. It would allocate the same carbon ration to each human being on the planet and allow them to trade it (Global Commons Institute, undated).

The paper on urban land transport used a comparative risk assessment method to compare the health effects, measured in DALYs, of urban land transport in two settings, London and Delhi (Woodcock et al, 2009). For each setting, a business-as-usual projection to 2030 was compared with introduction of lower carbon-emission motor vehicles, increased active travel and a combination of the two. The analysis focused on the direct effects of pollution, physical activity and kinetic energy. Evidence was available from a number of systematic reviews and a range of key assumptions were made explicit. The study looked at the distribution of travel patterns between different modes under the three scenarios. The majority of the DALYs saved in both cities were associated with active transport rather than with low-emission vehicles. The overall reduction in the burden of disease was estimated to be substantial. Reductions in disease burden of 10–20 per cent were estimated for ischaemic heart disease, cerebrovascular disease, dementia, breast cancer, diabetes and depression. In London, an increase in the percentage of road traffic crashes was expected as a result of a substantial increase in the number of pedestrians and cyclists. In Delhi, by contrast, reductions in road traffic crashes could be 27–69 per cent. Differences between the two cities were the result of differences in current rates of air pollution and motorized transport.

An overview paper summarizes the associations uncovered using the metric of 1000 DALYs saved per year per megatonne of CO_2 equivalent saved (Haines et al, 2009). In this metric, priority interventions for reducing GHG emissions and improving health include clean cooking stoves (in India), active transport, reduced production and consumption of meat, and low-carbon electricity production.

The authors describe some policy implications of their analysis. These include ensuring that strategies and technologies for GHG mitigation receive HIAs; taking account of the health co-benefits of GHG reduction; reducing inequities in access to clean energy sources; and promoting strategies and policies to reduce GHG emission in all professional work.

12.8 ENERGY SCARCITY

Energy scarcity is an associated and equally important concept as climate change, yet it has not received the same public or government attention. Its processes and its consequences are less well known but are likely to have a huge health impact (Hanlon and McCartney, 2008; McCartney and Hanlon, 2008; McCartney et al, 2008; Raffle, 2010).

Warning calls abound. For example, according to the International Energy Agency (2008), the world's energy system is at a crossroads and current global trends in energy supply and consumption are patently unsustainable. One recent report is entitled *The Oil Crunch, A Wake-up Call for the UK Economy* and was issued by some leading corporations (Branson et al, 2010). There are many documentaries that address the issue, for example 'A Crude Awakening: The Oil Crash' (Gelpke and McCormack, 2006). The evidence for a near-term peak in global oil production has been widely reviewed (Sorrell et al, 2009). There are a number of websites devoted to the topic (ASPO, 2008; Oil Drum, 2008). Energy scarcity is a driver for society to seek oil and gas in unconventional locations and through unconventional processes, such as tar sands

Table 12.7 *Examples of energy return on energy invested for different fuels*

Fuel	Energy return on energy invested
Conventional oil	10–100
Unconventional oil	1–4
Tar sand and shale	2, or less than 1
Biofuels from crop products	<1

and the bottom of deep seas and oceans. In doing so, it creates major pollution risks and increased GHG emissions.

This section explains the theory of peak oil and how it interacts with GHG emissions, and considers the implications for HIA. Before doing so, the related concept of energy return on energy invested must be explained.

12.8.1 Energy return on energy invested

One measure of the value of an energy resource is the energy return on energy invested (EROEI). This is the amount of usable acquired energy per unit of energy expended. For example, if we need to expend one barrel of oil in order to obtain ten barrels of usable oil the EROEI is ten.

The EROEI of some fuels is listed in Table 12.7. Conventional oil usually has a higher and better EROEI than unconventional oil. Unconventional oil is lower because it must be obtained from more difficult locations, or it is of poorer quality and requires increased processing. The extreme examples are tar sand and shale, as well as biofuels made from crop products. The EROEI of these may be less than one. There is evidence that 10 joules of energy are used in agriculture to produce each joule of food energy.

Coal has a lower EROEI than conventional oil and it appears to be more plentiful. However, quality and accessibility vary greatly and there are large GHG emissions associated with its combustion. There are proposals to capture the carbon emissions from power stations and bury them underground – carbon capture and storage (CCS). The total energy required for CCS is relatively high and will further reduce the EROEI.

CCS is only possible in stationary plants, such as electricity generators. But electricity is not suitable for powering most of our transport fleet as the energy density of batteries is much lower than that of liquid fuels. One option is to transform gas or coal into gasoline; however, the transport fleet would emit GHGs that could not be captured. Another option is to produce hydrogen for energy storage, but this also has significant challenges.

12.8.2 Hubbert's Peak

The future shape of the global oil production curve is not known with any accuracy but it could have many impacts on human health. There is good evidence, based on national production curves, that it will peak (Oil Drum, 2008). The shape is expected to look similar to Figure 12.3 and is named after Hubbert, who was one of the more famous advocates. The peak may occur within the next two decades, or it may have already occurred (Sorrell et al, 2009). There is a 30–40 year lag between discovery and production. Therefore, information about past discovery is used to predict future production. Discovery of large new conventional oilfields peaked 30–40 years ago.

According to the theory of peak oil, the curve has two components: gross and net (Oil Drum, 2008; Murphy, 2009). The gross curve, illustrated as a solid line, is slightly skewed and with a broad base. This represents the annual rate at which oil is produced at the well head. The net curve indicates the amount of available oil when the cost of production, including the EROEI, is included. This curve is narrower and more heavily skewed. During the ascent, the easy oil is produced first. During the descent, only the more difficult-to-extract oil remains. According to this theory, the descent based on net oil availability will be very rapid.

The timing of peak oil is partly determined by demand and this is, at present, likely to continue rising exponentially, ensuring an earlier peak. As supply and demand must remain equal, the price will have to increase. Increases in price are likely to trigger economic recession leading to reduction in demand. Recessions tend to increase health inequalities as people lose their employment and sometimes their homes. According to this theory, we can expect rapid changes in availability and extreme volatility in price. High prices will have impacts on the supply of heating, transport, food and basic services. Our civilization is not well prepared and proposals in preparation today are not designed for these stresses and may not be resilient. HIA practitioners cannot be experts on energy supply, but they can recommend that specialist studies are made.

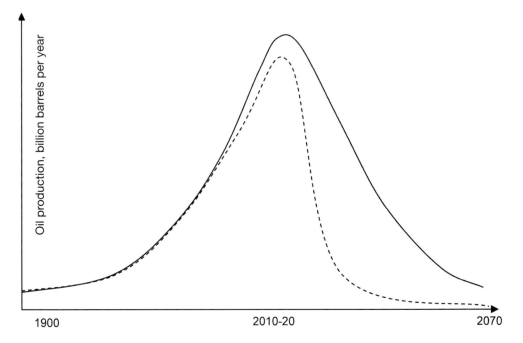

Figure 12.3 *Sketch of global gross (solid) and net (dashed) oil production*

12.8.3 The vicious spiral

At first sight it may seem that peak oil must reduce GHG emissions and so help to stabilize climate. However, the reverse is more likely. As conventional sources of fossil fuel become scarce, the price rises and there is an incentive to use unconventional sources. However, unconventional sources require a much higher level of GHG emission. Therefore, energy scarcity increases the rate of GHG emissions and accelerates climate change. In order to adapt to climate change an increased expenditure of energy is required to build flood defences, cope with extreme events and assure the supply of goods and services.

12.8.4 Alternative sources of energy

There is no space in this book to rehearse all the arguments surrounding the availability of alternative energy sources. In summary, there is no current technology available that can supply energy at the same intensity as fossil fuels. Renewable energy cannot fill the gap. Electricity cannot power long-distance road transport. Nuclear energy depends on uranium, which is a non-renewable resource. Many components of the nuclear energy life cycle are dependent on fossil fuels including mining, refining, transporting, as well as constructing and decommissioning nuclear facilities. Long-term storage of nuclear waste for time periods of the order of 100,000 years has an energy cost. All non-renewable resources have peaks in production. In most cases these are expected to occur within one generation (Heinberg, 2007).

At the time of writing, one new source of fossil fuel energy is emerging as a possible 'game changer': shale gas. This is both abundant and has lower carbon content than oil or coal. There are currently environmental, social and health impacts associated with its exploitation, but these may be overcome.

In conclusion, while the development of new sources of energy is essential, it would be imprudent to rely on that future possibility. Based on what we know, it seems likely that in the medium-term future, within the lifetime of most new proposals, our civilization will have to function with far less energy per capita than at present (Heinberg, 2007). This has implications for all current and future proposals, as well as the recommendations made to manage health impacts.

12.8.5 The energy gap and its implications

The conclusion above is represented graphically in Figure 12.4, using the UK as an example. The x axis measures time in years over the next 80 years or so. The y axis represents the percentage change required in net GHG emissions and the availability of fossil fuels. Within the next 40 years, emissions of GHGs must reduce by some 80–90 per cent. Fossil fuel availability is illustrated as diminishing along a similar gradient. At the same time, the proportion of primary energy provided by renewables is likely to increase from the current low of 1–2 per cent to, perhaps, 15 per cent by the year 2020 and 40 per cent by the year 2050. There is an energy gap because renewables do not replace the fossil fuels lost. This gap occurs within the lifetime of most proposals that are

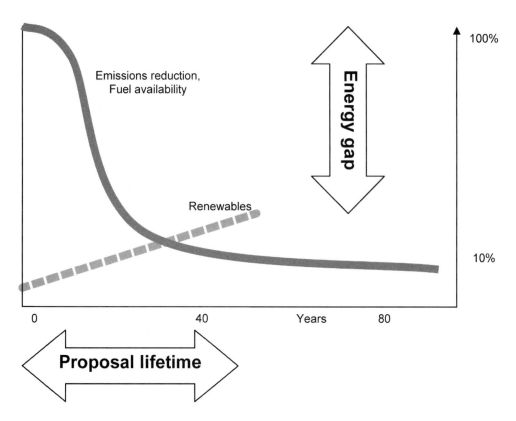

Figure 12.4 *Sketch of the potential energy gap in the UK*

under development today. This will affect the choice of proposals and the sustainability of health safeguards and mitigation measures recommended by HIAs. Therefore, these twin global cumulative driving forces are very important for the practice of HIA.

The health impacts can best be summarized as amplifying existing health risks and health inequalities on a global scale (Costello et al, 2009). The 'business as usual' scenario is likely to include increased inequality, reduced life expectancy, failure of medical services, war, starvation, mass migration and economic collapse. If this argument is accepted, then it follows that:

- All proposals should be scrutinized for their contribution to climate change and their dependency on scarce fossil fuel resources.
- All recommendations made in an impact assessment report should be energy efficient and minimize reliance on fossil fuels.

12.8.6 Models of change

The sections above have presented the evidence, as understood at the time of writing, regarding climate change and energy scarcity. While there is widespread debate about

Table 12.8 *Responses to knowledge about climate change and energy scarcity*

Emotion	Attitude	Practice
Fear	Denial	Inaction
	Despair	Inaction
Hope	Magical rescue	Inaction
	Change/transition	Action

these issues, the response exhibited by individuals, institutions and governments is very mixed. On a personal level, it is hard to accept. Neither the professional impact assessment community nor the decision-makers whom they inform have adapted to the implications. For example, motorways are still being built in Europe although targets for CO_2 emission reductions have been established. In order to construct persuasive and justifiable recommendations, those reactions must be analysed.

Knowledge, attitude and practice (KAP) is one model frequently used in public health in order to understand health-harming behaviour and to promote change. In the current case, knowledge is represented by our understanding of climate change and energy scarcity. Attitude and practice can be analysed using a simple model. There are many emotional responses to the knowledge, including fear and hope. These emotions, in turn, lead to attitudes such as those listed in Table 12.8. Denial is a common reaction and takes many forms. An alternative is despair; the facts are internalized but the situation seems hopeless and all action is worthless. Hope has two components. The first one may be called magical rescue, or 'deus ex machina'. The facts are accepted but it is assumed that science, technology, or God, will spring to the rescue. All three of these attitudes therefore lead to inaction by individuals and institutions.

The final attitude associated with the emotion of hope is change: the facts are accepted, there is great danger, but there is also opportunity.

An alternative model, based on social psychology research, suggests that our everyday activities are influenced by what we believe to be 'normal' behaviour (Schultz et al, 2007; Nolan et al, 2008; McAlaney et al, 2010; CHARM, undated). The majority of people simply wish to behave like their neighbours and to do whatever is considered normal. In this analysis, the objective is not to persuade people to abandon the norm and adopt a new and different way. Rather, the objective is to change what is believed to be the norm. The social norm approach attempts to change behaviour in socially desirable ways by telling people what other people do. This approach has been successfully employed in the contexts of alcohol and substance abuse, and sustainability issues such as electricity consumption, recycling and hotel towel reuse.

One approach to social change that, perhaps, captures both of these models has been popularly articulated by the Transition Town movement (Hopkins, 2008; Chamberlin, 2009). The practice of transition entails finding a way to make our lives, our towns and our proposals more resilient. Here, the objective is to find, or rediscover, ways to make life more rewarding, to increase happiness and to use less fossil fuel; to increase social capital so that communities can function in a more supportive manner; and to promote many of the determinants of health. In this scenario, goods and services will be more locally sourced and the community will be more able to cope with unexpected shocks and extreme events.

The theme of the 2010 IAIA conference was 'The Role of Impact Assessment in Transitioning to the Green Economy' (IAIA, undated). This provided an opportunity for participants to explore the many ways in which impact assessment can contribute to social change. My own paper explored the health impacts of the transport transition (Birley, 2010).

12.8.7 Associated issues

There are many associated issues that are outside the scope of this book. One of these is demographic trends and the total size of the human population. There is a growing literature that explores this linkage (see, for example, Bryant et al, 2009). There are debates about the kind of agriculture required to feed billions of people sustainably. Can we afford to continue eating meat? Global freshwater scarcity was mentioned in Chapter 9. Water is required for agriculture, industry and domestic consumption. Is there enough to go round and which use should have priority? The choices made to resolve these issues have health impacts.

12.9 CONCLUSIONS

This book has provided an introduction to the large and growing field of HIA. I have tried to pick up some of the threads, but there are many other threads that I will leave to others to describe.

As the subject evolves, one guiding principle could be that of *balance*. There should be a balance between the development of policy, processes, methods and tools. A good HIA method is only useful if there is a process in which to embed it. HIA cannot progress unless it enjoys policy support. There should also be a balance in the assessment of the environmental and social determinants of health. Both of these are required for human well-being. A pollution-free environment needs to be accompanied by increases in social equity and access to good food and physical activity. There should be a balance in the resources allocated for the different kinds of impact assessment. EIA is very important for the protection, preservation and enhancement of our physical environment. But it is of equal importance that human society and human health should be protected and enhanced. There should be a balance in the goals of development. Economic growth is of no value in itself. The value arises if it is sustainable and contributes to the health, well-being and happiness of all community groups.

Is a single, unified approach to HIA possible? It would have to be useful in low-, middle- and high-income countries, for policies as well as projects, and in both the private and public sectors. It would have to be capable of including a broad and holistic approach to human health. This is a controversial matter at the time of writing (Krieger et al, 2010). One view is that industry is unready to accept too much and a more limited set of objectives provides the best practical compromise. This book is based on the assumption that a holistic approach is possible, desirable and achievable.

This book started by describing the general policy context for HIA. It then discussed some features of the procedures, methods and tools that can be used. This was followed by descriptions of the application of HIA in three contrasting and diverse sectors.

The water resource development sector usually involves public utilities. It is particularly important in low-income economies and is frequently supported by international aid. A number of sub-sectors, including reservoirs and dams, have been responsible for the involuntary resettlement of over 40 million people. Water is an increasingly scarce global resource and it is the basis for irrigated agriculture and hence human food supply. Delivering clean drinking water is a challenge for urban development in low-income countries. Water is frequently a medium for the transmission of communicable diseases. In warm climates, it is associated with the transmission of vector-borne diseases and these affect many hundreds of millions of poor people. Climate change is altering water availability, causing both floods and droughts.

The housing and spatial planning sector has a mixture of public and private developers. Urbanization is taking place rapidly in all countries and over 50 per cent of the world's population now live in cities. The quality of the built environment has many impacts on the physical and mental well-being of the community. In low-income countries, urbanization is often unplanned. Where it is planned, it is often based on a model from high-income countries. In high-income countries, urban development is often designed around private car transport, and this is unsustainable and degrades the physical environment. The freedoms associated with private mobility give rise to congestion, pollution, severance and fear.

The extractive industry sector is often controlled by powerful, private, transnational corporations as well as governments. They are responsible for significant cumulative impacts at local, regional and global scales. They create dependencies and expectations in the communities in which they operate. They provide access to non-renewable resources that are increasingly scarce and expensive and increasingly located among vulnerable communities in low-income countries.

The question then arises, can an HIA practitioner have sufficient working knowledge to apply HIA in any sector where it is required? After trying to write convincingly about three sectors, I am left with the feeling that we cannot. Each sector has its own agenda, priorities, experiences and language. These, in turn, have consequences for public health. Proposals in each sector are shaped by policy. Some policy is enacted by national governments and some by global processes mediated by financial institutions. Some is accountable to the people and some is not. All of these have to be understood in order for HIA to be effective.

This diversity is united by a common thread of positive and negative health impacts and a common approach to assessing them. As a result of the diversity, some kind of specialization is inevitable. The HIA practitioner cannot be expected to know about all health determinants in all sectors. A better plan might be for public health policy to identify priority issues arising in each sector and to provide guidance and examples of how sectoral development impacts on those priorities. This might provide the building blocks that can be used in HIA to understand how specific proposals interact with those priorities.

For example, in high-income countries a current health priority is overweight and obesity. This is associated with a sedentary lifestyle and poor eating habits. This, in turn, is associated with passive travel and poor access to healthy foods. With knowledge of the health priority, spatial planning can be assessed according to the promotion of active travel and the location of healthy food outlets. Another example is the STIs associated with labour migration and large infrastructure development. They are determined by income and skill inequality, family separation and a construction workforce with an ethos that promotes unprotected and promiscuous sexual activity. Infrastructure development proposals can be assessed according to how they manage these issues.

The list of challenges remains large. We need better understanding of decision-making processes under uncertainty. We need better prioritization tools and better mechanisms for avoiding bias. We need an enabling environment, promoted by policy, with competency frameworks, research funding, capacity building and career paths that attract the new generation of practitioners that this book seeks to inform.

12.10 REFERENCES

APHO (Association of Public Health Observatories) (undated) The HIA Gateway, www.apho.org.uk/default.aspx?QN=P_HIA, accessed July 2009

ASPO (2008) Association for the Study of Peak Oil and Gas, www.peakoil.net, accessed June 2008

Birley, M. (2004) 'Health impact assessment in developing countries', in J. Kemm, J. Parry and S. Palmer (eds) *Health Impact Assessment: Concepts, Theory, Techniques and Applications*, Oxford University Press, Oxford

Birley, M. (2010) 'Health impacts of transport transition', www.birleyhia.co.uk/Publications/7_Birley_Health%20impacts%20transport%20transition.pdf, accessed 2010

Bollen, J., C. Brink, H. Eerens and T. Manders (2009) 'Co-benefits of climate change mitigation policies: Literature review and new results', www.pbl.nl/en/publications/2009/Co-benefits-of-climate-policy.html

Branson, R., I. Marchant, B. Souter, P. Dilley and J. Leggett (2010) *The Oil Crunch, A Wake-up Call for the UK Economy*. UK Industry Taskforce on Peak Oil & Energy Security, London, www.peakoiltaskforce.net

Bryant, L., L. Carver, C. D. Butler and A. Anage (2009) 'Climate change and family planning: Least-developed countries define the agenda', *Bulletin of the World Health Organization*, vol 87, no 11, pp805–884, www.who.int/bulletin/volumes/87/11/08-062562-ab/en/index.html

Carbon Disclosure Project (2008) 'Carbon disclosure project', www.cdproject.net, accessed July 2009

Chamberlin, S. (2009) *The Transition Timeline for a Local, Resilient Future*, Green Books, Totnes, www.greenbooks.co.uk

CHARM (undated) 'Using digital technologies for social norms', www.projectcharm.info, accessed July 2010

Chuengsatiansup, K. (2003) 'Spirituality and health: An initial proposal to incorporate spiritual health in health impact assessment', *Environmental Impact Assessment Review*, vol 23, no 1, pp3–15, www.sciencedirect.com/science/article/B6V9G-47BXFKW-1/2/6eda38a18715cb8bdb433cb2acf82991

Costello, A., M. Abbas, A. Allen, S. Ball, S. Bell, R. Bellamy, S. Friel, N. Groce, A. Johnson, M. Kett, M. Lee, C. Levy, M. Maslin, D. McCoy, B. McGuire, H. Montgomery, D. Napier, C. Pagel, J. Patel, J. A. de Oliveira, N. Redclift, H. Rees, D. Rogger, J. Scott, J. Stephenson, J. Twigg, J. Wolff and C. Patterson (2009) 'Managing the health effects of climate change: Lancet and University College London Institute for Global Health Commission', *The Lancet*, vol 373, no 9676, pp1693–733, www.ncbi.nlm.nih.gov/entrez/query.fcgi?cmd=Retrieve&db=PubMed&dopt=Citation&list_uids=19447250

Council of Europe (1950) 'The European Convention on Human Rights', www.hri.org/docs/ECHR50.html, accessed July 2009

Dannenberg, A. L., R. Bhatia, B. L. Cole, S. K. Heaton, J. D. Feldman and C. D. Rutt (2008) 'Use of health impact assessment in the U.S: 27 case studies, 1999–2007', *American Journal of Preventive Medicine*, vol 34, no 3, pp241–256, www.sciencedirect.com/science/article/b6vht-4rss76v-c/2/094f349ebbf8eb230e003d37cf2548a2

Davis, D. L. (1997) 'Short-term improvements in public health from global-climate policies on fossil-fuel combustion: An interim report', *The Lancet*, vol 350, no 9088, pp1341–1349, www.thelancet.com/journals/lancet/issue/vol350no9088/piis0140-6736(00)x0068-2

Department of Health (2008) 'Health effects of climate change in the UK 2008: An update of the Department of Health report 2001/2002', www.dh.gov.uk/en/publicationsandstatistics/publications/publicationspolicyandguidance/dh_080702, accessed October 2010

Edwards, P. and I. Roberts (2009) 'Population adiposity and climate change', *International Journal of Epidemiology*, vol 38, no 4, pp1137–1140, http://ije.oxfordjournals.org/cgi/content/abstract/dyp172v1

Findhorn Foundation (undated) 'Spiritual community, education centre, eco-village', www.findhorn.org/index.php?tz=-60, accessed March 2010

Gelpke, B. and R. McCormack (2006) 'A Crude Awakening: The Oil Crash', Lava Productions AG, Switzerland: 94 minutes

Global Commons Institute (undated) 'Contraction and convergence, climate justice without vengeance', www.gci.org.uk/index.html, accessed 2010

Global Humanitarian Forum (2009) *Human Impact Report Climate Change: The Anatomy of a Silent Crisis*, www.global-humanitarian-climate-forum.com, accessed February 2011

Gylfason, T. (2001), 'Natural resources and economic growth: From dependence to diversification, Group Meeting on Economic Diversification in the Arab World, United Nations Economic and Social Commission for Western Asia (UN-ESCWA) in cooperation with the Arab Planning Institute (API) of Kuwait, Beirut, Lebanon

Haines, A., A. J. McMichael, K. R. Smith, I. Roberts, J. Woodcock, A. Markandya, P. Armstrong, D. Campbell-Lendrum, A. D. Dangour, M. Davies, N. Bruce, C. Tonne, M. Barrett and P. Wilkinson (2009) 'Public health benefits of strategies to reduce greenhouse-gas emissions: overview and implications for policy makers', *The Lancet*, vol 374, no 9707, pp2104–2114, http://linkinghub.elsevier.com/retrieve/pii/S0140673609617591

Hanlon, P. and G. McCartney (2008) 'Peak oil: Will it be public health's greatest challenge?', *Public Health*, vol 122, no 7, pp647–652

Hansen, J., M. Sato, P. Kharecha, D. Beerling, R. Berner, V. Masson-Delmotte, M. Pagani, M. Raymo, D. L. Royer and J. C. Zachos (2008) 'Target atmospheric CO_2: Where should humanity aim?', *The Open Atmospheric Science Journal*, vol 2, pp217–231, doi:10.2174/1874282300802010217

Heinberg, R. (2007) *Peak Everything*, New Society Publishers, Vancouver, http://richardheinberg.com/

HM Government (2008) *Health is Global*, London, www.dh.gov.uk/prod_consum_dh/groups/dh_digitalassets/@dh/@en/documents/digitalasset/dh_088753.pdf, accessed 2010

Hopkins, R. (2008) *The Transition Handbook: From Oil Dependency to Local Resilience*, Green Books, Totnes, www.greenbooks.co.uk

IAIA (undated) International Association for Impact Assessment, www.iaia.org, accessed September 2010

International Energy Agency (2008) *World Energy Outlook*, www.iea.org/index.asp, accessed July 2008

IPCC (2007) 'Climate change 2007', www.ipcc.ch, accessed November 2007

Jobin, W. (2003) 'Health and equity impacts of a large oil project in Africa', *Bulletin of the World Health Organization*, vol 81, no 6, pp420–426, www.scielosp.org/scielo.php?script=sci_arttext&pid=S0042-96862003000600011&nrm=iso

Koenig, H., M. McCullough and D. Larson (2001) *Handbook of Religion and Health*, Oxford University Press, Oxford

Krieger, G. R., J. Utzinger, M. S. Winkler, M. J. Divall, S. D. Phillips, M. Z. Balge and B. H. Singer (2010) 'Barbarians at the gate: Storming the Gothenburg consensus', *The Lancet*, vol 375, no 9732, pp2129–2131, www.sciencedirect.com/science/article/b6t1b-50b9f8n-4/2/3c22bc4b16 e2600b721494ba52513a0f

Kristin, G. and T. J. Skone (2009) *An Evaluation of the Extraction, Transport and Refining of Imported Crude Oils and the Impact on Life Cycle Greenhouse Gas Emissions*, www.netl.doe.gov/energy-analyses/refshelf/detail.asp?pubID=227, accessed July 2009

The Lancet (2009) 'Health and climate change', www.thelancet.com/series/health-and-climate-change, accessed November 2009

Malone, M. J. (undated) 'Iban society', http://lucy.ukc.ac.uk/EthnoAtlas/Hmar/Cult_dir/Culture.7847, accessed October 2008

McAlaney, J., B. M. Bewick and J. Bauerle (2010) *Social Norms Guidebook: A Guide to Implementing the Social Norms Approach in the UK*, University of Bradford, University of Leeds, Department of Health, West Yorkshire, UK, www.normativebeliefs.org.uk, accessed 2010

McCartney, G. and P. Hanlon (2008) 'Climate change and rising energy costs: A threat but also an opportunity for a healthier future?', *Public Health*, vol 122, no 7, pp653–657

McCartney, G., P. Hanlon and F. Romanes (2008) 'Climate change and rising energy costs will change everything: A new mindset and action plan for 21st century public health', *Public Health*, vol 122, no 7, pp658–663

Milmo, C. (2007) '"Too late to avoid global warming," say scientists', *The Independent*, London, www.independent.co.uk/environment/climate-change/too-late-to-avoid-global-warming-say-scientists-402800.html, accessed July 2009

Mulonga.net (2005) 'The Valley Tonga', www.mulonga.net/index.php?option=com_content&task=view&id=213&Itemid=93, accessed March 2010

Murphy, D. (2009) 'The net Hubbert Curve: What does it mean?', http://tinyurl.com/l5gpcb, accessed May 2010

NICE (National Institute for Health and Clinical Excellence) (2008) *Social Value Judgements: Principles for the Development of NICE Guidance*, second edition, NICE, London, www.nice.org.uk/aboutnice/howwework/socialvaluejudgements/socialvaluejudgements.jsp

Nolan, J. M., P. W. Schultz, R. B. Cialdini, N. J. Goldstein and V. Griskevicius (2008) 'Normative social influence is underdetected', *Personality and Social Psychology Bulletin*, vol 34, no 7, pp913–923, http://psp.sagepub.com/cgi/content/abstract/34/7/913

Office of High Commissioner for Human Rights (1989) 'Convention on the Rights of the Child', www.ohchr.org, accessed February 2011

Oil Drum (2008) 'The Oil Drum: Discussions about energy and our future', www.theoildrum.com, accessed June 2008

Public Health Advisory Committee (2005) 'A guide to health impact assessment: A policy tool for New Zealand', Public Health Advisory Committee, Wellington, www.phac.health.govt.nz/moh.nsf/pagescm/764/$File/guidetohia.pdf, accessed February 2011

Quigley, R. J. and L. C. Taylor (2003) 'Evaluation as a key part of health impact assessment: The English experience', *Bulletin of the World Health Organization*, vol 81, no 6 pp415–419, www.scielosp.org/scielo.php?script=sci_arttext&pid=S0042-96862003000600010&nrm=iso

Raffle, A. E. (2010) 'Oil, health, and health care', *British Medical Journal*, vol 341 c4596, www.bmj.com/content/341/bmj.c4596.short

Royal College of Psychiatrists (2006) 'Spirituality and mental health', www.rcpsych.ac.uk/mentalhealthinfo/treatments/spirituality.aspx, accessed September 2009

Schultz, W., J. Nolan, R. Cialdini, N. Goldstein and V. Griskevicius (2007) 'The constructive, destructive, and reconstructive power of social norms', *Psychological Science*, vol 18, no 5, pp429–434, doi:10.1111/j.1467-9280.2007.01917.x

Scudder, T. (2005) *The Future of Large Dams: Dealing with Social, Environmental, Institutional and Political Costs*, Earthscan, London

Signal, L. (2008) 'Questions of spirituality: The challenge of assessing impacts on spiritual health', www.apho.org.uk/resource/item.aspx?RID=64331, accessed April 2010

Sorrell, S., J. Speirs, R. Bentley, A. Brandt and R. Miller (2009) 'An assessment of the evidence for a near-term peak in global oil production', www.ukerc.ac.uk/support/tiki-download_file.php?fileId=283, accessed October 2009

Stern, N. (2006) *Stern Review on the Economics of Climate Change*, www.hm-treasury.gov.uk/independent_reviews/stern_review_economics_climate_change/sternreview_index.cfm, accessed October 2006

Stroeve, J., M. Holland, W. Meier, T. Scambos and M. Serreze (2007) 'Models underestimate loss of Arctic Sea ice', http://nsidc.org/news/press/20070430_StroeveGRL.html, accessed July 2009

Sustainable Development Commission (2008) 'Health, place and nature: How outdoor environments influence health and well-being: A knowledge base', Sustainable Development Commission, London, www.sd-commission.org.uk/publications/downloads/Outdoor_environments_and_health.pdf

UK Government (2009) 'Early Day Motion 1250', http://edmi.parliament.uk/EDMi/EDMDetails.aspx?EDMID=38374&SESSION=899, accessed December 2009

WHO (World Health Organization) (1999) 'Amendments to the Constitution', http://apps.who.int/gb/archive/pdf_files/WHA52/ew24.pdf, accessed September 2009

Wismar, M., J. Blau, K. Ernst and J. Figueras (eds) (2007) 'The effectiveness of health impact assessment', European Observatory on Health Systems and Policies, Brussels www.euro.who.int, accessed February 2011

Woodcock, J., P. Edwards, C. Tonne, B. G. Armstrong, O. Ashiru, D. Banister, S. Beevers, Z. Chalabi, Z. Chowdhury, A. Cohen, O. H. Franco, A. Haines, R. Hickman, G. Lindsay, I. Mittal, D. Mohan, G. Tiwari, A. Woodward and I. Roberts (2009) 'Public health benefits of strategies to reduce greenhouse-gas emissions: Urban land transport', *The Lancet*, vol 374, no 9705, pp1930–1943, www.thelancet.com/series/health-and-climate-change#

World Bank (2001) 'Safeguard Policies, Operational Policy 4.12: Involuntary resettlement', http://web.worldbank.org/wbsite/external/projects/extpolicies/extsafepol/0,,menupk:584441~pagepk:64168427~pipk:64168435~thesitepk:584435,00.html, accessed October 2006

World Bank (2005) 'Operational Policy 4.10 – Indigenous peoples', http://web.worldbank.org/wbsite/external/projects/extpolicies/extopmanual/0,,contentmdk:20553653~menupk:64701637~pagepk:64709096~pipk:64709108~thesitepk:502184,00.html, accessed July 2010

For those seeking further information on HIA, there are a variety of sources. These are naturally fluid and may change substantially over time. The websites listed should provide appropriate starting points.

ORGANIZATIONS AND WEBSITES

The World Health Organization maintains an HIA Gateway at www.who.int/hia/en/. Several different departments of WHO are actively interested in HIA and these include Health Policy, Water and Sanitation and Public Health and Environment.

In England, an HIA Gateway is maintained by the Association of Public Health Observatories at www.apho.org.uk/default.aspx?qn=p_hia.

The International Association for Impact Assessment has an active HIA section and can be contacted at www.iaia.org.

In Australia, the University of New South Wales maintains an active website dedicated to building HIA capacity at www.hiaconnect.edu.au/. This provides links to activities in different parts of Australia as well as in New Zealand.

An HIA community knowledge Wiki site is maintained by Salim Vohra. See www.healthimpactassessment.info/.

A page on HIA is maintained on Wikipedia, see: http://en.wikipedia.org/wiki/health_impact_assessment.

A clearing house has been set up in the US at www.hiaguide.org, based in UCLA.

I maintain my own website at www.birleyhia.co.uk and can be contacted at martin@birleyhia.co.uk.

Healthy spatial planning in the US

Thanks to Michelle Marcus for the following two interesting websites:

Public Health Law and Policy at www.phlpnet.org/healthy-planning.
International City/County Management Association at http://icma.org/en/.

LISTSERVERS

The IAIA health section maintains an HIA listserver and this is available to subscribers through health@iaia.org.

In the UK, a monitored listserver is maintained on the academic Jiscmail system. Subscribers can access it through hianet@jiscmail.ac.uk. At the time of writing, the listserver was managed by staff at IMPACT, at the University of Liverpool. They can be contacted at www.ihia.org.uk.

In the Asia-Pacific region, the HIA listserver is hia-seao@explode.unsw.edu.au. It is managed by www.hiaconnect.edu.au.

In the US, the listserver is hia-usa@lists.onenw.org. It is managed by www.feetfirst. info/phbe/hia-usa-listserv.

BLOGS AND NEWSLETTERS

A regular HIA newsletter is issued in Australia, see www.hiaconnect.edu.au/hia_e-news. htm.

An IAIA HIA newsletter is produced and distributed by volunteers. The current editor is Ben Harris-Roxas.

A Health Impact Assessment Blog, maintained by the indefatigable Ben Harris-Roxas and Salim Vohra, seeks to provide the latest news and information about HIA worldwide. See http://healthimpactassessment.blogspot.com.

JOURNALS

The IAIA publishes a journal called *Impact Assessment and Project Appraisal* (*IAPA*). This carries occasional papers on HIA. More details can be found on the IAIA website at www.iaia.org.

The *EIA Review*, published by Elsevier, carries an increasing number of papers on HIA. More details can be found on the Elsevier website: www.elsevier.com.

Many public health journals carry occasional articles on HIA. Examples include:

- *Journal of Epidemiology and Community Health*;
- *Health Promotion International*;
- *British Medical Journal*;
- *NSW Public Health Bulletin*;
- *Bulletin of the World Health Organization*.

CONFERENCES

There are at least three annual conferences on HIA, in Europe, Southeast Asia and the US.

The annual conference of IAIA has an active HIA stream in which many papers are presented. Details are archived on the IAIA website.

The international HIA conference is hosted annually by a different organization within Europe. It has been held in the cities of Liverpool, Birmingham, Cardiff, Rotterdam and Dublin. The conference is currently in its 12th year and typically attracts about 200 people. The conference originated in Liverpool and was called the 'UK and Ireland HIA Conference'. Table 1 indicates where information about some of the past conferences is stored. The HIA Gateway now maintains an archive.

Table 1 *Details of some of the conferences on HIA in Europe*

Year	Location	Website
2011	Spain	www.hiainternationalconference.org
2009	Rotterdam	www.hia09.nl (no longer available)
2008	Liverpool	www.apho.org.uk/resource/item.aspx?RID=64038
2007	Dublin	www.publichealth.ie/internationalhiaconference
2006	Cardiff	www.wales.nhs.uk/sites3/page.cfm?orgid=522&pid=15502
1999–2005		Missing information
1998	Liverpool	The first conference in the series. Scanned copy at www.apho.org.uk

A number of conferences have been held in Southeast Asia in either Thailand or Australia. General information can be found at www.hiaconnect.edu.au. Table 2 indicates where information is archived.

Table 2 *Details of HIA conferences in Southeast Asia*

Year	Location	Website
2010	New Zealand	www.geography.otago.ac.nz
2008–2009	Thailand	www.hia2008chiangmai.com/home.php
2007	Australia	www.hiaconnect.edu.au/hia_events.htm#2007_conference

The first HIA in the Americas workshop was held in California, US during 2008. The proceedings are stored at: http://habitatcorp.com/whats_new/conference.html. The second HIA in the Americas was held in March 2010 and the proceedings are stored at www.hiacollaborative.org/hia-in-the-americas-march-2010-workshop.

As well as the list below, there are published glossaries of HIA and IA terms (Mindell et al, 2003; WHO, 2010a, 2010b; IAIA, undated).

GLOSSARY

Active travel	Walking, cycling and the use of public transport
Acute	Rapid onset, as opposed to chronic
AIDS	Acquired immune deficiency syndrome, caused by infection with the human immunodeficiency virus
Alzheimer's	A common type of senile dementia
Anaemia	A condition characterized by a low haemoglobin level in the blood
Anopheles (*An.*)	A genus of mosquitoes that transmit malaria
Anopheline	One of two groups into which mosquitoes are divided
Anorexia nervosa	An eating disorder characterized by refusal to maintain a healthy body weight and an obsessive fear of gaining weight, often coupled with a distorted self-image
Appraisal	Also called review or evaluation. Determining the quality of an HIA report
Arbovirus	An arthropod-borne virus
Arthropod	An animal group including insects, ticks and mites
Ascaris	A genus of large parasitic worms that infest the small and large intestines of humans and other animals, producing occasional symptoms. Also called roundworm. Found in temperate and tropical regions.
Authoritative evidence	Originating from an unbiased source such as a reference book
Baseline studies	Studies that include the health profile of existing communities before the proposal is implemented and the evidence in the literature concerning the health impacts of similar proposals
Bilharzia	See schistosomiasis
Biomedical model of health	Concerned with the presence or absence of disease

Body mass index	A comparative measure of body weight based on a person's weight and height
Bronchitis	A disease in which the linings of the bronchial tubes of the lungs are inflamed
Broncho-dilator	Substance that dilates bronchi, decreasing airway resistance and facilitating breathing
Business case	Captures the reasoning for initiating a project or task
Camp followers	Migrant communities attracted by construction sites
Carbon capture and storage	Capturing the carbon emissions from power stations and burying them underground
Cardiopulmonary	Pertaining to or affecting both the heart and the lungs and their functions
Cerebrovascular	Pertaining to the heart and blood vessels
Chronic	Developing slowly and persisting for a long period of time or constantly recurring
Communicable disease	Any disease that is transmitted from one animal to another via a host of agents, such as insects, foods and contaminated materials
Concurrent impact assessment	Undertaken during proposal implementation, construction or early operation
Contract and converge	Allocate the same carbon ration to each human being on the planet and allow them to trade it
Cost-benefit	Converts both the costs and the benefits to monetary units so that they can be compared
Cost effectiveness	Costs are measured in monetary units and effects are measured in some other unit
Counterfactual argument	An HIA is intended to change the design or operation of proposals. Therefore the outcomes are different to the outcomes that would have occurred if there had been no HIA. How then, can one determine whether the HIA changed the outcome?
Cretinism	A disorder with mental and physical symptoms. Associated with iodine deficiency
Culex (Cu.)	A genus of mosquitoes that transmit filariasis and viral encephalitis
Cultural relativism	The principle that a person's beliefs and activities should be understood in terms of their own culture
Cumulative impacts	The results of additive and aggregative actions producing impacts that accumulate incrementally or synergistically over time and space
Decommissioning	A late stage in the project life cycle, when a project is no longer operational, infrastructure is removed and the site is restored
Dengue	Acute tropical fever caused by a virus. Transmitted by certain mosquitoes
Diarrhoea	Persistent purging or looseness of bowels, commonly due to infection by microorganisms such as *Salmonella*

Direct impacts	Changes in health determinants that immediately result from a plan
Disability adjusted life years	A common metric with a unit of time that subtracts from expected duration of life the years lost due to disease-specific morbidity and mortality
Dose-response relationship	Also called exposure-response relationship, describes the change in effect on an organism caused by differing levels of exposure (or doses) to a stressor (usually a chemical) after a certain exposure time
E-learning	Training delivered through the Internet
Emphysema	Long-term, progressive disease of the lungs that primarily causes shortness of breath
Energy return on energy invested	Amount of usable acquired energy per unit of energy expended
Environmental impact assessment	Critical appraisal of the likely effects of a proposal or activity on the environment
Epidemic	The occurrence in a community or region of cases of an illness, specific health-related behaviour, or other health-related events clearly in excess of normal expectancy within a specific area and time period.
Epidemiological or risk transition	Relative change in the importance of communicable and non-communicable diseases with economic development
Epidemiology	The study of the distribution and determinants of health-related states or events in specified populations, and the application of this study to the control of health problems
Equator Principles	A set of environmental and social benchmarks for managing environmental and social issues in development project finance globally
Eudemonia	The highest human good, happiness
Filariasis	Disease caused by nematode parasites, can cause swellings to the limbs (elephantiasis)
Gastrointestinal infections	Caused by pathogens that have been ingested or have migrated by other means to the digestive tract
Gender	A social construct based on physiological sex
Green Papers	Reports issued as consultation documents to propose strategies that can be implemented in other legislation
Greenfield	Land that has not previously been built upon
Greenhouse gas	A gas in an atmosphere that absorbs and emits radiation within the thermal infrared range
Greenwash	Promoting oneself as behaving in an environmentally responsible manner without actually doing very much
Grey literature	Body of materials that cannot be found easily through conventional channels such as publishers, but which is frequently original and usually recent
Health	Health is not merely the absence of disease and infirmity, but a state of complete physical, mental, social and spiritual well-being

Health concerns	In this book, unstructured, non-prioritized, broad, vague or specific, and realistic or unrealistic concerns
Health determinant	The root causes of illness and well-being which may be physiological, social or environmental. Also referred to as risk factors and confounding factors
Health hazard	Potential sources of harm
Health impact assessment	A combination of procedures, methods and tools by which a policy, programme or project may be judged as to its potential effects on the health of a population, and the distribution of those effects within the population. HIA identifies appropriate actions to manage those effects
Health needs assessment	An investigation into the health status of the existing community, a rational prioritization of needs and a management plan for meeting those needs
Health outcomes	Medically defined states of disease and disability, as well as community-defined states of well-being
Health profile	A set of health indicators describing the state of health of a specific population
Health risk	Probability that a particular harm will affect a particular group of people in a particular setting
Health risk assessment	Concerned with the occupational health and safety of the future workforce associated with a proposal and, to some extent, with issues like explosions that could affect the peripheral community
Hermeneutics	Sets of practices or techniques used for the purpose of revealing intelligible meaning
HIV	Human immunodeficiency virus that causes the acquired immune deficiency syndrome (AIDS)
Host	An organism, on or in which a parasite lives and feeds
Impact (of a proposal)	A term indicating whether the project has had an effect on its surroundings in terms of: technical, economic, socio-cultural, health, institutional and environmental factors
Impact assessment	The process of identifying the future consequences of a current or proposed action
Incidence	The number of cases of a specified disease diagnosed or reported during a defined period of time, divided by the number of persons in the population in which they occurred
Indirect impacts	Changes in health determinants that are consequences of direct impacts
Inequity/inequality	Situation in which individuals in a society do not have equal social status
Infection	The invasion of the body by pathogenic microorganisms which multiply, causing disease
Interpretative	In this book, assumption that the observer imposes meaning and values and that social reality varies between groups. See also positivist

Ischaemic	Restriction in blood supply, generally due to factors in blood vessels
Joint venture	Agreement between a national government, investment companies and transnational corporations to create a national company for business purposes
Latent	Present or potential but not evident or active
Leishmaniasis	A disease caused by a protozoan parasite of the genus *Leishmania* that is transmitted from person to person by phlebotomine sandflies; also known as kala-azar and oriental sore
Licence to operate	Grant of permission to undertake an activity, subject to regulation or supervision
Lifeworld	A state of affairs in which the world is experienced. A world that subjects experience together
Line ministry	A government ministry or department responsible for/in charge of dealing with certain matters
Low-income country	The World Bank classifies countries into four income groups according to per capita gross national income. Low-income countries are below about US$1000. In this text, the term is used more loosely to separate low- and high-income countries
Malaria	A mosquito-borne disease caused by *Plasmodium* parasites. Most common are *falciparum* and *vivax* malaria
Micro-finance or micro-credit	The provision of financial services to low-income clients, including consumers and the self-employed, who traditionally lack access to banking and related services
Mind-mapping	Previously called brainstorming, a non-critical and creative process used by a group or individual to identify issues that may be relevant in a particular setting
Mitigate	Alleviate
Modal shift	A change in the proportion of people travelling in a particular way
Morbidity	The condition of illness or abnormality; the rate at which an illness occurs in a particular area or population
Mortality	The condition of being subject to death. The mortality (death) rate is the frequency or number of deaths in any specific region, age group, disease or other classification
Night storage	Stored at night for use during the day
Non-communicable disease	Cannot be spread from one person to another
Non-governmental organization (NGO)	Private organizations that pursue activities to promote the interests of the poor, protect the environment, provide basic social services, relieve suffering or undertake community development
Obesity	Medical condition in which excess body fat has accumulated to the extent that it may have an adverse effect on health

Onchocerciasis	Disease caused by a filarial nematode parasite, can cause blindness, also called river blindness
Ontology	Classification systems
Opistorchiasis	Parasitic disease sometimes acquired from eating inadequately cooked, infected fish. It is caused by *Opisthorchis viverrini*, a liver fluke. Causes chronic liver disease and can be fatal
Parasite	An organism that lives on or in another organism termed the host, and draws nourishment from it (giving the adjective parasitic)
Parasitaemia	A condition in which parasites are present in the body
Pathogen	An organism that causes disease. Most pathogens are microscopic in size
Planning gain	Where a developer agrees to provide additional benefits, usually for the community, as part of securing planning permission
PM_{10}	Particulates of size smaller than 10 microns
$PM_{2.5}$	Particulates of size smaller than 2.5 microns
Policy	A set of decisions which are oriented towards a long-term purpose
Positivist	In this book, assumption that there are natural laws governing the behaviour of large human groups and that these can be discovered through science. See also interpretative
Precautionary principle	In the absence of scientific consensus that an action is harmful, the burden of proof that it is not harmful falls on those taking the action
Prevalence	The number of people ill because of a particular disease at a particular time in a given population. Often expressed as a proportion of a population affected
Primary care trust	Local health departments in the UK with public health functions that were replaced by a new public health service in 2010
Primary data	Data collected solely for the purposes of the HIA itself
Programme	An array of projects which may proceed sequentially or concurrently
Project	A proposal, an organized undertaking, a special unit of work
Proponent	An institution, or person, that puts forward a proposal. In this text it is also a commissioner, decision-maker or manager
Proposal	A plan or suggestion, especially a formal or written one, put forward for consideration by others. In this book also a general term for policies, plans, programmes and projects
Prospective	Looking toward the future
Psychosocial	Refers to the psychological and social factors that influence mental health

Psychotic	Abnormal condition of the mind, involving loss of contact with reality
Quantitative risk assessment	A method that combines the dose-relationship with an exposure distribution in the population to estimate the disease burden attributable to a risk factor
Ramsar	A city in which a convention on wetland protection was signed, giving its name to the convention
Residual impact	Unwanted impact that remains after mitigation
Resilience	The set of factors associated with an individual or group that increases the probability that their well-being remains unaffected by change
Retrospective impact assessment	Undertaken after the proposal has become operational
Scabies	An infection of the skin by a mite called *Sarcoptes*
Schistosomiasis	A disease caused by infestation of the human body by trematode worms of the genus *Schistosoma*, characterized by the passing of blood in the urine or stool. Also called bilharzia
Scoping	Deciding what is to be included and what is to be excluded from the assessment
Screening	A process of sorting project proposals to ascertain the need for health impact assessment
Secondary data	Data collected for another purpose that is used in the HIA
Section 106 agreements (UK)	Legal agreements between a planning authority and a developer, or undertakings offered unilaterally by a developer, which ensure that certain extra works related to a development are undertaken
Sensitive receptors	People or other organisms that have a significantly increased sensitivity or exposure to pollutants
Severance	Separating a previously integrated community by a physical barrier such as a highway
Sexually transmitted infections	Diseases transmitted by human sexual contact, such as HIV/AIDS, gonorrhoea, syphilis and chlamydiasis
Snagging	Deficiencies in the quality of new homes that are identified by the occupants and which require correction by the developer
Social investment	Provision and use of finance to generate social as well as economic returns
Socio-environmental model of health	Concerned with the root economic, social, psychological and environmental causes of illness and well-being
Somatoform	Physical symptoms that cannot be explained by a medical condition
Spirituality	Identified with experiencing a deep-seated sense of meaning and purpose in life, together with a sense of belonging. It is about acceptance, integration and wholeness

Strategic environmental assessment	Environmental assessment as it applies to the development of policies, plans and programmes (rather than projects). It provides a systematic method of considering the likely effects on the environment of strategies, plans and programmes that set a broad-based context for future development activity. A systematic, objectives-led, evidence-based, proactive and participative decision-making support process for the formulation of sustainable policies, plans and programmes, leading to improved governance
Strategic health management	Incorporating workforce and community health considerations systematically into project planning and management
Stratification	In sampling, the population is first divided into groups or strata and then each group is sampled randomly
Super output area	A geographical area designed for the collection and publication of small area statistics (UK)
Sustainable	Potential for long-term maintenance of well-being, which has environmental, economic and social dimensions
Tangible evidence	Open to direct inspection
Terms of reference	The contractual agreement between client and consultant about the process, resources and content of the HIA
Testimonial evidence	Asserted by an informant
Tick	Eight legged arthropod, usually blood-feeders
Triangulation	Investigating a phenomenon using two or more different methods. Consistency adds weight to the conclusion
Trypanosomiasis	Parasitic disease transmitted by tsetse flies, also called sleeping sickness
Tuberculosis	Chronic and disabling disease of the lungs and less frequently other parts of the body, which is fatal if not treated
Varicella	Chickenpox, a highly contagious illness caused by primary infection with *varicella zoster* virus
Vector	An animal – often an insect – transmitting an infectious agent from an infected animal to another animal
VIP latrines	An improved design of a ventilated pit latrine
Vulnerability	The set of factors associated with an individual or group that increases their probability of experiencing a reduction in well-being associated with change
Well-being	Well-being is associated with quality of life and happiness
White Paper	Reports issued by government that lay out policy, or propose action, on topics of current concern
Willing to pay	Determines the value of anything by observing or asking people what they are prepared to pay
Zoonosis	An infectious disease transmissible, under natural conditions, to humans from other animals

ACRONYMS

ALARP	as low as reasonably practical
ASMs	artisanal and small-scale mines
BMI	body mass index
CAFE	Clean Air for Europe
CCS	carbon capture and storage
CD	communicable disease
CHETRE	Centre for Health Equity Training, Research and Evaluation in the University of New South Wales
CSDH	Commission on Social Determinants of Health
CSW	commercial sex worker
DALY	disability adjusted life year
DFID	UK Department for International Development, former name ODA
DPSEEA	driving forces, pressures, states, exposures, effects, actions
EC	European Commission
EHA	environmental health area
EHIA	environmental health impact assessment
EHS	environmental, health and safety
EIA	environmental impact assessment
EROEI	energy return on energy invested
ESHIA	environmental, social and health impact assessment
ESIA	environmental and social impact assessment
EU	European Union
FAO	Food and Agriculture Organization of the United Nations
GHG	greenhouse gas
GINI	a coefficient of inequality used by economists
GP	general practitioner, of medicine
GPS	global positioning system
HHSRS	Housing Health and Safety Rating System
HIA	health impact assessment
HiAP	health in all policies
HNA	health needs assessment
HRA	health risk assessment
HSE	health, safety and environment
HSSE	health, safety, security and environment
IA	impact assessment
IAIA	International Association for Impact Assessment
ICD	International Classification of Diseases
ICMM	International Council on Mining and Metals
IFC	International Finance Corporation
IMPACT	International Health Impact Assessment Consortium at Liverpool University
IPCC	Intergovernmental Panel on Climate Change

IPIECA	International Petroleum Industry Environmental Conservation Association
JV	joint venture
KAP	knowledge, attitude and practice
LDF	local development framework
MIGA	Multilateral Investment Guarantee Agency of the World Bank
NCD	non-communicable disease
nef	new economics foundation
NGO	non-governmental organization
NICE	National Institute for Health and Clinical Excellence
NIMBY	not in my backyard
NO_2	Nitrogen dioxide
NOx	mixtures of nitrogen oxides
O&G	oil and gas
OECD	Organisation for Economic Co-operation and Development
OGP	International Association of Oil and Gas Producers
OPIC	Overseas Private Investment Corporation
PCT	primary care trust in the UK (now local health authority)
PEEM	WHO/FAO/UNEP Panel of Experts on Environmental Management for Vector Control
PHASE	Promoting and supporting integrated approaches for health and sustainable development at the local level
PM	particulate matter, usually 10 micron or 2.5 micron in size
ppm	parts per million
PPS	policy planning statement
QALY	quality adjusted life year
QRA	quantitative risk assessment
RAM	risk assessment matrix
RSL	registered social landlord
RTPI	Royal Town Planning Institute
SEA	strategic environmental assessment
SIA	social impact assessment
SS	sensus strictu, used in taxonomy to distinguish closely related species
SMART	specific, measurable, achievable, realistic, timely
SOx	sulphur oxides
STI	sexually transmitted infection
TB	tuberculosis
ToR	terms of reference
UAE	United Arab Emirates
UNDP	United Nations Development Programme
UNECE	United Nations Economic Commission for Europe
UNEP	United Nations Environment Programme
VOC	volatile organic compound
WCD	World Commission on Dams
WHO	World Health Organization

REFERENCES

IAIA (undated) International Association for Impact Assessment, www.iaia.org, accessed September 2010

Mindell, J., E. Ison and M. Joffe (2003) 'A glossary for health impact assessment', *Journal of Epidemiology and Community Health*, vol 57, pp647–651, http://jech.bmjjournals.com/cgi/reprint/57/9/647.pdf

WHO (2010a) 'Glossary of terms used in health impact assessment', www.who.int/hia/about/glos/en/index.html, accessed September 2010

WHO (World Health Organization) (2010b) *Health Promotion Glossary*, www.who.int/hpr/nph/docs/hp_glossary_en.pdf, accessed September 2010